AF509946

GENERAL POST-NEWTONIAN ORBITAL EFFECTS

Orbital motions have always been used to test gravitational theories which, from time to time, have challenged the then-dominant paradigms. This book provides a unified treatment for calculating a wide variety of orbital effects due to general relativity and modified models of gravity, to its first and second post-Newtonian orders, in full generality. It gives explicit results valid for arbitrary orbital configurations and spin axes of the sources, without a priori simplifying assumptions on either the orbital eccentricity or inclination. These general results apply to a range of phenomena, from Earth's artificial satellites to the S-stars orbiting the supermassive black hole in the Galactic Centre to binary and triple pulsars, exoplanets, and interplanetary probes. Readers will become acquainted with working out a variety of orbital effects other than the time-honoured perihelion precession, designing their own space-based tests, performing effective sensitivity analyses, and assessing realistic error budgets.

LORENZO IORIO is qualified as Full Professor of Theoretical Physics and of Astrophysics at the Italian Ministry of University and Research. He earned his PhD from the University of Bari in 2001. His research focuses on gravitational physics, in particular, experimental/observational tests of general relativity and modified models of gravity. He is the author of more than 250 publications and is Editor-in-Chief of the journal *Universe*.

'This book provides a wonderful and very detailed guide for those interested in comparing observations with Einstein's theory and the many proposed alternatives. Written in a very readable and accessible manner, it is an indispensable guide to comparing theoretical gravitational predictions with the most recent data coming from celestial observations provided by satellites, space probes, and telescopes. I highly recommend it to anyone interested in a very practical handbook for comparing theory and observations.'

Jim Isenberg, Professor Emeritus, University of Oregon

'This scholarly book provides a comprehensive account of post-Newtonian orbital effects in gravitational systems. It is an authoritative contribution to modern relativistic celestial mechanics. Various gravitoelectric and gravitomagnetic effects of general relativity are treated in detail at the post-Newtonian level, while the last chapter of the book is devoted to modified gravity models. The presentation is clear and informative. This book is recommended to scientists working in astronomy and relativistic orbital mechanics.'

Bahram Mashhoon, Professor Emeritus of Physics, University of Missouri

'This is a self-contained text dealing with the main issue of any classical theory of gravity: orbital motion. The approach is very pedagogical. It is a precious toolkit to compare astronomical phenomenology with theories of gravity at any scale of astrophysical interest. The book is extremely useful for advanced undergraduate students as well as for PhD students in physics, astronomy, and mathematical physics. Furthermore, senior researchers working in the field can use it as a quick and comprehensive reference manual.'

Salvatore Capozziello, Full Professor, Università degli Studi di Napoli 'Federico II'

'Despite being mainly a theoretician, I find that this valuable book fills a gap in current literature since it sits at the interface between different fields which often do not communicate with one another. The approach and the methods developed in it can be straightforwardly extended from classical GR to alternative models of gravity, the orbital precessions of many of which are explicitly calculated. Interestingly, it explains how to potentially calculate the effect of any alternative gravity on several other observables such as astrometric angles, characteristic timescales, radial velocity, etc. I definitely recommend it to any serious student, researcher, and scholar involved in gravitational physics study.'

Sergei D. Odintsov, ICREA Research Professor, Institute of Space Sciences – CSIC, Spain

'An encyclopaedia of the 1pN and 2pN orbital effects, this book also explains approaches to testing GR in the said approximations. It is these practical applications which prove the great value of the post-Newtonian approximations, and which make this book an essential addition to the libraries of not only experts on GR and its applications, but also of the experts planning missions to giant planets. This excellent monograph provides a broad and up-to-date picture of post-Newtonian GR, as well as possible schemes of testing GR. I would recommend it to any graduate student or researcher working in the field of celestial mechanics and relativity.'

Michael Efroimsky, Astronomer, US Naval Observatory

GENERAL POST-NEWTONIAN ORBITAL EFFECTS

From Earth's Satellites to the Galactic Centre

LORENZO IORIO
Ministry of Education and Merit
(Ministero dell'Istruzione e del Merito)

CAMBRIDGE
UNIVERSITY PRESS

CAMBRIDGE
UNIVERSITY PRESS

Shaftesbury Road, Cambridge CB2 8EA, United Kingdom

One Liberty Plaza, 20th Floor, New York, NY 10006, USA

477 Williamstown Road, Port Melbourne, VIC 3207, Australia

314–321, 3rd Floor, Plot 3, Splendor Forum, Jasola District Centre, New Delhi – 110025, India

103 Penang Road, #05–06/07, Visioncrest Commercial, Singapore 238467

Cambridge University Press is part of Cambridge University Press & Assessment, a department of the University of Cambridge.

We share the University's mission to contribute to society through the pursuit of education, learning and research at the highest international levels of excellence.

www.cambridge.org
Information on this title: www.cambridge.org/9781009562874

DOI: 10.1017/9781009562911

© Lorenzo Iorio 2025

This publication is in copyright. Subject to statutory exception and to the provisions of relevant collective licensing agreements, no reproduction of any part may take place without the written permission of Cambridge University Press & Assessment.

When citing this work, please include a reference to the DOI 10.1017/9781009562911

First published 2025

A catalogue record for this publication is available from the British Library.

A Cataloging-in-Publication data record for this book is available from the Library of Congress.

ISBN 978-1-009-56287-4 Hardback

Online resources, including selected colour figures, for this publication at www.cambridge.org/9781009562874

Cambridge University Press & Assessment has no responsibility for the persistence or accuracy of URLs for external or third-party internet websites referred to in this publication and does not guarantee that any content on such websites is, or will remain, accurate or appropriate.

Intentio vero nostra est manifestare ... ea, quae sunt, sicut sunt.

– Frederick II of Hohenstaufen (Holy Roman Emperor)
De arte venandi cum avibus

Contents

Acknowledgements *page* xiii

1 Introduction 1

2 General Calculational Scheme 15
2.1 The Keplerian Picture 15
2.2 The pK Variations of the Keplerian Orbital Elements 19
2.2.1 The First-Order Effects 21
2.2.2 The Second-Order Effects 27
2.2.3 The Mixed, or Indirect Effects 28
2.3 The pK Corrections to the Keplerian Orbital Period 30
2.3.1 The Anomalistic Period 30
2.3.2 The Draconitic Period 31
2.3.3 The Sidereal Period 33
2.4 The pK Variations of a Generic Observable $\mathfrak{O}$ 36
2.4.1 The Radial Velocity of a Spectroscopic Binary 37
2.4.2 The Characteristic Timescales in Transiting Exoplanets 40
2.4.3 The Sky-Projected Spin-Orbit Angle of Transiting Exoplanets 46
2.4.4 The Variation of the Times of Arrival of Binary Pulsars 47
2.4.5 The Astrometric Angles RA and dec. 49
2.5 The pK Variations of the Two-Body Range and Range Rate 50
2.5.1 The Two-Body Range Shift 51
2.5.2 The Two-Body Range Rate Shift 51
2.5.3 How to Produce Time-Dependent Time Series 52

3 1pN Gravitoelectric Effects: Mass Monopole(s) 54
3.1 The 1pN Gravitoelectric Effects for a Test Particle 54
3.1.1 The Orbital Shifts of the Keplerian Orbital Elements 55
3.1.2 The Anomalistic Period 57
3.1.3 The Draconitic Period 58
3.1.4 The Sidereal Period 61
3.1.5 The Astrometric Angles RA and dec. 62
3.1.6 The Two-Body Range and Range Rate 63
3.2 The 1pN Gravitoelectric Effects for a Binary 64
3.2.1 The Orbital Shifts of the Keplerian Orbital Elements 65
3.2.2 The Anomalistic Period 67
3.2.3 The Draconitic Period 67
3.2.4 The Sidereal Period 68
3.2.5 The Radial Velocity of a Spectroscopic Binary 68
3.2.6 The Characteristic Timescales of Transiting Exoplanets 68
3.2.7 The Variation of the Times of Arrival of Binary Pulsars 72

4 2pN Gravitoelectric Effects: Mass Monopoles 73
4.1 The Direct Net Orbital Shifts 74
4.2 The Mixed Net Orbital Shifts 75
4.2.1 I: The 1pN Instantaneous Change of the Apsidal Line 75
4.2.2 II: The 1pN Instantaneous Orbital Shifts 76
4.2.3 The Total Mixed Net Orbital Shifts 77
4.3 The Total Net Orbital Shifts 77
4.4 The Total Net Orbital Precessions 78
4.5 Confrontation with Other Approaches in the Literature 80
4.5.1 The Damour–Deruelle Parameterization 80
4.5.2 The Calculation by Kopeikin and Potapov 82

5 1pN Gravitomagnetic Effects: Spin Dipole(s) 84
5.1 The Orbital Shifts of the Keplerian Orbital Elements 92
5.1.1 Some Special Orbital Configurations 94
5.1.2 The Contribution of Φ to the Mean Anomaly 96
5.1.3 Gravitomagnetic Ring Currents in Triple Systems 96
5.2 The Anomalistic Period 97

5.3 The Draconitic Period 98
5.4 The Sidereal Period 100
5.5 The Gravitomagnetic Clock Effect 103
5.6 The Radial Velocity of a Spectroscopic Binary 106
5.7 The Characteristic Timescales of Transiting Exoplanets 109
5.7.1 The Total Transit Duration δt_D 110
5.7.2 The Ingress/Egress Transit Duration $\delta t_{\mathrm{i/e}}$ 110
5.7.3 The Full Width at Half Maximum Primary Transit Duration δt_H 111
5.7.4 The Time of Inferior Conjunction t_{cj} 111
5.7.5 A Numerical Evaluation 112
5.8 The Sky-Projected Spin-Orbit Angle of Transiting Exoplanets 113
5.9 The Variation of the Times of Arrival of Binary Pulsars 114
5.10 The Astrometric Angles RA and dec. 115
5.11 The Two-Body Range and Range Rate 116

6 1pN Gravitomagnetic Effects: Spin Octupole 119
6.1 The Averaged Orbital Precessions 120
6.1.1 Some Special Orbital Configurations 121

7 Newtonian Effects: Mass Quadrupole(s) 123
7.1 The Orbital Shifts of the Keplerian Orbital Elements 126
7.1.1 Some Special Orbital Configurations 127
7.1.2 The Contribution of Φ to the Mean Anomaly 129
7.2 The Anomalistic Period 129
7.3 The Draconitic Period 130
7.4 The Sidereal Period 132
7.5 The Radial Velocity of a Spectroscopic Binary 135
7.6 The Characteristic Timescales of Transiting Exoplanets 137
7.6.1 The Total Transit Duration δt_D 138
7.6.2 The Ingress/Egress Transit Duration $\delta t_{\mathrm{i/e}}$ 138
7.6.3 The Full Width at Half Maximum Primary Transit Duration δt_H 139
7.6.4 The Time of Inferior Conjunction t_{cj} 140
7.6.5 A Numerical Evaluation 140
7.7 The Sky-Projected Spin-Orbit Angle of Transiting Exoplanets 141
7.8 The Variation of the Times of Arrival of Binary Pulsars 141
7.9 The Astrometric Angles RA and dec. 142
7.10 The Two-Body Range and Range Rate 144

8 1pN Gravitoelectric Effects: Mass Quadrupole 146
8.1 The Direct Effects 147
8.1.1 The Net Orbital Precessions 148
8.2 The Mixed Effects 150
8.2.1 The Net Orbital Shifts 150
8.2.2 The Net Orbital Precessions 154

9 pK Tidal Effects: Distant 3rd Body 155
9.1 Tidal Orbital Effects in a Kinematically Non-rotating Frame 155
9.1.1 The Newtonian Effects 158
9.1.2 The 1pN Gravitoelectric Effects 158
9.1.3 The 1pN Gravitomagnetic Effects 159
9.2 Tidal Orbital Effects in a Kinematically Rotating Frame 160
9.2.1 The Impact of the de Sitter–Fokker Precession 160
9.2.2 The Impact of the Spin of the Distant Body 162

10 Modified Models of Gravity: Orbital Precessions 165
10.1 Power-Law Modified Potentials 165
10.2 Yukawa Modified Potential 168
10.3 Logarithmic Potential 169
10.4 Once-Per-Revolution (1-cpr) Accelerations 169
10.5 Constant and Uniform Acceleration 171
10.6 Tidal-Type Matrix Acceleration 172
10.7 Dark Matter Distributions 172
10.7.1 Exponential Mass Density Profile 172
10.7.2 Power-Law Mass Density Profile 173
10.8 Lorentz-Violating Gravitomagnetic Acceleration 175

Appendix A List of Acronyms and Abbreviations 177

Appendix B Notations and Definitions 180
B.1 Some Constants of Nature and Astronomical Quantities 180
B.2 Parameters of a Mass-Energy Source and Spacetime Variables 181
B.3 Relevant Mathematical Functions and Notations 182
B.4 Relevant Physical Parameters of the Central Body 182
B.5 Relevant Physical Parameters of a Binary System 184
B.6 Relevant Physical and Orbital Parameters of a Test Particle 185
B.7 Relevant Parameters of a Hierarchical Triple System 189
B.7.1 Physical and Orbital Parameters of the Third Body 189

Appendix C pK Disturbing Functions from the Spacetime Metric 190
C.1 The 1pN Gravitoelectric Lagrangian per Unit Mass 191
C.2 The 1pN Lense–Thirring Lagrangian per Unit Mass 192
C.3 The Newtonian J_2 Lagrangian per Unit Mass 193

Appendix D Spin-Orbit Orientation and Lorentz-Violating Coefficients 194
D.1 Equatorial Orbits: General Spin Orientation 195
D.2 Polar Orbits: General Spin Orientation 195
D.3 Known Spin Orientation 196
D.4 Lorentz-Violating Gravitomagnetic Coefficients 196

Appendix E Coefficients of the Lense–Thirring Orbital Shifts 197
E.1 The Inclination 197
E.2 The Longitude of the Ascending Node 197
E.3 The Argument of Pericentre 198

Appendix F Coefficients of the Newtonian J_2 Orbital Shifts 199
F.1 The Semimajor Axis 199
F.2 The Eccentricity 199
F.3 The Inclination 200
F.4 The Longitude of the Ascending Node 200
F.5 The Argument of Pericentre 201
F.6 The Mean Anomaly at Epoch 201
F.7 $\Phi(t)$ 202

Appendix G Coefficients of the Total Mixed Net Orbital Shifts of the Order of J_2/c^2 203
G.1 The Semimajor Axis 203
G.2 The Eccentricity 203
G.3 The Inclination 204
G.4 The Longitude of the Ascending Node 204
G.5 The Argument of Pericentre 205
G.6 The Mean Anomaly at Epoch 205

Appendix H The Coefficients of the Tidal Orbital Precessions 207
H.1 The General Expressions of the Net Orbital Precessions 207
H.1.1 The Eccentricity 207
H.1.2 The Inclination 207
H.1.3 The Longitude of the Ascending Node 208
H.1.4 The Argument of Pericentre 208
H.1.5 The Mean Anomaly at Epoch 208

H.2 The Net 1pN Gravitoelectric Tidal Coefficients 209
H.3 The Net Gravitomagnetic Tidal Coefficients 209

Appendix I Coefficients of the Power-Law Mass Density Profile Orbital Precessions 212
I.1 The Disturbing Function 212
I.2 The Argument of Pericentre 212
I.3 The Mean Anomaly at Epoch 213

Appendix J Numerical Values of Relevant Astronomical Parameters 214

Appendix K Post-Keplerian Orbital Effects: Numerical Values 217
K.1 Solar System's Planets 217
K.1.1 The Inclination 217
K.1.2 The Longitude of the Ascending Node 218
K.1.3 The Argument of Perihelion 219
K.1.4 The Mean Anomaly at Epoch 220
K.1.5 The Current Level of Accuracy in Determining the Orbital Precessions 220
K.2 The Spacecraft Juno around Jupiter 220
K.3 The Earth's Geodetic Satellite LAGEOS 222
K.4 The Double Pulsar 224
K.5 Triple Pulsars 225
K.6 The Star S4716 around Sgr A* 227
K.6.1 The Net Shifts per Orbit of the Orbital Elements 228
K.6.2 The Net Shift per Orbit of the Radial Velocity Semiamplitude 229

Appendix L A Cursory Overview on Some Proposed Spacecraft-Based Missions 230
L.1 HERO 230
L.2 IORIO 231
L.3 EURO 231
L.4 LETSGO 231
L.5 ELXIS 232
L.6 Other Proposed Space-Based Missions 232

References 234
Author Index 269
Subject Index 270

Acknowledgements

I gratefully thank Michael Efroimsky for his long-standing encouragement to write a book like the present one. I wish to thank also my dear friend Grazia Demaio for her constant and joyful presence, and the unconditional support she has always given me.

1
Introduction

The feeblest of the four fundamental interactions governing the natural world is gravitation.[1]

The General Theory of Relativity[2] (GTR) is the formulation of gravitation set out by Albert Einstein in 1915 (Einstein, 1915c,d,a) and completed one year later (Einstein, 1916). It is the simplest possible gravitational theory compatible with his Special Theory of Relativity (STR) (Einstein, 1905). For contemporary comprehensive expositions of GTR, see, for example, Fok (1959), Synge (1960), Weinberg (1972), Hawking and Ellis (1973), Wald (1984), Stephani (1990), Cheng (2009), Padnanabhan (2010), Ohanian and Ruffini (2013), Zee (2013), Misner et al. (2017), Carroll (2019), Thorne and Blandford (2021), Schutz (2022), and Kenyon (2023). Some recent review articles, which appeared in the literature on the occasion of its last centenary, are, for example, Blandford (2015), Iorio (2015a), and Debono and Smoot (2016).

The time-honoured Law of Universal Gravitation proposed by Isaac Newton at the end of the seventeenth century in his immortal book *Philosophiæ Naturalis Principia Mathematica* (Newton, 1687; Chandrasekhar, 1995) describes it by means of a mysterious – remarkably, for Newton himself – force acting instantaneously between two or more material bodies, even if mutually separated in empty space by distances r much larger than their characteristic sizes D; as such, it benefits from the properties of the forces established by the three Newtonian laws of dynamics.

Instead, GTR adopts a completely different conceptual framework. According to it, gravitation is no longer best understood as a force, being, instead, a manifestation of the curvature, in a very specific sense, of a four-dimensional pseudo-Riemannian

[1] From the adjective *grăvis, e* ('heavy') and the noun *grăvĭtas, ātis*, ('weight, heaviness').
[2] From *Allgemeine Relativitätstheorie*.

Lorentzian manifold[3] known as spacetime (Oloff, 2023) with respect to the so-called 'flat' version of the spacetime employed by STR. Stated differently, the Einsteinian picture replaced the Newtonian concept of gravitational force with the notion of deformation of the chronogeometric[4] structure of spacetime (Damour, 2007) due to *all* forms of energy weighing it; as such, GTR can be defined as a chronogeometrodynamic theory of gravitation (Torretti, 1991). Indeed, the weight force on the Earth, which Newton unified with the agent determining the course of the heavens in the framework of his Universal Gravitation, is just an illusion due to the fact that we are born, live continuously, and die on the surface of our planet.[5] Actually, what we perceive as weight is not due to gravitation, but to the reaction force, of non-gravitational nature, exerted on our bodies by any physical surface we rest on; a chair, a floor, a bed. What kills us when we fall from a building is not gravity, but the non-gravitational reaction force by the ground. Indeed, if we are in free fall, that is, if we move subjected only to gravity and no[6] forces act on us, all the different parts of our body proceed with the same acceleration,[7] and we are not torn apart as would occur if gravity acted differently on bodies of diverse composition. Thus, as long as the regime of free fall continues, we are weightless, and the gravity seems to have been cancelled in our neighbourhood; for us, all things go as predicted by STR, we would obtain always the value of c in any experiment aimed at measuring the speed of light, and the worldlines of non-interacting, electrically neutral material objects appear as just straight in our freely falling experimental setup. It can be said that we are in a *local* (in both the spatial and temporal sense) inertial reference frame. It is one aspect of the so-called Equivalence Principle (EP).[8] In fact, such a removal of gravitation is not exact, being dictated by how

[3] According to differential geometry, a differentiable manifold is said to be pseudo-Riemannian (Benn and Tucker, 1987; Bishop and Goldberg, 1980) if it is endowed with a metric tensor that is everywhere nondegenerate, thus relaxing the requirement of positive-definiteness characterizing the Riemannian manifolds. A n_{d}-dimensional Lorentzian manifold is a special case of a pseudo-Riemannian manifold whose metric signature is $(1,\ n_{\mathrm{d}}-1)$.

[4] From Χρόνος, 'Chronos', the personification of Time, not to be confused with Κρόνος, 'Kronos', the Titan father of Zeus, corresponding also to the Roman deity Saturn.

[5] From πλανήτης, -ου, ὁ, meaning 'wanderer', composed by the verb πλανάω ('I wander') and the masculine agent noun suffix -της.

[6] If gravity were a force, here one would have to prefix the adjective 'other' to the word 'forces'.

[7] The tale according to which Galilei experimentally proved it by dropping objects of different weights from the Leaning Tower of Pisa (Drake, 1978) is, in all likelihood, apocryphal (Adler and Coulter, 1978; Segre, 1989; Crease, 2006).

[8] So far, one has only talked about bodies whose self-gravity is negligible in holding their constituent parts together, and whose free fall is not affected by their reciprocal gravitational interaction. Such a weak version of the EP (Nobili and Anselmi, 2018) has been recently tested to a relative accuracy of $\simeq 10^{-15}$ (Touboul et al., 2022a) in the spaceborne experiment Micro-Satellite à traînée Compensée pour l'Observation du Principe d'Equivalence (MicroSCOPE) (Touboul et al., 2022b) with two objects made of platinum and titanium alloys, respectively, kept in free fall around the Earth inside a spacecraft which shielded them from any potentially disturbing non-gravitational influences. As shown by analyses of the motions of the Earth and the Moon in the field of the Sun with the Lunar Laser Ranging (LLR) technique (Williams et al., 2012; Müller et al., 2019; Biskupek et al., 2021) and, more recently, of the binary pulsar-white dwarf PSR

uniform the gravitational field is on the scale of our body and of the things that free fall in our vicinity along with us. The more uniform the field is, or the smaller our neighbourhood is, the more accurate the absence of gravity is. In any case, free falling non-interacting objects left to themselves will sooner or later move, more or less rapidly, towards or apart from each other because of the unavoidable non-uniformity of the gravitational field in which they all fall together. That is *not* an illusion, and there is *no* way of *wholly* removing such a state of affairs: it is the true essence of gravitation for Einstein (Taylor and Wheeler, 1992). In Newtonian language, one would explain the aforementioned pattern in terms of residual, or differential, gravitational forces, commonly dubbed tidal since they are the analogue of the lunar gravitational pulls which, varying from one end to the other over the entire extension of the terrestrial globe, raise the tides on it. Instead, in the language of spacetime, the worldlines of such objects 'tidally' driven towards or apart from each other no longer appear straight, being curved. Since, as remarked before, this is the key feature of gravity, in the Einsteinian framework it is said that gravity is a manifestation of the curvature of spacetime and GTR relies upon the EP. Thus, GTR is, at the same time, a theory of space and time, and of gravitation as well; furthermore, light and free massive particles move along geodesics of a curved spacetime, which are the generalization of straight lines taking place when gravity is absent. Their equation is

$$\frac{d^2x^\sigma}{d\lambda^2} = -\Gamma^\sigma_{\ \upsilon\iota}\frac{dx^\upsilon}{d\lambda}\frac{dx^\iota}{d\lambda},\ \sigma = 0, 1, 2, 3, \tag{1.1}$$

where λ is some affine[9] parameter which, in the case of a massive body, coincides with its proper time τ, while

$$\Gamma^\sigma_{\ \upsilon\iota} := \frac{1}{2}g^{\sigma\kappa}\left(\frac{\partial g_{\kappa\upsilon}}{\partial x^\iota} + \frac{\partial g_{\kappa\iota}}{\partial x^\upsilon} - \frac{\partial g_{\upsilon\iota}}{\partial x^\kappa}\right),\ \sigma, \upsilon, \iota = 0, 1, 2, 3 \tag{1.2}$$

are the Christoffel symbols of the second kind (Weinberg, 1972; Bishop and Goldberg, 1980; Misner et al., 2017); $g^{\sigma\lambda}$ is the inverse of the spacetime metric tensor $g_{\sigma\lambda}$. In terms of the temporal coordinate $x^0 := ct$, the equations of motion for a test particle retrievable from Equation (1.1) for $\lambda \rightarrow \tau$ and $\sigma = 1, 2, 3$, can be written as follows (Weinberg, 1972; Brumberg, 1991):

J0337+1715 (Ransom et al., 2014; Shao, 2016) in the field of another distant white dwarf, searching for violations of the EP in terms of the Nordtvedt effect (Nordtvedt, 1968b,a), the EP holds also in its stronger version, according to which the mutual gravitational attraction among bodies along with their own self-gravity is taken into account as well, to the $\simeq 10^{-4}$ (Hofmann and Müller, 2018) and $\simeq 10^{-6}$ (Archibald et al., 2018; Voisin et al., 2020) levels, respectively. The challenges of testing the EP in different regimes, including also the quantum realm in which it is not obvious that the former is valid, are reviewed in Tino et al. (2020).

9 From *affinis, e*, 'bordering on, adjacent, contiguous'.

$$\frac{d^2x^i}{dx^{0\,2}} = -\Gamma^i_{\ \sigma\lambda}\frac{dx^\sigma}{dx^0}\frac{dx^\lambda}{dx^0} + \Gamma^0_{\ \sigma\lambda}\frac{dx^\sigma}{dx^0}\frac{dx^\lambda}{dx^0}\frac{dx^i}{dx^0},\ i = 1, 2, 3. \tag{1.3}$$

On the other hand, another crucial aspect of the EP consists of the fact that gravity can also be *emulated*, to a certain extent, by adopting an accelerated reference frame. Indeed, the motions of material objects referred to the latter are characterized by accelerations which depend neither on the mass nor on the composition of the former ones, which is just the distinctive trait of the gravitational interaction itself. Such a feature, together with STR, allows one to predict a number of peculiar phenomena pertaining to the propagation of electromagnetic waves and the motion of material objects which are unknown to the Newtonian gravitational picture. Suffice it to think about the Coriolis acceleration affecting a moving particle with respect to a rotating reference frame and the corresponding gravitomagnetic counterpart arising in GTR since the latter has to fulfil the Lorentz symmetry (Jantzen et al., 1992b; Schmid, 2023).

Since GTR is a *relativistic* theory of gravitation, and in STR *all* forms of energy are equivalent to mass, for Einstein, the source of gravitation, that is, of the spacetime curvature, is made by *several more* entities than for Newton and his scalar potential $\mathcal{U}$ *alone*. That is, a material body gravitates not only because it possesses its own rest energy, but also because it is compressed or dilated, or because it is distorted by internal stresses, and even if it moves. All that is encoded by the symmetric energy-momentum tensor $\mathsf{T}_{\sigma\lambda},\ \sigma, \lambda = 0,\ 1,\ 2,\ 3$ (Provost, 2017; d'Inverno and Vickers, 2022). Thus, there is *no* longer just a *single* gravitational potential sourced only by the matter density ρ, as in the Newtonian scheme, but now there are *ten* generally different quantities playing the role of gravitational potentials: the independent components of the symmetric spacetime metric tensor. The way the distribution of matter and energy actually deforms the spacetime ultimately determining the metric tensor is established by Einstein's field equations (Fok, 1959; Synge, 1960; Weinberg, 1972; Hawking and Ellis, 1973; Wald, 1984; Stephani, 1990; Cheng, 2009; Padnanabhan, 2010; Ohanian and Ruffini, 2013; Zee, 2013; Misner et al., 2017; Carroll, 2019; Thorne and Blandford, 2021; Schutz, 2022; Kenyon, 2023),

$$\mathsf{R}_{\sigma\lambda} - \frac{1}{2}\mathsf{g}_{\sigma\lambda}\mathsf{R} = \kappa_g \mathsf{T}_{\sigma\lambda},\ \sigma, \lambda = 0, 1, 2, 3, \tag{1.4}$$

which represent a set of complicated nonlinear partial differential equations. In Equation (1.4), $\mathsf{R}_{\sigma\lambda}$ is the Ricci curvature tensor of the spacetime, defined by contracting two indices of the Riemann tensor (Weinberg, 1972; Bishop and Goldberg, 1980; Parker and Christensen, 1994b; Misner et al., 2017; Schutz, 2022),

$$\mathsf{R}^{\epsilon}_{\ \sigma\psi\lambda} := \frac{\partial \Gamma^{\epsilon}_{\ \lambda\sigma}}{\partial x^{\psi}} - \frac{\partial \Gamma^{\epsilon}_{\ \psi\sigma}}{\partial x^{\lambda}} + \Gamma^{\epsilon}_{\ \psi\chi}\Gamma^{\chi}_{\ \lambda\sigma} - \Gamma^{\epsilon}_{\ \lambda\chi}\Gamma^{\chi}_{\ \psi\sigma},\ \epsilon, \sigma, \psi, \lambda = 0, 1, 2, 3, \tag{1.5}$$

in the following way (Weinberg, 1972; Bishop and Goldberg, 1980; Parker and Christensen, 1994a; Misner et al., 2017):

$$\mathsf{R}_{\sigma\lambda} := \mathsf{R}^{\epsilon}{}_{\sigma\epsilon\lambda},\ \sigma, \lambda = 0, 1, 2, 3. \tag{1.6}$$

Furthermore,

$$\mathsf{R} := \mathsf{g}^{\mu\nu}\mathsf{R}_{\mu\nu} \tag{1.7}$$

is the trace of the Ricci tensor, and κ_g is Einstein's gravitational constant (Adler et al., 1975). Nonetheless, if the characteristic motions of the system at hand are quite slow, and the gravitational fields are weak and almost static, the general relativistic field equations reduce to just the Poisson equation

$$\nabla^2 \mathcal{U} = 4\pi G\rho \tag{1.8}$$

for the potential $\mathcal{U}$ of the Newtonian theory. Such a correspondence fixes the value of Einstein's gravitational constant entering Equation (1.4) to[10]

$$\kappa_g := \frac{8\pi G}{c^4}, \tag{1.9}$$

where G is Newton's constant of gravitation. In view of its tensorial nature, if $\mathsf{R}^{\epsilon}{}_{\sigma\psi\lambda},\ \epsilon, \sigma, \psi, \lambda = 0, 1, 2, 3$ vanishes in a given coordinate system, it is zero in *all* other coordinates as well; in this case, gravity is effectively absent even if the spacetime appears *formally* curved in some coordinates; they would refer to a merely accelerated reference frame. Indeed, the geodesic deviation equation, known also as Jacobi equation (Chicone and Mashhoon, 2002) in differential geometry, which expresses the tidal forces, that is, the true manifestation of gravity, within the GTR framework, is proportional just to the Riemann tensor (Wald, 1984; Ohanian and Ruffini, 2013; Carroll, 2019).

Of course, GTR is not limited only to providing a different theoretical scheme to frame and reproduce the same phenomena described by the Newtonian one. The Einsteinian theory is much richer than Newton's Universal Gravitation, predicting a whole set of new phenomena. Indeed, GTR is able to treat motions occurring in gravitational fields so intense – in the sense that their gravitational potentials are close to the speed of light squared c^2 – that they accelerate bodies to speeds close to c itself and bend the path of electromagnetic waves in unparalleled ways undergoing also exceptionally rapid variations in time and from a point in space to another nearby one. The most spectacular – and expensive, as well as long-lasting – tests of GTR, recently performed by large international teams after several decades, undoubtedly come from such strong regimes. Suffice it to think about the gravitational waves (Cervantes-Cota et al., 2016) emitted in the end-of-life stages of

[10] With such a choice, each component of $\mathsf{T}_{\sigma\lambda}$ has the dimensions of energy density, that is, energy per volume, or, equivalently, pressure.

binary black holes (BHs) (LIGO Scientific Collaboration and Virgo Collaboration, 2016) and neutron stars (LIGO Scientific Collaboration and Virgo Collaboration, 2017), detected so far by the Laser Interferometer Gravitational-wave Observatory (LIGO) and Virgo facilities, or the shadows of the supermassive black holes (SMBHs) at the centre of the supergiant elliptical galaxy Messier 87 (M87) (Event Horizon Telescope Collaboration, 2019) and in Sgr A* at the Galactic Centre (GC) (Event Horizon Telescope Collaboration, 2022) imaged by the Event Horizon Telescope (EHT) collaboration (Doeleman et al., 2009). In such domains, Newton fails miserably.

The first approximation of GTR to the next order to the purely Newtonian one, in which new terms in the equations of motion appear, is named post-Newtonian (pN); see, for example, Damour (1987), Asada and Futamase (1997), Blanchet (2003), Blanchet (2006), Futamase and Itoh (2007), Will (2018), and references therein. It is a computational scheme for solving the GTR field equations relying upon the assumptions that the characteristic speeds of the bodies under consideration are smaller than c and that the gravitational fields inside and around them are weak. Nonetheless, as pointed out by Will (2011b), such a framework turned out to be notably effective in describing also certain strong field and fast motion systems such as compact binaries made of at least one dense neutron star and inspiralling pairs of BHs emitting gravitational waves; the reasons for that are largely unknown (Will, 2011b). Thus, putting the pN approximation to the test in as many different scenarios and at the highest order of approximation as possible is of paramount importance to gain ever increasing confidence in it and extrapolating the validity of its effects to their counterparts in stronger regimes. In principle, such pN tests have the benefit that, if, on the one hand, the expected signals of interest have smaller magnitude with respect to the corresponding ones in the strong field regime, on the other hand, the knowledge of the competing features of motion of classical origin is relatively better, and the impact of their mismodelling can be more accurately assessed with respect to less accessible astrophysical scenarios whose environments are, generally, less reliably known. Furthermore, the measurement techniques routinely used, or under development, for tracking solar system's artificial or natural bodies like, for example, LLR, Satellite Laser Ranging (SLR) (Coulot et al., 2011), and Planetary Laser Ranging (PLR) (Dirkx et al., 2019) are becoming more and more accurate, allowing, in principle, one to detect increasingly smaller features of motion. As if that weren't enough, the technological efforts needed to measure such tiny effects could be useful one day in other, unsuspected fields. Last but not least, a somewhat opportunistic approach may be more easily followed by exploiting existing or planned missions directed to different goals, with a remarkable gain of time and money. In its technical realm of validity, the pN approximation has been, or is currently being, tested only to the first post-Newtonian (1pN) order,[11] since its 2pN effects are deemed too small to be currently measurable. Moreover, the tests done or currently underway largely refer to the mass monopole and, to a much lesser extent, the spin dipole moments of the source, namely its mass M and angular momentum $\boldsymbol{J}$. In particular, the perihelion[12] precessions of Mercury (Shapiro et al., 1972; Shapiro, 1990), of other inner planets of the solar system (Anderson et al., 1978, 1993), and of the asteroid Icarus (Shapiro et al., 1968, 1971) were measured long ago. More recently, Earth's geodetic satellites[13] (Pearlman et al., 2019), tracked with the SLR technique, were used (Lucchesi and Peron, 2010, 2014). Finally, the perinigricon[14] shift of the S star S2 in the field of the SMBH in Sgr A* was recently measured as well (GRAVITY Collaboration et al., 2020). Furthermore, the periastron[15] advance of a two-body system of comparable masses M_{A} and M_{B} was measured with different binary radiopulsars (Weisberg and Taylor, 1984; Stairs, 2003; Champion et al., 2004; Weisberg and Taylor, 2005; Kramer et al., 2006). As far as the 1pN orbital[16] effects induced by the angular momentum $\boldsymbol{J}$ of the primary, known collectively as the Lense–Thirring (LT) effect (Lense and Thirring, 1918; Mashhoon et al., 1984), are concerned, tests have been underway with SLR geodetic satellites since 1996 (Ciufolini et al., 1996). Some aspects of them, like their realistic accuracy, are currently being debated; see, for example, Renzetti (2013b) and references therein. So far, the only uncontroversial test of another 1pN feature due to the Earth's angular momentum is the one performed with the Gravity Probe B (Everitt, 1974) (GP-B) mission which measured the Pugh–Schiff precessions (Pugh, 1959; Schiff, 1960) of four spaceborne gyroscopes to a $\simeq 19\%$ accuracy (Everitt et al., 2011, 2015), despite the fact that for many decades it was assumed that the final accuracy would be around 1% (Everitt, 1974; Everitt et al., 2001). Actually, to the 1pN level, other dynamical effects arise induced by mass and spin multipole moments of higher order (Soffel and Han, 2019).

In this book, extensive use is made of the Keplerian orbital elements (Brouwer and Clemence, 1961; Soffel, 1989; Brumberg, 1991; Klioner and Kopeikin, 1994; Bertotti et al., 2003; Roy, 2005; Kopeikin et al., 2011; Poisson and Will, 2014; Soffel and Han, 2019). They are the semimajor axis a, the eccentricity e, the inclination I, the longitude of the ascending node Ω, the argument of pericentre[17] ω, and

[11] It can be formulated to yield field equations for just two potentials (Soffel and Brumberg, 1991).

[12] From περί (+ accusative), meaning 'around, near, about, from', and Ἥλιος, -ου, ὁ, 'Hēlios', the god of the Sun.

[13] From *sătellĕs, ĭtis*, meaning 'attendant upon a distinguished person', 'lifeguard'. For a discussion of the word satellite, its origin and its use in astronomy, see Sparavigna (2016).

[14] From περί (+ accusative), meaning 'around, near, about, from', and *nĭger, gra, grum* ('black').

[15] From περί (+ accusative), meaning 'around, near, about, from', and ἄστρον, -ου, τό ('celestial body, star').

[16] From *orbis, is*, 'a ring, circle, re-entering way, circular path, hoop, orbit'.

[17] From περί (+ accusative), meaning 'around, near, about, from', and κέντρον, -ου, τό, meaning, among other things, 'stationary point of a pair of compasses', 'centre (of a circle)'.

the mean anomaly at epoch[18] η. Such a choice, which, by no means, should be deemed obligatory since other orbital parameterizations also exist (Bond and Janin, 1981; Gurfil, 2004; Efroimsky, 2005; Kopeikin et al., 2011; Gurfil and Efroimsky, 2022; Pogossian, 2022), is motivated by their immediately intuitive meaning which greatly helps in visualizing the effects described with them. Furthermore, they are easy to use in order to suitably design space-based experiments and preliminarily assessing the impact of other competing dynamical effects of classical origin.

However, nowadays, *actual* tests of dynamical features of motion are usually performed *differently*. Large datasets are reduced in the following way. Highly detailed mathematical models of (a) the dynamics of the moving bodies, including pN effects X_{pN} to a certain degree of completeness (b) the propagation of the electromagnetic waves between the Earth's stations and the (re)transmitting/reflecting artificial or natural bodies of interest (c) the measurement devices, all containing several key parameters $\mathfrak{p}$ characterizing the physical and orbital features of the system's components at hand (masses, initial positions and velocities, bias of transponders, etc.), are fitted to huge amounts of data. The latter consist of measurements of the directly observable quantities[19] $\mathfrak{O}$. In such grand fits (Nordtvedt, 2000), $\mathfrak{p}$ are estimated in a least-square way[20] along with their errors and reciprocal correlations, all stored in the covariance matrix. Finally, time series of post-fit residuals[21] are produced by subtracting the *measured* values of the observables $\mathfrak{O}$ from their *analytical* counterparts calculated with the previously estimated values of $\mathfrak{p}$. In order to *realistically* assess the accuracy of the parameter(s) of interest, different data sets and background reference models can be used, and the resulting values $\mathfrak{p}$ are confronted with each other. In principle, such post-fit residuals should account for, among other things, all the mismodelled – or even unmodelled – dynamics. Thus, if they are statistically compatible with *zero*, there is the temptation to straightforwardly compare them to their analytically predicted counterparts in order to infer *upper bounds* on X_{pN} if the latter is not included in the dynamical models fit to the observations. Furthermore, should the post-fit residuals be considered *different from zero* at a statistically significant level, one would be likely tempted to claim a *measurement* of the unmodelled effect X_{pN} of interest. This is a widely adopted practice in the literature. Actually, *great care* is needed

18 There is not a symbol commonly adopted for it in the literature. Suffice it to say that, for example, η is used by Milani et al. (1987), while in the notation by Brumberg (1991) the mean anomaly at epoch is l_0. Furthermore, Kopeikin et al. (2011) denote it as $\mathcal{M}_0$, while Bertotti et al. (2003) adopt ϵ'.

19 The Keplerian orbital elements do not belong to them, being computed from observations through some intermediate steps.

20 Recently, the Bayesian approach also has been gaining ground (Mariani et al., 2023).

21 It is possible to produce time-dependent 'residuals' of the Keplerian orbital elements (Lucchesi and Balmino, 2006; Lucchesi, 2007) only when the spacecraft motion proceeds steady and seamlessly, without interruptive orbital manoeuvres needed for, for example, pointing an antenna towards the Earth.

in proceeding as just outlined, especially when the expected size of the pN signal one is interested in is *not much larger* than the measurement errors[22] (Fienga and Minazzoli, 2024). Indeed, if $X_{\rm pN}$ is *not* modelled, its possible signature may be more or less *absorbed* in some of – or all – the parameters $\mathfrak{p}$ estimated in the fit, like, for example, the initial conditions. Thus, it would be partially or totally *removed* from the post-fit residuals. In this case, one would infer artificially *too tight* constraints on (some of the parameters of) $X_{\rm pN}$, when, instead, the real impact of the latter on the system's dynamics actually is *larger*. Furthermore, if the post-fit residuals produced *without modelling* $X_{\rm pN}$ are significantly *different from zero*, it may be that their resulting anomalous pattern is *not* due to $X_{\rm pN}$ at all, as one would hope, being, instead, caused by some fortunate *mutual partial cancellation* of completely different effects leaving a signature which, by chance, has just the characteristics of $X_{\rm pN}$ one is looking for. Then, the *correct* way to proceed consists of *explicitly* modelling the pN feature of motion $X_{\rm pN}$ one wants to test and *simultaneously estimating* the parameter(s) $\mathfrak{p}_{X_{\rm pN}}$ characterizing it[23] along with all the other ones. Then, one can compare the post-fit residuals produced in this way with, say, those generated without modelling $X_{\rm pN}$ at all to see if significant differences, larger than the measurement error level, can be spotted. Finally, the errors of $\mathfrak{p}_{X_{\rm pN}}$ along with their correlations with the other simultaneously estimated parameters in the covariance matrix obtained just by modelling $X_{\rm pN}$ are to be inspected. See Section K.3 for a discussion of a case in which this standard approach is, for some reasons, disregarded.

A clarification is in order when one talks about *tests* of pN gravity. Let $\mathfrak{B}$ be the theoretical prediction of a certain pN effect, namely an analytical formula usually containing, among other things, one or more parameter(s) characterizing the physical properties of the environment in which the former takes place; they could be, for example, the masses and some other relevant physical quantities (angular momenta, multipole moments) of, say, a two-body system. Let it be assumed that there is an agreement, within the experimental errors, between $\mathfrak{B}$ and a corresponding measured or observed quantity. Then, one can correctly speak of a *genuine test* of the effect under consideration only if the parameters entering $\mathfrak{B}$ are known *independently from that very same effect*; for example, they could have been previously determined by exploiting different, even non-dynamical, features. Conversely, if the theory at hand is widely accepted in the common knowledge at the time, $\mathfrak{B}$

22 The scope of data reductions is to finally produce post-fit residuals as small as the measurement errors.

23 A widely adopted set of parameters usually estimated in pN gravity tests are those belonging to the so-called parametrized post-Newtonian (PPN) formalism (Will, 2018), among which $\beta_{\rm PPN}$ and $\gamma_{\rm PPN}$, both equal to 1 in GTR, are those that attract the greatest interest. The PPN scheme can be applied to all metric gravitational theories, namely, those relying upon the EP. The speed of light c remains constant in it, and the metric tensor $g_{\sigma\lambda}$ is always assumed symmetric.

and the corresponding measured value can be used just to measure or constrain the parameter(s) entering the former.

The same considerations hold also for the plethora of long-range, or infrared, modified models of gravity (Brax et al., 2004; Nojiri and Odintsov, 2007; De Felice and Tsujikawa, 2010; Maartens and Koyama, 2010; Capozziello and de Laurentis, 2011; Skordis, 2011; Clifton et al., 2012; Ferraro, 2012; de Rham, 2014; Capozziello et al., 2015; Ruggiero and Radicella, 2015; Cai et al., 2016; Joyce et al., 2016; Maggiore, 2017; Mashhoon, 2017; Kobayashi, 2019; Roshan and Mashhoon, 2022) that have been continually churned out mainly since the accelerated cosmic expansion was discovered in 1998 (Riess et al., 1998; Perlmutter et al., 1999; Riess, 2000; Astier and Pain, 2012; Schmidt, 2012) and, more recently, since the issue of the Hubble tension gained prominence (Cervantes-Cota et al., 2023; Hu and Wang, 2023; Vagnozzi, 2023; Capozziello et al., 2024). Also the puzzle of nonbaryonic dark matter at galactic scales (Merrifield, 2005; Garrett and Duda, 2011; Bullock and Boylan-Kolchin, 2017; Wechsler and Tinker, 2018; de Martino et al., 2020) prompted the birth of several alternative theoretical frameworks among which the most prominent one is, perhaps, the MOdified Newtonian Dynamics (MOND) paradigm (Milgrom, 1983a,b,c; Sanders and McGaugh, 2002; Bekenstein, 2009; Famaey and McGaugh, 2012; Milgrom, 2014; Bugg, 2015; McGaugh, 2015; Banik and Zhao, 2022). For epistemological discussions about the MOND/dark matter debate, see Duerr and Wolf (2023). Another model put forth to cope with, among other things, the dark matter issue is the Scalar Tensor Vector Gravity (STVG), or MOdified Gravity (MOG) (Brownstein and Moffat, 2006a,b; Moffat, 2006; Moffat and Toth, 2009; Harikumar, 2022). For a comparison between MOND and MOG and other less known theories trying the explain the same phenomenology, see Pascoli (2024), and references therein. Recently, also the Modified General Relativity (MGR) paradigm popped up (Nash, 2019; Das and Sur, 2022; Nash, 2023). A further theoretical scenario arising in the framework of the long-lasting attempts to find a consistent quantum theory of gravity is the effective field theory called[24] Standard Model Extension (SME) (Kostelecký, 2004; Kostelecký and Potting, 2005, 2009). Among other things, it encompasses local Lorentz violations in the gravity sector which may manifest themselves to a pN level with several phenomena including also orbital effects (Bailey and Kostelecký, 2006). For a recent review of modern tests of Lorentz invariance, see, for example, Mattingly (2005), and references therein. Another theoretical scheme encompassing violations of the Lorentz symmetry is the Einstein–Æther theory, a generally covariant theory of gravity coupled to a dynamical, unit timelike vector field that breaks the aforementioned symmetry (Jacobson and Mattingly, 2004;

[24] Here, the reference is to the Standard Model of elementary particles and fields (Gouttenoire, 2023).

Eling et al., 2006; Jacobson, 2008). Reliably testing such proposed modifications of the currently known laws of gravitation in local systems with, for example, orbital motions is of paramount importance in order to gain knowledge on them *independently* of the very same effects for which they were introduced which, otherwise, would remain their sole, ad hoc justification.

This book, in the wake of the meritoriously celebrated texts by[25] Soffel (1989), Brumberg (1991), and Soffel and Han (2019), treats the effect of pN and alternative gravity on different quantities (Keplerian orbital elements, astrometric angles RA and decl., radial velocity of spectroscopic binaries, variation of the times of arrival in binary pulsars, characteristic timescales and sky-projected spin-orbit angles in transiting exoplanets,[26] two-body range and range rate) within a unified and uniform calculational scheme for arbitrary orbital geometries and generic orientations of the spin axes of the sources of the gravitational field in space. It mainly adopts the language of celestial mechanics, being aimed at the widest possible audience of readers typically working on celestial mechanics, astronomy, and astrodynamics in astronomical observatories, laser-ranging stations, and data centres. Spatially isotropic or harmonic coordinates[27] are adopted (Soffel and Brumberg, 1991). Furthermore, the coordinate time t is used to calculate temporal intervals; they coincide with those obtained by an observer comoving with the orbiting particle in terms of its proper time[28] τ up to corrections of the order of $\mathcal{O}\left(1/c^4\right)$.

The book is organized as follows.

The general scheme needed to calculate the desired post-Keplerian[29] (pK) orbital effects is outlined in Chapter 2. In it, after an overview of the Keplerian picture for a restricted two-body system in Section 2.1, the pK variations of the osculating Keplerian orbital elements are treated in Section 2.2; the first-order shifts in the perturbing acceleration are worked out in Section 2.2.1, while the second-order ones are dealt with in Section 2.2.2. The mixed effects arising when two pK accelerations enter simultaneously the equations of motion are the subject of Section 2.2.3. The methods for calculating the pK corrections to various characteristic orbital

25 To a different level, see also O'Leary (2021).

26 From ἐκ- (ἐξ- before a vowel), meaning, among other things, 'out of, forth from; outside of, beyond', and the adjectival form ἔξω ('outer, external', or 'foreign'). With reference to our solar system, an exoplanet is, then, a planet outside of it.

27 As explained by Brumberg (2010), in order to effectively cope with the problem of the coordinate-dependent quantities in relativistic celestial mechanics and astrometry, in 1991 the International Astronomical Union (IAU) recommended to adopt one specific type of coordinates once and forever: the harmonic coordinates, determined by four specific non-tensorial differential relations to be added to the tensorial field equations of GTR (Fok, 1959; Weinberg, 1972; Brumberg and Kopeikin, 1989b; Damour et al., 1991).

28 The coordinate and the proper times coincide, up to corrections of higher order in $1/c$, when the orbiter is quite distant from the source of the gravitational field.

29 Here, by post-Keplerian (pK) I mean dynamical features arising from *any* acceleration, Newtonian or not, different from the simple Newtonian inverse-square one. Then, in *this* sense of the term pK, the classical acceleration due to, say, the primary's oblateness is pK.

temporal intervals are presented in Section 2.3: they are the anomalistic (Section 2.3.1), draconitic (Section 2.3.2) and sidereal (Section 2.3.3) periods, which all coincide with each other in the Keplerian case. Section 2.4 illustrates how to calculate the pK shifts of a generic observable quantity for which an analytical model can be given; the cases treated are (a) the radial velocity of a spectroscopic binary in Section 2.4.1, (b) some characteristic timescales in transiting exoplanets in Section 2.4.2, (c) the rate of change of the sky-projected spin-orbit angle for such kinds of exoplanets, dealt with in Section 2.4.3, (d) the variation of the times of arrival (TOAs) of binary pulsars in Section 2.4.4, and (e) the astrometric angles RA and dec. in Section 2.4.5. Finally, the pK shifts of the two-body range and range-rate are calculated in Section 2.5.

Chapter 3 is devoted to the calculation of various 1pN gravitoelectric features of motion for a test particle (Section 3.1) and a binary system of bodies with comparable masses (Section 3.2): the Keplerian orbital elements in Section 3.1.1 (test particle) and Section 3.2.1 (binary system), the anomalistic (Section 3.1.2 for a test particle and Section 3.2.2 for a binary system), draconitic (Section 3.1.3 for a test particle and Section 3.2.3 for a binary system), and sidereal (Section 3.1.4 for a test particle and Section 3.2.4 for a binary system) orbital periods, RA and dec. (Section 3.1.5), the two-body range and range rate (Section 3.1.6), the radial velocity (Section 3.2.5), the characteristic timescales of transiting exoplanets (Section 3.2.6), and the TOAs of binary pulsars (Section 3.2.7).

The 2pN gravitoelectric orbital precessions of a binary system are calculated in Chapter 4.

The 1pN LT acceleration, sourced by the source's spin dipole moment(s) and dubbed also as 'gravitomagnetic', is treated in Chapter 5 along with several features of motion induced by it: the Keplerian orbital elements (Section 5.1), the anomalistic (Section 5.2), draconitic (Section 5.3), and sidereal (Section 5.4) orbital periods, the gravitomagnetic clock effect (Section 5.5), the radial velocity (Section 5.6), the characteristic timescales of transiting exoplanets (Section 5.7), the sky-projected spin-orbit angle (Section 5.8), the TOAs of binary pulsars (Section 5.9), RA and dec. (Section 5.10), and the two-body range and range rate (Section 5.11).

Other 1pN gravitomagnetic orbital precessions, due to the spin octupole moment of the central body, are dealt with in Chapter 6.

Several Newtonian features of motion due to the quadrupole mass moment(s) of the source are the subject of Chapter 7: the Keplerian orbital elements (Section 7.1), the anomalistic (Section 7.2), draconitic (Section 7.3), and sidereal (Section 7.4) orbital periods, the radial velocity (Section 7.5), the characteristic timescales of transiting exoplanets (Section 7.6), the sky-projected spin-orbit angle (Section 7.7), the TOAs of binary pulsars (Section 7.8), RA and dec. (Section 7.9), and the two-body range and range rate (Section 7.10).

The 1pN orbital precessions of the order of $\mathcal{O}\left(J_2/c^2\right)$ are calculated for a test particle in Chapter 8.

Newtonian and pN tidal orbital precessions of a test particle orbiting a primary induced by a distant third body are calculated in Chapter 9. In particular, in Section 9.1, the impact of the pN precessions of the axes of the reference frame comoving with the two-body system in geodesic motion in the spacetime of the third body is omitted, being, instead, treated in Section 9.2.

The orbital precessions induced by some categories of popular modified models of gravity are treated in Chapter 10: they are due to power-law (Section 10.1), Yukawa-like (Section 10.2), $1/r$ (Section 10.3), empirical once-per-revolution (Section 10.4), constant (Section 10.5), and tidal-like (Section 10.6) extra-accelerations. The effects of some dark matter distributions are the subject of Section 10.7. Models encompassing violations of the Lorentz symmetry in the gravitational sector are treated as well (Section 10.8).

Appendix A collects a list of acronyms and abbreviations.

Notations and definitions are listed in Appendix B.

In Appendix C, it is shown how to calculate pK Lagrangians, to be used as disturbing functions in the Lagrange equations for the variations of the Keplerian orbital elements, from the spacetime metric tensor.

Appendix D presents some useful coefficients accounting for the various spin-orbit configurations.

Appendix E contains the coefficients entering the LT instantaneous shifts of the orbital elements.

The coefficients of the instantaneous orbital shifts due to the Newtonian J_2 acceleration are listed in Appendix F.

Appendix G collects the coefficients of the total net mixed orbital shifts of the order of $\mathcal{O}\left(J_2/c^2\right)$.

Appendix H displays the explicit expressions of the coefficients of the orbital precessions of tidal origin.

The coefficients of the averaged disturbing functions of the power-law and exponential dark matter density profiles along with those of the resulting orbital precessions can be found in Appendix I.

In Appendix J, numerical values for the relevant physical parameters of some major bodies of the solar system (the Sun, the Earth, and Jupiter) are provided along with those of the double pulsar.

Appendix K contains the numerical values of the several pK orbital effects calculated for various natural and artificial bodies in the solar system and outside it: the Sun's planets (Section K.1), the spacecraft Juno[30] around Jupiter (Section K.2), the

[30] From *Iūnō, ōnis*, Roman deity, wife of Jupiter.

Earth's Laser GEOdynamics Satellite (LAGEOS) (Section K.3), the double pulsar PSR J0737–3039 (Section K.4), the triple pulsars (Section K.5), and the star S4716 in the GC (Section K.6).

Appendix L reviews some space-based missions aimed at testing several pN orbital effects recently proposed by the author: Highly Elliptical Relativity Orbiter (HERO) (Section L.1), In-Orbit Relativity Iuppiter Observatory, or IOvis Relativity In-Orbit Observatory (IORIO) (Section L.2), Elliptical Uranian Relativity Orbiter (EURO) (Section L.3), LEnse–Thirring Sun–Geo Orbiter (LETSGO) (Section L.4), and ELXIS (Section L.5). Further missions proposed by other authors are presented in Section L.6.

2

General Calculational Scheme

2.1 The Keplerian Picture

Consider a gravitationally bound restricted two-body system $\mathcal{S}$ in which a test particle, that is, an uncharged and nonspinning object of negligible mass, moves in the gravitational field generated by an isolated, massive body. It means that $\mathcal{S}$ is, actually, freely falling in some external gravitational field whose action reduces just to residual tidal effects, assumed negligible[1] throughout the extension of $\mathcal{S}$; thus, the latter defines, to a sufficiently high level of approximation, a local inertial reference frame $\mathcal{K}$ with respect to which the course of the particle is studied. It is assumed that $\mathcal{K}$ is both kinematically and dynamically nonrotating (Brumberg and Kopeikin, 1989a), that is, no Coriolis and centrifugal inertial forces appear (dynamically nonrotating), and the general relativistic de Sitter–Fokker (de Sitter, 1916b; Schouten, 1918; Fokker, 1921) and Pugh–Schiff (Pugh, 1959; Schiff, 1960) precessions, which would naturally alter the orientation of the reference axes simply because $\mathcal{K}$ moves being parallel transported (Fermi, 1922; Levi-Civita, 1926; Synge, 1927) along the geodesic worldline of M in a deformed external spacetime, are corrected for (kinematically nonrotating).[2]

Let the orbital motion of the satellite be affected by a pK acceleration $\boldsymbol{A}$ whose magnitude is small with respect to the Newtonian monopole term given by

$$\boldsymbol{A}_{\mathrm{N}} = -\frac{\mu}{r^2}\,\hat{\boldsymbol{r}}. \tag{2.1}$$

In Equation (2.1), $\mu := GM$ is the standard gravitational parameter of the primary, defined as the product of its mass M by the Newtonian constant of gravitation G, r

1 When such an approximation is untenable and the external tidal field is due to a distant body, the orbital dynamics within $\mathcal{S}$ is affected in various pK ways described in Chapter 9.

2 Alternatively, it can be assumed that the motion is studied over timescales much shorter than those characterizing the aforementioned general relativistic precessions. When the tidal effects within $\mathcal{S}$ can no longer be deemed as negligible, specific orbital effects arise if the external field is due to a massive body; they are studied in Section 9.1. If, instead, $\mathcal{K}$ is kinematically rotating, further orbital perturbations, solely of pN origin, have an impact on the dynamics within $\mathcal{S}$; see Section 9.2.

is the instantaneous distance between the orbiter and the source, and $\hat{\boldsymbol{r}} := \boldsymbol{r}/r$ is the unit vector of the position vector $\boldsymbol{r}$ of the particle with respect to the body acting as centre of force. In what follows, $\boldsymbol{A}$ can be either Newtonian or pN. For example, if the central body experiences departures from spherical symmetry, the first even zonal harmonic coefficient J_2 of the multipolar expansion of the Newtonian part of its gravitational potential, accounting for its oblateness,[3] induces a pK acceleration $\boldsymbol{A}^{J_2}$ of classical origin; its orbital effects are treated in Chapter 7. On the other hand, to the 1pN level, pK accelerations of the order of $\mathcal{O}\left(1/c^2\right)$ arise originating from the source's mass and spin multipole moments. The 1pN gravitoelectric orbital effects induced by the mass monopole and quadrupole moments are treated in Chapter 3 and Chapter 8, respectively, while the 1pN gravitomagnetic ones due to the spin dipole and octupole moments are worked out in Chapter 5 and Chapter 6, respectively. In general, $\boldsymbol{A}$ can be due to some modified model of gravity as well; the orbital effects due to some alternative theories of gravity are the subject of Chapter 10. Furthermore, it does not necessarily have to be of gravitational origin, as it could well also be caused by nonconservative forces. Be that as it may, it is assumed that $\boldsymbol{A}$ can be explicitly modelled, that is, an analytical expression of it can be given, as in the aforementioned examples.

If $\boldsymbol{A}$ were absent, the test particle would move along a Keplerian ellipse[4] of constant shape and size, and whose orientation in space is fixed. It is parameterized in terms of the Keplerian orbital elements.

The size of the ellipse is fixed by a; it determines also the mean motion $n_{\mathrm{K}} := \sqrt{\mu/a^3}$ related to the orbital period $T_{\mathrm{K}} := 2\pi/n_{\mathrm{K}}$ needed for the test particle to complete a full orbital revolution with respect to *any* fixed direction in space.

The shape of the orbit is determined by e in such a way that $0 \leq e < 1$. The value $e = 0$ corresponds to a circle, while $e \lesssim 1$ gives a highly eccentric ellipse.

The semilatus rectum is defined as $p := a\left(1 - e^2\right)$.

The inclination of the orbit to the fundamental plane $\{x, y\}$ of the reference frame adopted is given by I, with $0 \leq I < 180^\circ$. The motion is called prograde if $0 \leq I < 90^\circ$, while it is dubbed retrograde if $90^\circ < I < 180^\circ$. About the reference plane, in the case of our solar system, it often coincides with the Earth's equatorial plane at some reference epoch, and the reference x axis points towards the Vernal Equinox ♈ at the same epoch (Ma et al., 1998; Charlot et al., 2020); the ecliptic plane at some reference epoch may be used as well (Capitaine and Soffel, 2015).[5] In studying astronomical binary systems like, for example, exoplanets, the fundamental plane is assumed to be coincident with the plane of the

3 From *oblātus, a, um*, made of *ob-* ('in front of', 'before'), and *lātus, a, um* ('broad', 'wide').

4 From ἔλλειψις, -εως, ἡ, meaning 'a falling short', 'defect', and ἐλλείπω, meaning, among other things, 'I fall short of', 'I am in want of', 'I lack'. Such a conic section was first named so by Apollonius of Perga because its cutting plane makes a smaller angle with the base of the cone than that made by the side of the latter.

5 Another possible choice may be the solar system's invariable plane (Souami and Souchay, 2012).

sky, and the reference z axis is directed along the line of sight towards or away from the observer depending on the conventions followed by the authors at hand (Kaplan, 2015).

The angle Ω, between 0 and 360°, is counted in the fundamental plane from the reference x axis to the line of nodes, that is, the intersection between the orbital and the $\{x, y\}$ planes, towards the ascending node; the latter, marked as ☊, is the point on the line of nodes where the test particle crosses the fundamental plane from below. The descending node is denoted as ☋.

The angle ω, between 0 and 360° in the orbital plane, is subtended by the arc of trajectory from the ascending node to the pericentre.

The longitude of pericentre $\varpi := \Omega + \omega$, which, in general, is a broken angle, provides the pericentre position with respect to the reference x axis when the orbital plane lies in the fundamental one.

The three angles I, Ω, ω, corresponding to the Euler angles characterizing the configuration of a rigid body with a fixed point, determine the orientation of the orbital plane in space.

The angle η is the value of the mean anomaly[6] at a reference instant t_0 to which the orbital elements are referred; t_0 does not necessarily coincide with the time of passage at pericentre t_{p}; for the relation connecting η and t_{p}, see Equation (2.3).

The instantaneous position of the test particle along its orbit is reckoned by the true anomaly $f\,(t)$, counted from the position of the pericentre on the line of apsides.[7]

If the instantaneous location of the orbiter is reckoned relative to the line of nodes or to the reference x axis, the argument of latitude $u\,(t) := \omega + f\,(t)$ and the true longitude $l\,(t) := \varpi + f\,(t)$ are used, respectively. It should be noted that, actually, $l\,(t)$ is a broken or dogleg angle since Ω and $u\,(t)$ lie in different planes; it has a clear meaning only when the orbital plane coincides with the fundamental one, that is, for $I = 0$, in which case it is just the real longitude of the test particle as measured with respect to the x axis.

Another time-dependent fast variable used to locate the position of the orbiter along its path is the mean anomaly $\mathcal{M}\,(t)$ defined as

$$\mathcal{M}\,(t) := n_{\mathrm{K}}\left(t - t_{\mathrm{p}}\right) = \eta + n_{\mathrm{K}}\,(t - t_0)\,. \tag{2.2}$$

From Equation (2.2), it turns out that the mean anomaly at epoch is proportional to the time of passage at pericentre as per

6 From ἀνωμαλία, -ας, ἡ, made of the privative prefix ἀ- and ὁμαλός, ή, όν ('average', 'regular'). In astronomical contexts, the word 'anomaly' was used since ancient times to indicate irregularities in motions of celestial objects. Geminus Astronomicus (*Gem.* 1.20; cf. *Ptol. Alm.* 3.3), deals with ἀνωμαλία τῆς κινήσεως ('irregularity of motion'). Plutarch, in his *Lives* (*Plut. AEm.* 17), mentions ἀνωμαλίαι ἐκλειπτικαί ('irregularities of the Moon's orbit').

7 From ἁψίς, -ῖδος, ἡ, meaning, among other things, 'arch', 'vault'.

$$\eta = n_{\rm K}\left(t_0 - t_{\rm p}\right). \tag{2.3}$$

As $f(t)$, also $\mathcal{M}(t)$ is measured from the pericentre in such a way that it is equal to 0 when the test particle is at the point of closest approach, 180° at the apocentre[8] and 360° after a full orbital revolution. Contrary to f, the mean anomaly does not refer to any physical objects, except at pericentre or apocentre, or for a circular orbit, being just a convenient uniform measure of how far around its orbit the particle has progressed since pericentre. Basically, it tracks the position of a fictitious particle describing a full revolution along a putative circular orbit in the same temporal interval $T_{\rm K}$ as the real particle on its true elliptical path. The explicit relation between $f(t)$ and the time t is given through the mean anomaly by Brouwer and Clemence (1961, p. 77, equation (74)):

$$f(t) = \mathcal{M}(t) + 2\sum_{s=1}^{\infty}\frac{1}{s}\left\{\mathcal{J}_s(se) + \sum_{j=1}^{\infty}\beta^j\left[\mathcal{J}_{s-j}(se) + \mathcal{J}_{s+j}(se)\right]\right\}\sin s\mathcal{M}(t), \tag{2.4}$$

where $\beta := \left(1 - \sqrt{1-e^2}\right)/e$, and $\mathcal{J}_s(se)$ is the Bessel function of the first kind of order s. From a practical point of view, $s \leq s_{\max}, j \leq j_{\max}$ where $s_{\max}, j_{\max}$ are set by the desired accuracy level.

A further fast angular variable which can be used to track the position of the test particle along its orbit is the eccentric anomaly $E(t)$, defined as

$$\mathcal{M}(t) := E(t) - e\sin E(t). \tag{2.5}$$

Also $E(t)$ does not refer to any physical moving objects, being just another convenient tool for tracking the progress of the test particle as it advances along its orbit. Its connection with t is given by Brouwer and Clemence (1961, p. 76, equation (70)) and Murray and Dermott (1999, p. 39, equation (2.80)):

$$E(t) = \mathcal{M}(t) + 2\sum_{s=1}^{\infty}\frac{1}{s}\mathcal{J}_s(se)\sin s\mathcal{M}(t). \tag{2.6}$$

The time-dependent position and velocity vectors $\boldsymbol{r}$ and $\boldsymbol{v}$ can be conveniently expressed in terms of the Keplerian orbital elements as

$$\boldsymbol{r}(t) = r(t)\left[\hat{\boldsymbol{l}}\cos u(t) + \hat{\boldsymbol{m}}\sin u(t)\right], \tag{2.7}$$

$$\boldsymbol{v}(t) = \sqrt{\frac{\mu}{p}}\left\{-\hat{\boldsymbol{l}}\left[e\sin\omega + \sin u(t)\right] + \hat{\boldsymbol{m}}\left[e\cos\omega + \cos u(t)\right]\right\}. \tag{2.8}$$

In Equations (2.7)–(2.8),

[8] From ἀπό (+ genitive), meaning 'away from', and κέντρον, -ου, τό, meaning, among other things, 'stationary point of a pair of compasses', 'centre (of a circle)'.

$$\hat{\boldsymbol{l}} = \begin{cases} \cos\Omega \\ \sin\Omega \\ 0 \end{cases} \tag{2.9}$$

is a unit vector directed along the line of nodes towards the ascending node, while

$$\hat{\boldsymbol{m}} = \begin{cases} -\cos I \sin\Omega \\ \cos I \cos\Omega \\ \sin I \end{cases} \tag{2.10}$$

is another unit vector lying in the orbital plane and directed perpendicularly to $\hat{\boldsymbol{l}}$ such that their cross product $\hat{\boldsymbol{l}} \times \hat{\boldsymbol{m}}$ is parallel to the unit vector $\hat{\boldsymbol{h}}$ of the orbital angular momentum, defined in Equation (2.25). Furthermore, the distance entering Equation (2.7) is

$$r(t) = \frac{a\left(1 - e^2\right)}{1 + e\cos f(t)}. \tag{2.11}$$

Figure 2.1 shows the orbital configuration of a test particle going along a Keplerian ellipse arbitrarily oriented in space around a primary with mass M and angular momentum $\boldsymbol{J}$ directed along a generic direction.

2.2 The pK Variations of the Keplerian Orbital Elements

A pK disturbing acceleration $\boldsymbol{A}$ causes the trajectory to change over time; since the former is small with respect to Equation (2.1), its action can be treated with the standard perturbative methods of celestial mechanics.

A very popular way of looking at how things go on is assuming that, at every instant, the actual trajectory followed by the moving particle can be approximated by an osculating ellipse,[9] whose Keplerian orbital elements are slightly different with respect to those parameterizing the ellipse which 'kissed' the trajectory in the previous instant, and so on. Thus, it can be admitted that, under the action of $\boldsymbol{A}$, the Keplerian orbital elements do vary slightly over time, producing a series of osculating ellipses, which, from moment to moment, approximate the real path traveled by the satellite.

In the following, pK perturbed trajectories due to secular rates of change of all the Keplerian orbital elements taken individually one at a time are shown along with their osculating Keplerian ellipses approximating them at the initial instant of time t_0.

[9] From *ōsculor ātus, ārī* ('to kiss'). Moreover, *ōscŭlum, ī* ('little mouth', 'pretty mouth', 'sweet mouth') comes from *ōs, ōris* ('mouth') and the diminutive suffix *-culum, ī*.

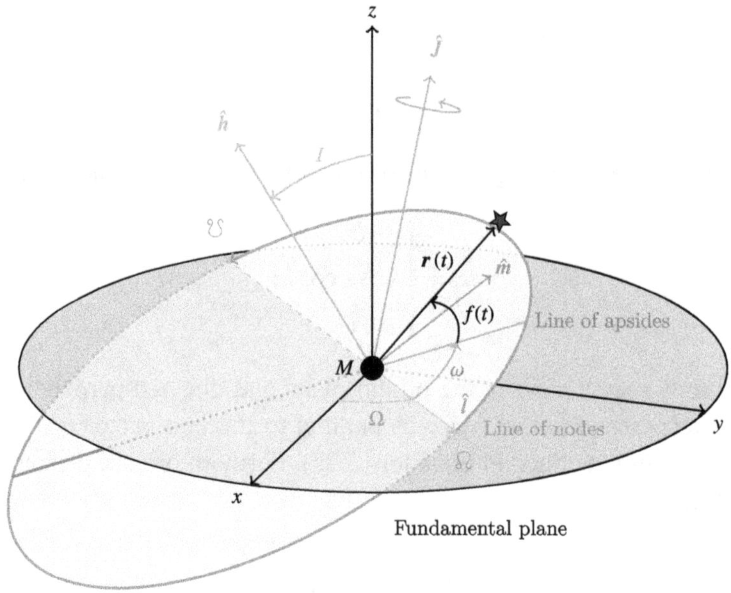

Figure 2.1 Keplerian ellipse followed by a test particle moving around a spinning primary of mass M and angular momentum $\boldsymbol{J}$ arbitrarily oriented in space. As an example, they are represented by an S star of spectral class B (Eckart and Genzel, 1996; Ali et al., 2020) and an SMBH like the one in Sgr A* at the GC (Eckart et al., 2002; Ghez et al., 2008; Genzel et al., 2010), respectively. The generic values $e = 0.4$, $I = 30°$, $\Omega = 45°$, $\omega = 50°$ were adopted for the orbital configuration, while the satellite is at $f = 60°$. In this particular example, the orientation of $\boldsymbol{J}$, parameterized as $\hat{J}_x = \cos\alpha_J \cos\delta_J$, $\hat{J}_y = \sin\alpha_J \cos\delta_J$, $\hat{J}_z = \sin\delta_J$, is set by $\alpha_J = 150°$, $\delta_J = 75°$. The unit vectors, which should all be the same length, are not drawn to scale purely for display reasons.

The impact of a decrease of the semimajor axis a is the subject of Figure 2.2. Figure 2.3 shows the distortion induced by a (negative) rate of change of the eccentricity e. The effect of a (positive) secular rate of change of the inclination I is illustrated in Figure 2.4 for a circular, polar orbit.

Figure 2.5 is devoted to the impact of a (positive) secular precession of the node Ω on a circular, polar orbit as well.

Figure 2.6 depicts a perturbed path obtained by assuming that only the pericentre experiences a (retrograde) secular precession.

A secular rate of the mean anomaly at epoch η does alter neither the shape nor the orientation of the orbit, merely accelerating or decelerating the motion of the test particle along its Keplerian ellipse. Thus, it is not worth dedicating a specific figure to this effect. Figure 2.7 shows a pK trajectory overall influenced by secular perturbations affecting all the Keplerian orbital elements.

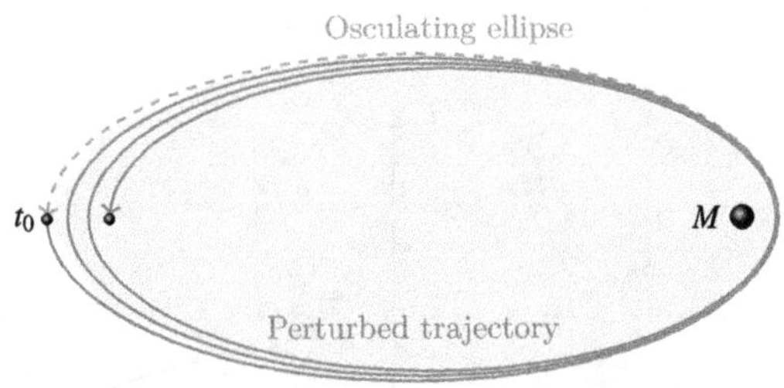

Figure 2.2 Perturbed trajectory (continuous curve) and the osculating Keplerian ellipse (dashed curve) at the initial instant of time t_0 characterized by $e = 0.9$, $I = 0$, $\Omega = 0$, $\omega = 90^\circ$, $\eta = 180^\circ$. In this example, it is assumed that a undergoes a (negative) secular rate of change large enough for better visualizing its effect. The motion takes place within the orbital plane during a time span equal to three Keplerian orbital periods $T_{\rm K}$.

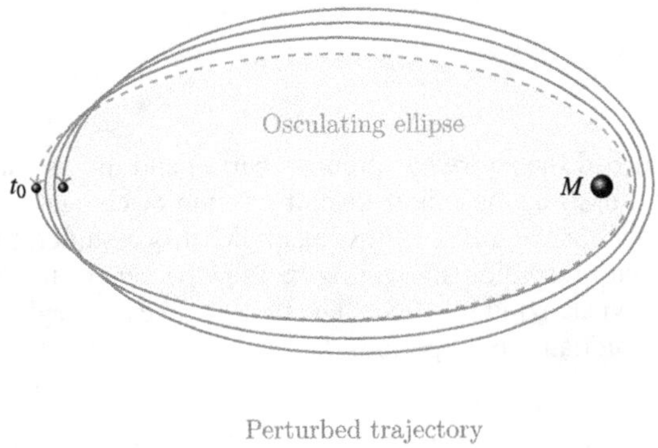

Figure 2.3 Perturbed trajectory (continuous curve) and the osculating Keplerian ellipse (dashed curve) at the initial instant of time t_0 characterized by $e = 0.9$, $I = 0$, $\Omega = 0$, $\omega = 90^\circ$, $\eta = 180^\circ$. In this example, it is assumed that e undergoes a (negative) secular rate of change large enough for better visualizing its effect. The motion takes place within the orbital plane during a time span equal to three Keplerian orbital periods $T_{\rm K}$.

Also other parameterizations of the orbital motion can be used, employing, for example, nonosculating elements (Bond and Janin, 1981; Gurfil, 2004; Efroimsky, 2005; Kopeikin et al., 2011; Gurfil and Efroimsky, 2022; Pogossian, 2022); they will not be treated in the present book.

2.2.1 The First-Order Effects

If κ is any of the six Keplerian orbital elements, its finite variation $\Delta\kappa\,(f)$ occurring in the time interval corresponding to a change of the true anomaly from its initial value f_0 to a generic one $f_0 < f < f_0 + 2\pi$ can be calculated, to the first order in A, as

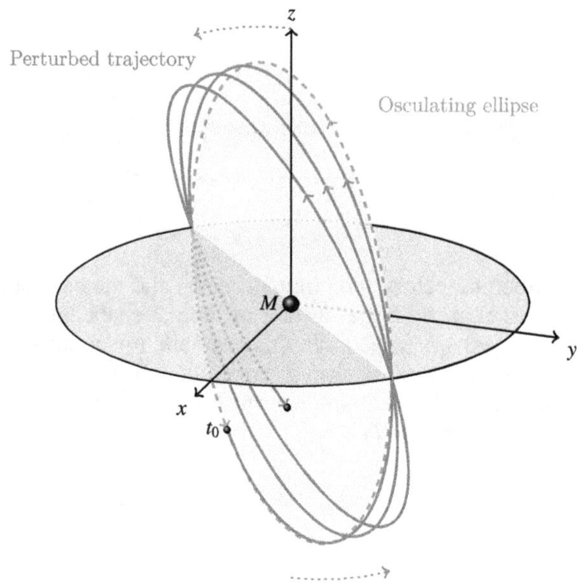

Figure 2.4 Perturbed trajectory (continuous curve) and the osculating Keplerian ellipse (dashed curve) at the initial instant of time t_0 characterized by $e = 0$, $I = 90°$, $\Omega = 45°$, $u_0 = 230°$. In this example, it is assumed that I undergoes a (positive) secular precession amounting to a few percent of the Keplerian mean motion for better visualizing its effect. The motion takes place during a time span equal to three Keplerian orbital periods T_K.

$$\Delta\kappa\,(f) = \int_{f_0}^{f} \frac{d\kappa}{df'}df' = \int_{f_0}^{f} \frac{d\kappa}{dt}\frac{dt}{df'}df', \tag{2.12}$$

where $d\kappa/dt$ are the right-hand sides of the equations for the variations of the Keplerian osculating elements in the Euler–Gauss form (Brouwer and Clemence, 1961; Soffel, 1989; Brumberg, 1991; Bertotti et al., 2003; Roy, 2005; Kopeikin et al., 2011; Poisson and Will, 2014; Soffel and Han, 2019)

$$\frac{da}{dt} = \frac{2}{n_K\sqrt{1-e^2}}\left[eA_r\sin f + \left(\frac{p}{r}\right)A_\tau\right], \tag{2.13}$$

$$\frac{de}{dt} = \frac{\sqrt{1-e^2}}{n_K a}\left\{A_r\sin f + A_\tau\left[\cos f + \frac{1}{e}\left(1-\frac{r}{a}\right)\right]\right\}, \tag{2.14}$$

$$\frac{dI}{dt} = \frac{1}{n_K a\sqrt{1-e^2}}A_h\left(\frac{r}{a}\right)\cos u, \tag{2.15}$$

$$\frac{d\Omega}{dt} = \frac{1}{n_K a\sin I\sqrt{1-e^2}}A_h\left(\frac{r}{a}\right)\sin u, \tag{2.16}$$

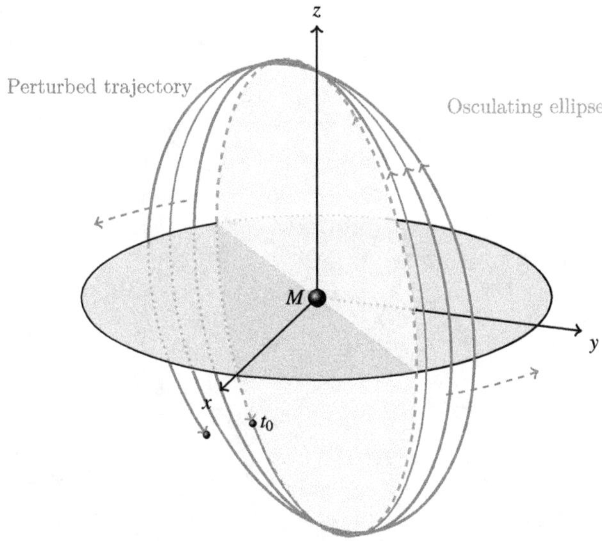

Figure 2.5 Perturbed trajectory (continuous curve) and the osculating Keplerian ellipse (dashed curve) at the initial instant of time t_0 characterized by $e = 0$, $I = 90°$, $\Omega = 45°$, $u_0 = 230°$. In this example, it is assumed that Ω undergoes a (positive) secular precession amounting to a few percent of the Keplerian mean motion for better visualizing its effect. The motion takes place during a time span equal to three Keplerian orbital periods T_{K}.

$$\frac{d\omega}{dt} = \frac{\sqrt{1-e^2}}{n_{\mathrm{K}} a e}\left[-A_r \cos f + A_\tau \left(1 + \frac{r}{p}\right)\sin f\right] - \cos I \frac{d\Omega}{dt}, \tag{2.17}$$

$$\frac{d\eta}{dt} = -\frac{2}{n_{\mathrm{K}} a} A_r \left(\frac{r}{a}\right) - \frac{(1-e^2)}{n_{\mathrm{K}} a e}\left[-A_r \cos f + A_\tau \left(1 + \frac{r}{p}\right)\sin f\right], \tag{2.18}$$

evaluated onto the Keplerian ellipse given by Equation (2.11), and

$$\frac{dt}{df} = \frac{r^2}{\sqrt{\mu p}}. \tag{2.19}$$

In Equations (2.13)–(2.18),

$$A_r := \boldsymbol{A} \cdot \hat{\boldsymbol{r}}, \tag{2.20}$$

$$A_\tau := \boldsymbol{A} \cdot \hat{\boldsymbol{\tau}}, \tag{2.21}$$

$$A_h := \boldsymbol{A} \cdot \hat{\boldsymbol{h}} \tag{2.22}$$

are the projections of $\boldsymbol{A}$ onto the radial, transverse and out-of-plane (or normal)[10] unit vectors $\hat{\boldsymbol{r}}$, $\hat{\boldsymbol{\tau}}$, $\hat{\boldsymbol{h}}$ of a moving local orbital frame attached to the test particle.

[10] From *norma, -ae* ('a square, employed by carpenters, masons, etc., for making right angles').

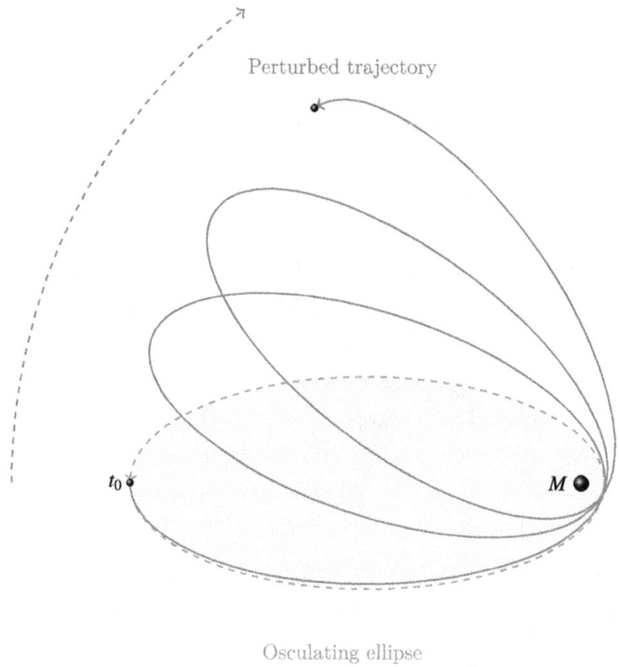

Figure 2.6 Perturbed pK trajectory (continuous curve) and the osculating Keplerian ellipse (dashed curve) at the initial instant of time t_0 characterized by $e = 0.9, I = 0, \Omega = 0, \omega = 90^\circ, \eta = 180^\circ$. In this example, it is assumed that only ω undergoes a (negative) secular precession amounting to a few percent of the Keplerian mean motion for better visualizing its effect. The motion takes place within the orbital plane during a time span equal to three Keplerian orbital periods T_K.

The radial unit vector, lying in the orbital plane and directed from the primary to the test particle, is

$$\hat{\boldsymbol{r}} = \begin{cases} \cos\Omega\cos u - \cos I\sin\Omega\sin u \\ \sin\Omega\cos u + \cos I\cos\Omega\sin u \ . \\ \sin I\sin u \end{cases} \tag{2.23}$$

The transverse unit vector, lying in the orbital plane perpendicularly to $\hat{\boldsymbol{r}}$, is

$$\hat{\boldsymbol{\tau}} = \begin{cases} -\cos\Omega\sin u - \cos I\sin\Omega\cos u \\ -\sin\Omega\sin u + \cos I\cos\Omega\cos u \ . \\ \sin I\ \cos u \end{cases} \tag{2.24}$$

The normal unit vector, directed perpendicularly to the orbital plane along the orbital angular momentum, is

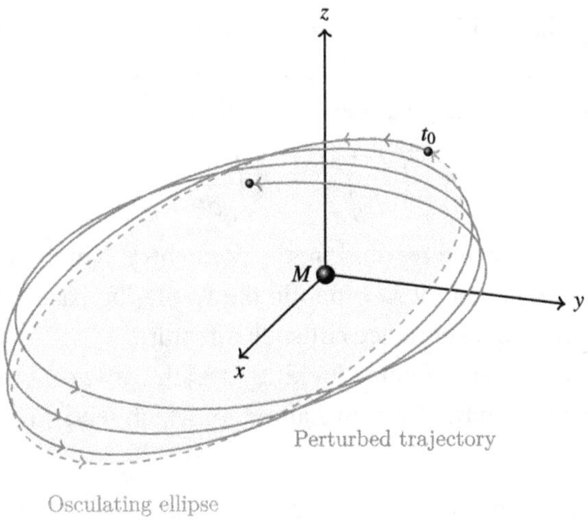

Figure 2.7 Perturbed trajectory (continuous curve) and its osculating Keplerian ellipse (dashed curve) at the initial instant of time t_0 characterized by $e = 0.4$, $I = 30°$, $\Omega = 45°$, $\omega = 50°$, $\eta = 60°$. It is assumed that *all* the Keplerian orbital elements undergo secular rates of change amounting to a few percent of the Keplerian mean motion for a better visualization of their impact. The motion covers three Keplerian orbital periods $T_{\rm K}$.

$$\hat{\boldsymbol{h}} = \begin{cases} \sin I \sin \Omega \\ -\sin I \cos \Omega \, . \\ \cos I \end{cases} \tag{2.25}$$

For them, the relation $\hat{\boldsymbol{\tau}} = \hat{\boldsymbol{h}} \times \hat{\boldsymbol{r}}$ holds. It is important to note that, in the calculation of the order of $\mathcal{O}(A)$, all the Keplerian orbital elements are kept *fixed* to their values at t_0 in the integrand of Equation (2.12)

The instantaneous first-order shifts of the Keplerian orbital elements given by Equation (2.12) are the building blocks for calculating the impact of $\boldsymbol{A}$ on other quantities closely related to observables. Furthermore, they have relevance in themselves whenever the characteristic timescales of the system(s) under consideration are much longer than any realistically conceivable observational time span; then, only relatively short arcs are accessible to observation, since the average effects are not available.

The net shift per orbit $\overline{\Delta\kappa}$ can straightforwardly be obtained by the replacement $f \rightarrow f_0 + 2\pi$ in the expression resulting from Equation (2.12). Depending on the resulting positive or negative sign, $\overline{\Delta\kappa}$ is an advance or a retardation, respectively.

To the first order in A, the precession of κ averaged over one orbital revolution $\langle d\kappa/dt \rangle$ can be obtained simply by taking the ratio of $\overline{\Delta\kappa}$ to the Keplerian orbital period $T_{\rm K}$. As per the meaning of Equation (2.18), from Equation (2.3) it turns out that, by assuming that neither M nor a vary so that n_{K} stays constant, the rate of change of the mean anomaly at epoch is proportional to the opposite of the pace of variation of the time of passage at pericentre, that is,

$$\left\langle \frac{d\eta}{dt} \right\rangle = -n_{\mathrm{K}} \left\langle \frac{dt_{\mathrm{p}}}{dt} \right\rangle. \tag{2.26}$$

Thus, if η *increases* after one revolution, the pericentre, which in the meantime has generally moved, is reached *earlier* than in the Keplerian case, and vice versa; see Section 2.3.1 for some consequences of such a feature.

An effective calculational scheme yielding straightforwardly $\langle d\kappa/dt \rangle$ relies upon the planetary equations in the form of Lagrange, which read[11] (Brumberg, 1991, p. 11, Equation (1.144))

$$\left\langle \frac{da}{dt} \right\rangle = \frac{2}{n_{\mathrm{K}} a} \frac{\partial \langle \mathfrak{R} \rangle}{\partial \eta}, \tag{2.27}$$

$$\left\langle \frac{de}{dt} \right\rangle = \frac{1-e^2}{n_{\mathrm{K}} a^2 e} \frac{\partial \langle \mathfrak{R} \rangle}{\partial \eta} - \frac{\sqrt{1-e^2}}{n_{\mathrm{K}} a^2 e} \frac{\partial \langle \mathfrak{R} \rangle}{\partial \omega}, \tag{2.28}$$

$$\left\langle \frac{dI}{dt} \right\rangle = \frac{\cot I}{n_{\mathrm{K}} a^2 \sqrt{1-e^2}} \frac{\partial \langle \mathfrak{R} \rangle}{\partial \omega} - \frac{\csc I}{n_{\mathrm{K}} a^2 \sqrt{1-e^2}} \frac{\partial \langle \mathfrak{R} \rangle}{\partial \Omega}, \tag{2.29}$$

$$\left\langle \frac{d\Omega}{dt} \right\rangle = \frac{\csc I}{n_{\mathrm{K}} a^2 \sqrt{1-e^2}} \frac{\partial \langle \mathfrak{R} \rangle}{\partial I}, \tag{2.30}$$

$$\left\langle \frac{d\omega}{dt} \right\rangle = -\frac{\cot I}{n_{\mathrm{K}} a^2 \sqrt{1-e^2}} \frac{\partial \langle \mathfrak{R} \rangle}{\partial I} + \frac{\sqrt{1-e^2}}{n_{\mathrm{K}} a^2 e} \frac{\partial \langle \mathfrak{R} \rangle}{\partial e}, \tag{2.31}$$

$$\left\langle \frac{d\eta}{dt} \right\rangle = -\frac{1-e^2}{n_{\mathrm{K}} a^2 e} \frac{\partial \langle \mathfrak{R} \rangle}{\partial e} - \frac{2}{n_{\mathrm{K}} a} \frac{\partial \langle \mathfrak{R} \rangle}{\partial a} - \frac{3}{n_{\mathrm{K}} a^2} \langle \boldsymbol{v} \cdot \nabla_{v} \mathfrak{R} \rangle. \tag{2.32}$$

In Equations (2.27)–(2.32), $\mathfrak{R}$, known as disturbing function, is any pK correction $L^{\mathrm{pK}}(\boldsymbol{r}, \boldsymbol{v}, t)$ (Brumberg, 1991) to the pointlike Newtonian Lagrangian per unit mass

$$L_{\mathrm{N}} = \frac{v^2}{2} + \frac{\mu}{r}; \tag{2.33}$$

it is the *opposite* of any extra-pK potential energy per unit mass. Using Equations (2.27)–(2.32) to calculate $\langle d\kappa/dt \rangle$ implies performing the average of only one function,[12] namely $\mathfrak{R}$, whose inspection can immediately give an insight of which

[11] See also, for example, Brouwer and Clemence (1961); Kaula (2000); Murray and Dermott (1999); Bertotti et al. (2003); Capderou (2005); Roy (2005); Xu (2008); Kopeikin et al. (2011); Gurfil and Seidelmann (2016); and Soffel and Han (2019) where such a topic is treated at varying levels of completeness, especially as far as η is concerned.

[12] In fact, if L^{pK} depends explicitly on the velocity $\boldsymbol{v}$, also $\langle \boldsymbol{v} \cdot \nabla_{v} \mathfrak{R} \rangle$ has to be computed for the rate of change of mean anomaly at epoch, as per Equation (2.32).

averaged orbital changes one can expect. This proves particularly useful for several modified models of gravity, treated in Chapter 10, for which a spherically symmetric extra-potential is usually inferred from the spacetime metric coefficient g_{00}. Since, in view of Equation (2.11), $\langle\mathfrak{R}\rangle$, in this case, can only contain at most a and e, from Equations (2.27)–(2.32) it can straightforwardly be predicted that only ω and η may experience nonvanishing net shifts per orbit. In Appendix C, it is explained how to obtain a pK correction L^{pK}, relativistic or not, to Equation (2.33) that can be used in Equations (2.27)–(2.32).

2.2.2 The Second-Order Effects

The Second-Order Orbital Shifts

In order to calculate the shift of κ to the second order in A, Equation (2.19) must be replaced in Equation (2.12) with

$$\frac{dt}{df} \simeq \frac{r^2}{\sqrt{\mu p}}\left[1 + \frac{r^2}{\sqrt{\mu p}}\left(\frac{d\omega}{dt} + \cos I\frac{d\Omega}{dt}\right)\right]; \tag{2.34}$$

indeed, it is (Egorov, 1958; Taratynova, 1959; Mioc and Radu, 1979; Brumberg, 1991; Bertotti et al., 2003; Poisson and Will, 2014)

$$\frac{df}{dt} = \frac{\sqrt{\mu p}}{r^2}\left[1 - \frac{r^2}{\sqrt{\mu p}}\left(\frac{d\omega}{dt} + \cos I\frac{d\Omega}{dt}\right)\right]. \tag{2.35}$$

Equations (2.34)–(2.35) account for the fact that, actually, the apsidal line does instantaneously vary as the satellite goes along its perturbed trajectory. Thus, the contribution of Equation (2.12) to the shift of κ to the second order in A is

$$\int_{f_0}^{f}\left(\frac{d\kappa}{dt}\right)\frac{r^4}{\mu p}\left(\frac{d\omega}{dt} + \cos I\frac{d\Omega}{dt}\right)\Bigg|_{\mathrm{K}} df'. \tag{2.36}$$

Indeed, $d\kappa/dt$ is of the order of $\mathcal{O}(A)$, being any of Equations (2.13)–(2.18), and $d\omega/dt + \cos I d\Omega/dt$, coming from Equations (2.16)–(2.17), is of the order of $\mathcal{O}(A)$ as well. The subscript $_{\mathrm{K}}$ in Equation (2.36) and in the following means that the content to which it is appended has to be calculated onto the unperturbed Keplerian ellipse of Equation (2.11).

Furthermore, one has also to account for the fact that, actually, all the Keplerian orbital elements instantaneously change during the time it takes the satellite to complete a full orbital revolution. Thus, one has also to compute

$$\int_{f_0}^{f}\sum_{\kappa}^{a,e,\ldots\eta}\frac{\partial\left(d\kappa/df'\right)}{\partial\kappa}\Delta\kappa\left(f'\right)\Bigg|_{\mathrm{K}} df', \tag{2.37}$$

where $\Delta\kappa\,(f)$ are the instantaneous variations of the order of $\mathcal{O}\,(A)$ calculated according to Equation (2.12). In taking the partial derivatives entering Equation (2.37), the Keplerian expression of Equation (2.19) has to be used in $d\kappa/df = (d\kappa/dt)\,(dt/df)$; thus, $\partial\,(d\kappa/df)\,/\partial\kappa$ is of the order of $\mathcal{O}\,(A)$, like $\Delta\kappa\,(f)$, and their product entering Equation (2.37) is just of the order of $\mathcal{O}\left(A^2\right)$.

As a result, the complete orbital shift of κ to the second order in A $\Delta\kappa^{(2)}\,(f)$ consists of the sum of two terms: Equation (2.36) and Equation (2.37), that is,

$$\Delta\kappa^{(2)}\,(f) = \int_{f_0}^{f}\left[\frac{d\kappa}{dt}\frac{r^4}{\mu p}\left(\frac{d\omega}{dt} + \cos I\frac{d\Omega}{dt}\right) + \sum_{\kappa}^{a,e,\ldots\eta}\frac{\partial\left(d\kappa/df'\right)}{\partial\kappa}\Delta\kappa\left(f'\right)\right]_{\mathrm{K}} df'. \tag{2.38}$$

Also in this case, the net shift per orbit $\overline{\Delta\kappa}^{(2)}$ is obtained with the replacement $f \to f_0 + 2\pi$ in the expression resulting from Equation (2.38).

The preceding scheme is used in Sections 4.2–4.3 to calculate the contributions of the 1pN gravitoelectric acceleration itself to the total 2pN gravitoelectric orbital shifts.

The Averaged Second-Order Orbital Precessions

Care is required in correctly calculating the full precession $\left\langle d\kappa^{(2)}/dt\right\rangle$ of κ to the second order in A. The mere division of $\overline{\Delta\kappa}^{(2)}$ by the Keplerian orbital period T_{K} is not enough; indeed, one has also to expand to the second order in A the ratio of $\overline{\Delta\kappa}$ of the order of $\mathcal{O}\,(A)$ to the pK perturbed orbital period $T^{\mathrm{pK}} := T_{\mathrm{K}} + \Delta T^{\mathrm{pK}}$. To this aim, it should be noted that, according to Equation (2.34), the latter has to be meant as the apsidal period, namely the temporal interval between two consecutive passages at the pericentre. See Section 2.3.1 for the details on how to calculate ΔT_{ano}.

The preceding considerations are used in Section 4.4 to calculate the total 2pN gravitoelectric averaged orbital precessions.

2.2.3 The Mixed, or Indirect Effects

The Mixed, or Indirect Orbital Shifts

If the pK acceleration $\boldsymbol{A}$ perturbing the otherwise Keplerian motion of the test particle is made of the sum of two pK accelerations $\boldsymbol{A}^{\mathrm{I}}$ and $\boldsymbol{A}^{\mathrm{II}}$, assuming that both are small in such a way that their sum is still much smaller than Equation (2.1), their simultaneous presence in the equations of motion induces mixed effects of the order of $\mathcal{O}\left(A^{\mathrm{I}}A^{\mathrm{II}}\right)$. It is just the case of, for example, the features of motion of the order of $\mathcal{O}\left(J_2/c^2\right)$, treated in Chapter 8 whose *complete* calculation requires taking into account *also* the contributions of either the 1pN gravitoelectric or the Newtonian quadrupolar accelerations (see Section 8.2).

They can be calculated by considering the instantaneous variations of all the Keplerian orbital elements occurring during the time interval needed to describe a full orbital revolution due to *both* the accelerations. That is, the mixed variation of κ due to $\boldsymbol{A}^{\mathrm{I}}$ and $\boldsymbol{A}^{\mathrm{II}}$ can be worked out as

$$\Delta\kappa_{\mathrm{mix}}^{(\mathrm{I-II})}(f) = \sum_{j=1}^{4} \mathfrak{K}_j^{\mathrm{mix}}, \tag{2.39}$$

with

$$\mathfrak{K}_1^{\mathrm{mix}} := \int_{f_0}^{f} \sum_{\kappa}^{a,e,\ldots\eta} \frac{\partial\left(d\kappa/df'\right)^{\mathrm{I}}}{\partial\kappa} \Delta\kappa^{\mathrm{II}}\left(f'\right)\Big|_{\mathrm{K}} df', \tag{2.40}$$

$$\mathfrak{K}_2^{\mathrm{mix}} := \int_{f_0}^{f} \sum_{\kappa}^{a,e,\ldots\eta} \frac{\partial\left(d\kappa/df'\right)^{\mathrm{II}}}{\partial\kappa} \Delta\kappa^{\mathrm{I}}\left(f'\right)\Big|_{\mathrm{K}} df', \tag{2.41}$$

$$\mathfrak{K}_3^{\mathrm{mix}} := \int_{f_0}^{f} \left(\frac{d\kappa}{dt}\right)^{\mathrm{I}} \frac{r^4}{\mu p} \left(\frac{d\omega}{dt} + \cos I \frac{d\Omega}{dt}\right)^{\mathrm{II}}\Bigg|_{\mathrm{K}} df', \tag{2.42}$$

$$\mathfrak{K}_4^{\mathrm{mix}} := \int_{f_0}^{f} \left(\frac{d\kappa}{dt}\right)^{\mathrm{II}} \frac{r^4}{\mu p} \left(\frac{d\omega}{dt} + \cos I \frac{d\Omega}{dt}\right)^{\mathrm{I}}\Bigg|_{\mathrm{K}} df'. \tag{2.43}$$

The suffixes ^{I} and ^{II} in Equations (2.40)–(2.43) mean that the terms which they are appended to have to be calculated with the accelerations $\boldsymbol{A}^{\mathrm{I}}$ and $\boldsymbol{A}^{\mathrm{II}}$, respectively. Note that Equation (2.38) is equal just to half of Equation (2.39), along with Equations (2.40)–(2.43), with $\boldsymbol{A}^{\mathrm{I}} = \boldsymbol{A}^{\mathrm{II}} = \boldsymbol{A}$. The net mixed shift per orbit $\left\langle\Delta\kappa_{\mathrm{mix}}^{(\mathrm{I-II})}\right\rangle$ is obtained straightforwardly with the replacement $f \rightarrow f_0 + 2\pi$ in Equations (2.39)–(2.43).

The Averaged Mixed, or Indirect Orbital Precessions

The averaged mixed precession $\left\langle d\kappa_{\mathrm{mix}}^{(\mathrm{I-II})}/dt\right\rangle$ can be obtained by adding the expansion to the required order of

$$\frac{\overline{\Delta\kappa}^{\mathrm{II}}}{T^{\mathrm{I}}} + \frac{\overline{\Delta\kappa}^{\mathrm{I}}}{T^{\mathrm{II}}} \tag{2.44}$$

to $\overline{\Delta\kappa}_{\mathrm{mix}}^{(\mathrm{I-II})}/T_{\mathrm{K}}$. In Equation (2.44), $\overline{\Delta\kappa}^{\mathrm{I}}$ and $\overline{\Delta\kappa}^{\mathrm{II}}$ are the first-order net shifts per orbit due to $\boldsymbol{A}^{\mathrm{I}}$ and $\boldsymbol{A}^{\mathrm{II}}$, respectively, while T^{I} and T^{II} are the apsidal periods calculated with $\boldsymbol{A}^{\mathrm{I}}$ and $\boldsymbol{A}^{\mathrm{II}}$, respectively; how to compute the latter is explained in Section 2.3.1.

2.3 The pK Corrections to the Keplerian Orbital Period

From a theoretical point of view, various time intervals T characterizing different cyclic patterns of the orbital motion can be defined when a pK acceleration $\boldsymbol{A}$ acts on a satellite. They are the amounts of time elapsed between two successive passages of the latter at some directions which, in the Keplerian case, are all fixed; in this case, all such periods coincide with the Keplerian one T_{K}. Instead, a perturbing acceleration breaks such a degeneracy, and the aforementioned temporal intervals generally differ one from each other.

2.3.1 *The Anomalistic Period*

The anomalistic period T_{ano} is defined as the time interval between two successive instants when the real position of the test particle coincides with the pericentre position on the corresponding orbit. Among other things, it is required to calculate the averaged orbital precessions to the second order in A (Section 2.2.2), and when two pK accelerations are present (Section 2.2.3). It can be calculated as follows (Zhongolovich, 1960; Mioc and Radu, 1979; Iorio, 2016b):

$$T_{\mathrm{ano}} = T_{\mathrm{K}} + \Delta T_{\mathrm{ano}} = \int_0^{2\pi} \left(\frac{dt}{df}\right) df, \tag{2.45}$$

where dt/df, when a pK acceleration $\boldsymbol{A}$ is present, is given by Equation (2.34). The true anomaly f enters Equation (2.45) as a fast variable of integration just because the line of apsides is involved in the definition of anomalistic period. In order to obtain the full correction ΔT_{ano} of the order of A to the Keplerian orbital period, the contribution of the second term of Equation (2.34) to Equation (2.45) is not enough. Indeed, also the partial derivatives of Equation (2.19) with respect to a and e, multiplied by the finite variations from f_0 to f of the same orbital elements, have to be taken; in this way, one fully accounts for the fact that the Keplerian orbital elements vary instantaneously as the satellite goes along its trajectory. Thus, it is finally obtained:

$$\Delta T_{\mathrm{ano}} = \int_0^{2\pi} \left\{ \frac{3}{2}\sqrt{\frac{a\left(1-e^2\right)^3}{\mu}} \frac{\Delta a\left(f\right)}{\left(1+e\cos f\right)^2} - \sqrt{\frac{a^3\left(1-e^2\right)}{\mu}} \frac{\left[3e + \left(2+e^2\right)\cos f\right]}{\left(1+e\cos f\right)^3} \Delta e\left(f\right) + \frac{r^4}{\mu p}\left(\frac{d\omega}{dt} + \cos I \frac{d\Omega}{dt}\right) \right\}_{\mathrm{K}} df. \tag{2.46}$$

In Iorio (2016b), a variant of the preceding calculation can be found; in Equation (2.19), the semilatus rectum p is adopted as independent variable along with the eccentricity e, and simpler expressions for the partial derivatives of Equation (2.19) are obtained. The resulting expressions for calculating ΔT_{ano} turn out to be

$$\Delta T_{\mathrm{ano}} = \int_0^{2\pi} \left\{ \frac{3}{2}\sqrt{\frac{p}{\mu}} \frac{\Delta p\,(f)}{(1+e\cos f)^2} - 2\sqrt{\frac{p^3}{\mu}} \frac{\cos f\,\Delta e\,(f)}{(1+e\cos f)^3} + \frac{r^4}{\mu p}\left(\frac{d\omega}{dt} + \cos I \frac{d\Omega}{dt}\right) \right\}_{\mathrm{K}} df. \tag{2.47}$$

The first-order variation $\Delta p\,(f)$ of the semilatus rectum can be calculated from Taratynova (1959) and Mioc and Radu (1979):

$$\frac{dp}{df} = \frac{2r^3 A_\tau}{\mu}. \tag{2.48}$$

In the end, both Equation (2.46) and Equation (2.47) give the same result.

The presence or not of the pK anomalistic correction ΔT_{ano} to the orbital period can be intuitively explained as follows. According to Equation (2.26), the rate of change of the mean anomaly at epoch η is proportional to the *opposite* of the pace of variation of the time of passage at pericentre t_{p}. Thus, should η *increase*, the crossing of the pericentre position would be *anticipated* with respect to the Keplerian case since t_{p} would *decrease*, and vice versa. In this case, the variation of η would result in an orbit-by-orbit *advance* or *delay* of the passages at the pericentre. As will be shown, while the 1pN gravitoelectric acceleration due to the mass monopole(s) does induce a *negative* rate of η, the gravitomagnetic LT one leaves the mean anomaly at epoch unchanged. Furthermore, several modified models of gravity, inducing radial pK accelerations dependent only on r, secularly change both ω and η. Also the Newtonian acceleration raised by the primary's oblateness J_2 affects, among other things, also η.

2.3.2 The Draconitic Period

For a perturbed trajectory, the draconitic[13] period T_{dra} is defined as the time interval between two successive instants when the real position of the test particle coincides with the ascending node position on the corresponding osculating ellipse.

[13] This adjective originally referred to the passage of the Moon at its ascending node, when an eclipse occurs. Indeed, the ancient Greeks thought that, during an eclipse, our natural satellite was swallowed up by a dragon (δέρκομαι, meaning literally 'I stare', from which δράκων, -οντος, ὁ, 'which stares') hiding near the nodes of the lunar orbit (Capderou, 2005).

It can be calculated as follows (Mioc and Radu, 1977; Iorio, 2016b):

$$T_{\rm dra} = T_{\rm K} + \Delta T_{\rm dra} = \int_0^{2\pi} \left(\frac{dt}{du}\right) du, \tag{2.49}$$

where dt/du, when $\boldsymbol{A}$ is present, can be obtained as follows. From the definition of u and Equation (2.35), it straightforwardly turns out (Ochocimskij et al., 1959; Mioc and Radu, 1977):

$$\frac{du}{dt} = \frac{\sqrt{\mu p}}{r^2}\left(1 - \frac{r^2 \cos I}{\sqrt{\mu p}}\frac{d\Omega}{dt}\right). \tag{2.50}$$

Thus, it is

$$\frac{dt}{du} \simeq \frac{r^2}{\sqrt{\mu p}} + \frac{r^4 \cos I}{\mu p}\frac{d\Omega}{dt}. \tag{2.51}$$

Note that $d\Omega/dt$ is already expressed in terms of u, as per Equation (2.16).

By using the nonsingular orbital elements[14] (Mioc and Radu, 1977; Montenbruck et al., 2006)

$$k := e \sin\omega, \tag{2.52}$$

$$q := e \cos\omega, \tag{2.53}$$

Equation (2.11) can be rewritten as

$$r = \frac{p}{1 + q\cos u + k \sin u} \tag{2.54}$$

in which p, q, k enter as independent variables.

By proceeding as in Section 2.3.1, it can be obtained (Mioc and Radu, 1977; Iorio, 2016b)

$$\Delta T_{\rm dra} = \int_0^{2\pi} \left\{ \frac{3}{2}\sqrt{\frac{p}{\mu}} \frac{\Delta p\,(u)}{(1 + q\cos u + k\sin u)^2} - 2\sqrt{\frac{p^3}{\mu}} \frac{\cos u \Delta q\,(u) + \sin u \Delta k\,(u)}{(1 + q\cos u + k\sin u)^3} + \frac{r^4 \cos I}{\mu p}\frac{d\Omega}{dt} \right\}_{\rm K} du. \tag{2.55}$$

[14] They are the components of the eccentricity vector (Taff, 1985), an alternative formulation of the Laplace–Runge–Lenz vector (Goldstein, 1980). In the context of pulsar astronomy, they are also known as first and second Laplace–Lagrange parameters ϵ_1, ϵ_2 (Lorimer and Kramer, 2005).

The first-order variations $\Delta p\,(u)$, $\Delta q\,(u)$, and $\Delta k\,(u)$ entering Equation (2.55) can be obtained by integrating the following expressions (Mioc and Radu, 1977):

$$\frac{dp}{du} = \frac{2r^3 A_\tau}{\mu}, \tag{2.56}$$

$$\frac{dq}{du} = \frac{r^2 \sin u A_r}{\mu} + \frac{r^2\left[rq + (r+p)\cos u\right] A_\tau}{\mu} + \frac{\cot I r^3 k \sin u A_h}{\mu p}, \tag{2.57}$$

$$\frac{dk}{du} = -\frac{r^2 \cos u A_r}{\mu} + \frac{r^2\left[rk + (r+p)\sin u\right] A_\tau}{\mu} - \frac{\cot I r^3 q \sin u A_h}{\mu p} \tag{2.58}$$

from u_0 to u.

As far as the actual measurability of the draconitic period in some astronomical scenario of interest is concerned, it was demonstrated (Amelin, 1966; Kassimenko, 1966; Zhongolovich, 1966) that it is possible to measure it, for an artificial Earth's satellite,[15] as the ratio of the difference of the times of passages of the sub-satellite point through a chosen parallel for two following epochs to the number of satellite revolutions corresponding to this difference. The accuracy reached at that time seems to be of the order of $\simeq 10^{-4}$ s (Kassimenko, 1966); it is plausible that it could be improved by orders of magnitude with the most recent techniques currently available.

2.3.3 The Sidereal Period

In general, both the line of nodes and the line of apsides do vary over time because of some pK acceleration(s). Thus, it may be useful to look at a characteristic orbital timescale involving the crossing of some fixed reference direction in space; the sidereal period $T_{\rm sid}$,[16] defined as the time interval between two successive instants when the real position of the test particle lies on a given reference direction, is right up our alley.

For an orbit arbitrarily inclined to the fundamental plane, the sidereal period can be calculated as

$$T_{\rm sid} = T_{\rm K} + \Delta T_{\rm sid} = \int_0^{2\pi} \left(\frac{dt}{d\phi}\right) d\phi, \tag{2.59}$$

where $\phi\,(t)$ is the azimuthal angle reckoned from the reference x axis in the fundamental plane; when the latter is assumed to be coincident with the Earth's equatorial

[15] In their analyses, Amelin (1966), Kassimenko (1966), and Zhongolovich (1966) used the Soviet satellite 1960 ε 3.

[16] From the adjective *sīdĕralīs, e*, ('of or belonging to the stars').

plane at some reference epoch, $\phi(t)$ is the right ascension (RA) $\alpha(t)$ of the celestial body of interest. From

$$x(t) = r(t)\left[\cos\Omega\cos u(t) - \cos I\sin\Omega\sin u(t)\right], \tag{2.60}$$

$$y(t) = r(t)\left[\sin\Omega\cos u(t) + \cos I\cos\Omega\sin u(t)\right], \tag{2.61}$$

one obtains $\phi(t)$ as

$$\phi(t) = \arctan\left[\frac{y(t)}{x(t)}\right]; \tag{2.62}$$

it is a function of the generally varying $I(t)$, $\Omega(t)$, and $u(t)$, namely $\phi(t) = \phi(I(t), \Omega(t), u(t))$.

Since the ongoing calculation is to the first order in the pK acceleration, the differential $d\phi$ in Equation (2.59) can be written as

$$d\phi \simeq \left(\frac{\partial\phi}{\partial u}\right) du. \tag{2.63}$$

The integrand of Equation (2.59) can be obtained as

$$\frac{dt}{d\phi} = \frac{1}{\frac{d\phi}{dt}} = \frac{1}{\frac{\partial\phi}{\partial I}\frac{dI}{dt} + \frac{\partial\phi}{\partial\Omega}\frac{d\Omega}{dt} + \frac{\partial\phi}{\partial u}\frac{du}{dt}} = \frac{1}{\frac{du}{dt}\frac{\partial\phi}{\partial u}\left[1 + \frac{\partial u}{\partial\phi}\left(\frac{\partial\phi}{\partial I}\frac{dI}{du} + \frac{\partial\phi}{\partial\Omega}\frac{d\Omega}{du}\right)\right]}. \tag{2.64}$$

Thus, to the first order in the pK acceleration, the integral of Equation (2.59) can be approximated as

$$\begin{aligned} T_{\mathrm{sid}} &\simeq \int_0^{2\pi} \frac{dt}{du}\left[1 - \frac{\partial u}{\partial\phi}\left(\frac{\partial\phi}{\partial I}\frac{dI}{du} + \frac{\partial\phi}{\partial\Omega}\frac{d\Omega}{du}\right)\right] du \\ &= \int_0^{2\pi}\left(\frac{dt}{du}\right) du - \int_0^{2\pi} \frac{1}{\frac{\partial\phi}{\partial u}}\left(\frac{\partial\phi}{\partial I}\frac{dI}{du} + \frac{\partial\phi}{\partial\Omega}\frac{d\Omega}{du}\right)\left(\frac{dt}{du}\right) du. \end{aligned} \tag{2.65}$$

The first term in Equation (2.65) is nothing but the draconitic period, and can be calculated to the order $\mathcal{O}(A)$ as outlined in Section 2.3.2. The second term in Equation (2.65) is a correction to the former

$$\Delta T_{\mathrm{sid\,II}} := -\int_0^{2\pi} \frac{1}{\frac{\partial\phi}{\partial u}}\left(\frac{\partial\phi}{\partial I}\frac{dI}{du} + \frac{\partial\phi}{\partial\Omega}\frac{d\Omega}{du}\right)\left(\frac{dt}{du}\right)_{\mathrm{K}} du, \tag{2.66}$$

taking into account the fact that, in general, the orbital plane is displaced by the pK acceleration; indeed, the rates of I and Ω enter it. In Equation (2.66), dt/du is intended to be calculated onto the unperturbed Keplerian ellipse in order to keep the calculation to the first order in A.

If the orbital plane coincides with the fundamental one, the previously outlined calculational strategy may lead to analytical expressions for $T_{\rm sid}$ which, for some pK accelerations, are singular in $I = 0$. In that case, the sidereal period can be straightforwardly calculated by means of the true longitude l as follows (Iorio, 2016b):

$$T_{\rm sid} = T_{\rm K} + \Delta T_{\rm sid} = \int_0^{2\pi} \left(\frac{dt}{dl}\right) dl, \tag{2.67}$$

in close analogy with Sections 2.3.1 to 2.3.2. It should be recalled that l is generally a dogleg angle since Ω and u lie in different planes; it is the true longitude of the test particle actually moving along its real orbit only if $I = 0$. When a perturbing acceleration $\boldsymbol{A}$ enters the equations of motion, dt/dl can be obtained in the following way. From the definition of l and Equation (2.34), it is

$$\frac{dl}{dt} = \frac{\sqrt{\mu p}}{r^2}\left[1 + \frac{2r^2 \sin^2(I/2)}{\sqrt{\mu p}} \frac{d\Omega}{dt}\right]. \tag{2.68}$$

Then, it can be written

$$\frac{dt}{dl} \simeq \frac{r^2}{\sqrt{\mu p}} - \frac{2r^4 \sin^2(I/2)}{\mu p} \frac{d\Omega}{dt}. \tag{2.69}$$

The sine of the argument of latitude entering Equation (2.16) for $d\Omega/dt$ can be written in terms of l as $\sin(l - \Omega)$.

By introducing the nonsingular equinoctial elements (Broucke and Cefola, 1972)

$$\widetilde{q} := e \cos \varpi, \tag{2.70}$$

$$\widetilde{k} := e \sin \varpi, \tag{2.71}$$

Equation (2.11) can be rewritten as

$$r = \frac{p}{1 + \widetilde{q}\cos l + \widetilde{k}\sin l} \tag{2.72}$$

in which p, $\widetilde{q}$, $\widetilde{k}$ enter as independent variables.

By proceeding as in Sections 2.3.1 to 2.3.2, one obtains (Iorio, 2016b)

$$\Delta T_{\rm sid} = \int_0^{2\pi} \left\{ \frac{3}{2}\sqrt{\frac{p}{\mu}} \frac{\Delta p(l)}{\left(1 + \widetilde{q}\cos l + \widetilde{k}\sin l\right)^2} - 2\sqrt{\frac{p^3}{\mu}} \frac{\cos l \Delta\widetilde{q}(l) + \sin l \Delta\widetilde{k}(l)}{\left(1 + \widetilde{q}\cos l + \widetilde{k}\sin l\right)^3} - \frac{2r^4 \sin^2(I/2)}{\mu p}\frac{d\Omega}{dt} \right\}_{\rm K} dl. \tag{2.73}$$

The first-order variations $\Delta p\,(l)$, $\Delta\widetilde{q}\,(l)$ and $\Delta\widetilde{k}\,(l)$ entering Equation (2.73) can be obtained by integrating the following expressions (Iorio, 2016b):

$$\frac{dp}{dl} = \frac{2r^3 A_\tau}{\mu}, \tag{2.74}$$

$$\frac{d\widetilde{q}}{dl} = \frac{r^2 \sin l A_r}{\mu} + \frac{r^2\left[r\widetilde{q} + (r+p)\cos l\right] A_\tau}{\mu} - \frac{\tan\left(I/2\right) r^3 \widetilde{k} \sin\left(l-\Omega\right) A_h}{\mu p}, \tag{2.75}$$

$$\frac{d\widetilde{k}}{dl} = -\frac{r^2 \cos l A_r}{\mu} + \frac{r^2\left[r\widetilde{k} + (r+p)\sin l\right] A_\tau}{\mu} + \frac{\tan\left(I/2\right) r^3 \widetilde{q} \sin\left(l-\Omega\right) A_h}{\mu p} \tag{2.76}$$

from l_0 to l. If the orbital plane is aligned with the fundamental one, Equation (2.73) and Equations (2.75)–(2.76) have to be calculated with $I = 0$.

It is generally expected that if the orbital plane stays constant in space, that is, if neither the nodes, when defined, nor the orbit's projection onto the fundamental plane change over time, the sidereal period coincides with the draconitic one since the line of nodes is a fixed direction in space.

2.4 The pK Variations of a Generic Observable $\mathfrak{O}$

Let $\mathfrak{O}$ be some generally time-dependent observable quantity for the system at hand like, for example, the radial velocity, one out of the two astrometric angles RA and dec., a characteristic time scale of a transiting exoplanet, or the ratio of the projection onto the line of sight of the barycentric orbit of an emitting pulsar in a binary system to the speed of light c.

If the former can be analytically modelled in terms of some explicit function $F\,(a,\ e,\ I,\ \Omega,\ \omega,\ \eta,\ f)$ of the Keplerian orbital elements, its instantaneous pK change $\Delta F\,(f)$ due to $\boldsymbol{A}$ can be calculated, to the order of $\mathcal{O}\,(A)$, as

$$\Delta F\,(f) = \sum_{\kappa}^{a,e,\ldots,\eta} \frac{\partial F}{\partial \kappa}\Delta\kappa\,(f) + \frac{\partial F}{\partial f}\Delta f\,(f). \tag{2.77}$$

In Equation (2.77), the instantaneous pK variations of all the Keplerian orbital elements, apart from the true anomaly f, are computed, to the first order in A, by means of Equations (2.12)–(2.19). The pK change $\Delta f\,(f)$ of the true anomaly f encompasses the variation $\Delta\mathcal{M}\,(f)$ of the mean anomaly $\mathcal{M}$, whose calculation requires care, as shown in the following equations. According to Equation (A.6) of Casotto (1993), the variation of f can be written as

$$\Delta f(f) = \left(\frac{a}{r}\right)\left[\sin f\left(1+\frac{r}{p}\right)\Delta e(f) + \sqrt{1-e^2}\left(\frac{a}{r}\right)\Delta\mathcal{M}(f)\right]. \tag{2.78}$$

While $\Delta e(f)$ can be straightforwardly worked out according to Equation (2.12) calculated with Equation (2.14) and Equation (2.19), $\Delta\mathcal{M}(f)$ is more complicated to evaluate; see, for example, Brumberg (1991) and Sobel (1989). As shown in Iorio (2017), the instantaneous variation of the mean anomaly can be obtained, for example, as

$$\Delta\mathcal{M}(f) = \Delta\eta(f) + \Phi(f), \tag{2.79}$$

where the contribution due to the change Δn_{K} of the mean motion n_{K} induced by the variation of the semimajor axis a is

$$\Phi(f) := \int_{t_0}^{t} \Delta n_{\mathrm{K}}\left(t'\right) \mathrm{d}t' = -\frac{3}{2}\frac{n_{\mathrm{K}}}{a}\int_{f_0}^{f} \Delta a\left(f'\right)\frac{dt}{df'}\,\mathrm{d}f'. \tag{2.80}$$

The shifts $\Delta\eta(f)$ in Equation (2.79) and $\Delta a(f)$ in Equation (2.80) can be calculated as per Equation (2.12) with Equation (2.13) and Equation (2.18), along with Equation (2.19) for dt/df.

The net change per orbital revolution $\overline{\Delta F}$ is obtained with the substitution $f \rightarrow f_0 + 2\pi$ in the final expression of Equation (2.77); it turns out that, in general, $\overline{\Delta F}$ may depend on f_0.

2.4.1 The Radial Velocity of a Spectroscopic Binary

A very common observable in studies of binary systems made of two bodies A and B with comparable masses M_{A} and M_{B} is the radial velocity V (Lindegren and Dravins, 2003) of the member for which a spectroscopically determined light curve is available (Struve and Huang, 1958). In this case, one speaks of a single-lined spectroscopic binary (SB1); if both light curves are measured, the system is named double-lined spectroscopic binary (SB2). In exoplanets (Mason, 2008; Seager, 2011; Kitchin, 2012; Deeg and Belmonte, 2018; Perryman, 2018), which have been attracting a growing interest for some years as possible tools to test GTR and modified gravity (Adams and Laughlin, 2006c,a,b; Iorio, 2006a; Jordán and Bakos, 2008; Pál and Kocsis, 2008; Jordán and Bakos, 2009; Ragozzine and Wolf, 2009; Iorio and Ruggiero, 2010; Damiani and Lanza, 2011; Fukui et al., 2011; Iorio, 2011b,a; Eibe et al., 2012; Kane et al., 2012; Li, 2012; Zhao and Xie, 2013; Xie and Deng, 2014; Vargas dos Santos and Mota, 2017; Blanchet et al., 2019; Marzari and Nagasawa, 2019; Ruggiero and Iorio, 2020; Antoniciello et al., 2021; Gou et al., 2021; Kozak and Wojnar, 2021), the only light curve accessible to observations is that of the parent star (Lovis and Fischer, 2010; Hara and Ford, 2023). The same occurs for the S stars (Eckart and Genzel, 1996; Gezari et al.,

2002; Gillessen et al., 2009, 2017; Nishiyama et al., 2018; Peißker et al., 2020, 2022) revolving about the SMBH in Sgr A* at the GC (Eckart et al., 2002; Ghez et al., 2008; Genzel et al., 2010, 2024) in the sense that the light curve accessible to observations is, obviously, theirs.

The radial velocity[17] V is intended to be the projection onto the line of sight, usually assumed coincident with the z axis of a reference frame whose fundamental plane is tangential to the Celestial Sphere at the position of the binary's barycentre,[18] of the sum of the constant barycentre's velocity, or systemtic velocity, $\boldsymbol{V}^0$ and the barycentric velocity $\boldsymbol{v}$ of the visible binary's member which, as seen in the previous examples, is either a star hosting an exoplanet or a S star in Sgr A*. Thus, it can be written

$$V = V_z^0 + \dot{z}. \tag{2.81}$$

From Equation (2.8), along with Equations (2.9)–(2.10), it turns out

$$V - V_z^0 = \frac{m_c}{M_b}\frac{n_K a \sin I}{\sqrt{1-e^2}}\left(\cos u + e\cos\omega\right), \tag{2.82}$$

where m_c is the mass of the unseen companion, namely either an exoplanet p or the SMBH in Sgr A*, and M_b is the total mass of the system; the rest of the orbital parameters refer to the relative motion of one about the other. The pK instantaneous shifts of Equation (2.82) can be calculated according to Equation (2.77) by means of the following partial derivatives[19]

$$\frac{\partial V}{\partial a} = -\frac{\mu_c\left(e\cos\omega + \cos u\right)\sin I}{2\sqrt{\mu_b a^3\left(1-e^2\right)}}, \tag{2.83}$$

$$\frac{\partial V}{\partial e} = \frac{\mu_c\left(\cos\omega + e\cos u\right)\sin I}{\sqrt{\mu_b a\left(1-e^2\right)^3}}, \tag{2.84}$$

$$\frac{\partial V}{\partial I} = \frac{\mu_c\left(e\cos\omega + \cos u\right)\cos I}{\sqrt{\mu_b a\left(1-e^2\right)}}, \tag{2.85}$$

$$\frac{\partial V}{\partial \omega} = -\frac{\mu_c\left(e\sin\omega + \sin u\right)\sin I}{\sqrt{\mu_b a\left(1-e^2\right)}}, \tag{2.86}$$

[17] See, for example, Chapter 18 of the online book *Celestial Mechanics* by J. B. Tatum, retrievable at www.astro.uvic.ca/~tatum/celmechs.html, accessed on 20 January 2024.

[18] From βαρύς, εῖα, ύ, 'heavy'.

[19] In obtaining Equation (2.83), the Keplerian mean motion n_K is considered as an explicit function of a.

$$\frac{\partial V}{\partial f} = -\frac{\mu_c \sin u \sin I}{\sqrt{\mu_b a \left(1 - e^2\right)}} \tag{2.87}$$

and Equation (2.78).

In Equation (2.82), the semiamplitude of the radial velocity curve is defined as

$$K := \frac{m_c}{M_b} \frac{n_K a \sin I}{\sqrt{1 - e^2}}; \tag{2.88}$$

it is one of the directly estimated parameters in exoplanet studies (Eastman et al., 2019). Its nonvanishing partial derivatives are

$$\frac{\partial K}{\partial a} = -\frac{\mu_c \sin I}{2\sqrt{\mu_b a^3 \left(1 - e^2\right)}}, \tag{2.89}$$

$$\frac{\partial K}{\partial e} = \frac{m_c}{M_b} \frac{n_K a e \sin I}{\left(1 - e^2\right)^{3/2}}, \tag{2.90}$$

$$\frac{\partial K}{\partial I} = \frac{m_c}{M_b} \frac{n_K a \cos I}{\sqrt{1 - e^2}}. \tag{2.91}$$

About the current and forthcoming accuracy in radial velocity measurements in extrasolar planets, according to Crass et al. (2021), for nearly a decade, precision instruments and surveys devoted to implementing the Extreme Precision Radial Velocity (EPRV) technique[20] have been unable to routinely reach accuracies of less than roughly 1 m/s (Fischer et al., 2016; Dumusque et al., 2017). However, modern spectrographs promise measurement precision of $\simeq 0.2-0.5$ m/s for an ideal target star (Gilbertson et al., 2020), or even down to the $\simeq 0.01$ m/s level (Matsuo et al., 2022). Furthermore, the next generation of EPRV instruments, which have been designed to have instrumental accuracies approaching 0.1 m/s with a few cm/s stability over many year, have recently been developed: the Echelle SPectrograph for Rocky Exoplanets and Stable Spectroscopic Observations (ESPRESSO) (Pepe et al., 2010), the EXtreme PREcision Spectrograph (EXPRES) (Jurgenson et al., 2016), M dwarf Advanced Radial velocity Observer Of Neighbouring eXoplanets (MAROON-X) (Seifahrt et al., 2018), and NEID (Schwab et al., 2016).

As far as the S stars are concerned, the uncertainty in the radial velocity of the massive, young main sequence star S2 of spectral class B, also known as S0–2 (Ghez et al., 2003; Martins et al., 2008; Habibi et al., 2017), was recently pushed down to the $\simeq$ 7 km/s level for the best data, with a median error of 12.3 km/s (GRAVITY Collaboration, 2019), from measurements collected over

[20] Its main goal is to detect potentially habitable Earthlike rocky planets orbiting at about 1 au from solar-type stars (Crass et al., 2021).

several years[21] with the SINgle Faint Object Near-IR Investigation (SINFONI) instrument,[22] now decommissioned.

2.4.2 The Characteristic Timescales in Transiting Exoplanets

Transiting exoplanets (Haswell, 2010) are discovered by monitoring the reduction in the collected electromagnetic radiation as they pass in front of their parent star in what is called a primary transit, or primary eclipse[23] (Winn, 2011). Most of them are seen edge-on, namely, $I \simeq 90^\circ$ (Mason, 2008). Since it is the combined flux from both the star and the planet p which is actually measured, a further reduction of the former occurs also when the planet passes behind the star in what is called a secondary transit, or secondary eclipse (Winn, 2011). In both cases, some characteristic timescales are measured along with their variations as a powerful tool to reveal transiting – and even non-transiting – planets (Kipping et al., 2011; Nesvorný et al., 2012; Fabrycky et al., 2012; Steffen et al., 2012a,b; Nesvorný et al., 2013): the total transit duration δt_D, the ingress/egress[24] transit duration $\delta t_{i/e}$, the full width at half maximum primary transit duration δt_H, and also the times of inferior and superior conjunctions[25] $t_{\rm cj}$ (Eastman et al., 2019).

When the primary eclipse occurs, the transit starts at the first instant of contact $t_{\rm I}$, when the planetary disk, moving towards the star, becomes externally tangent to the stellar one. Then, at the second instant of contact $t_{\rm II}$, the two disks are internally tangent with the planet's disk superimposed on the stellar one. At the third instant of contact $t_{\rm III}$, the planet's disk begins to leave the stellar one becoming internally tangent to it. The transit ends at the fourth instant of contact $t_{\rm IV}$ when the planetary disk, moving away from the stellar one, becomes externally tangent to it.

The collected data records, spanning months or even years, usually cover a huge number of transits $N_{\rm tr}$ since most of the detected transiting exoplanets are close to their parent stars. As an example, K2–137b (Smith et al., 2018), discovered in 2017, is only 0.0058 au from its star and has an orbital period $T_{\rm K}$ as short as 4.3 hr; in principle, more than $N_{\rm tr} = 10{,}000$ transits are available nowadays for it. About the actual level of uncertainty in measuring, say, δt_D, in a specific system, the case of[26] HD 286123b (Brahm et al., 2018; Yu et al., 2018), known also as K2–232b,

[21] The orbital period of S2 is about 16 yr (Gillessen et al., 2009; Boehle et al., 2016).

[22] It was an adaptive optics assisted near-infrared integral field spectrometer mounted to the Very Large Telescope (VLT) run by the European Southern Observatory (ESO) (Eisenhauer et al., 2003).

[23] From ἔκλειψις, -εως, ἡ, 'disappearance, abandonment', formed by ἐκ- ('out') and λείπω ('to leave, depart, disappear').

[24] From *in-grĕdĭor, essus*, 'to go into, to enter', and *ē-grĕdĭor, gressus*, 'to leave, exit'.

[25] From *coniūnctiō, ōnis*, 'a connecting, uniting, union, agreement'.

[26] It is a sub-Jovian planet orbiting a Sunlike star in 11.6 d, corresponding to a $a = 0.0991$ au, along an elliptical orbit with $e = 0.2555$.

can be fruitfully analysed. The *formal*, statistical accuracy in measuring its total transit duration amounts to (Yu et al., 2018)

$$\sigma_{\delta t_D} \simeq 0.0003\,\mathrm{d} = 28\,\mathrm{s} \tag{2.92}$$

over $N_{\mathrm{tr}} = 7$ since its orbital period is $T_{\mathrm{K}} = 11.16\,\mathrm{d}$, and it was monitored during 80 d (Yu et al., 2018). It should likely not be unrealistic to expect some future improvements in $\sigma_{\delta t_D}$; indeed, it generally gets smaller with the number of transits N_{tr} as $\sqrt{N_{\mathrm{tr}}}$. Thus, for $N_{\mathrm{tr}} = 100$, $\sigma_{\delta t_D}$ should reduce just by a factor of 10. If one assumed a continuous monitoring over, say, 10 yr, corresponding to $N_{\mathrm{tr}} \simeq 330$, $\sigma_{\delta t_D}$ would improve by a factor of 18 corresponding to a measurement uncertainty as little as 1.6 s. Furthermore, if it were possible to observe $N_{\mathrm{tr}} = 1{,}000$ transits over about 30 yrs, the experimental accuracy would improve by a factor of 32 reaching the 0.8 s level. However, it must be remarked that the quoted errors are, in fact, just the *formal*, statistical ones; they do not account for several sources of systematics like, e.g., confusing time standards, neglecting star spots, and neglecting clouds. According to the online NASA Exoplanet Archive available at https://exoplanetarchive.ipac.caltech.edu,[27] the best current accuracy in measuring the total transit duration of transiting exoplanets is of the order of

$$\sigma_{\delta t_D} \simeq 0.0004\,\mathrm{hr} = 1.4\,\mathrm{s}. \tag{2.93}$$

However, Equation (2.93) refers to planets moving along circular orbits. As far as the other timescales of HD 286123b are concerned, the *formal* uncertainty in measuring $\delta t_{\mathrm{i/e}}$ is (Yu et al., 2018)

$$\sigma_{\delta t_{\mathrm{i/e}}} \simeq 0.00020 - 0.00047\,\mathrm{d} = 17 - 40\,\mathrm{s}, \tag{2.94}$$

while it is (Yu et al., 2018)

$$\sigma_{\delta t_H} \simeq 0.0002\,\mathrm{d} = 17\,\mathrm{s}. \tag{2.95}$$

The reported *formal* uncertainty in measuring t_{cj} is, for HD 286123b (Yu et al., 2018),

$$\sigma_{t_{\mathrm{cj}}} \simeq 0.00004\,\mathrm{d} = 3.6\,\mathrm{s} \tag{2.96}$$

after $N_{\mathrm{tr}} = 7$ transits. After 3 yr and $N_{\mathrm{tr}} = 100$ transits, the measurement error should be reduced down to $\simeq 0.3$ s. If it were possible to observe $N_{\mathrm{tr}} = 330$ transits over 10 yr, the error should become as little as $\simeq 0.2$ s. About the realistic obtainable accuracy, the same caveats pointed out for δt_D hold also here.

In the following discussion, only the timescales characterizing the primary transits are treated since the ones corresponding to the secondary eclipses are, in general, less accurately measured (Yu et al., 2018).

[27] Accessed 12 February 2024.

The Total Transit Duration δt_D

The transit is viewed in the plane of the sky, assumed as reference plane $\{x, y\}$ of an astrocentric coordinate system whose reference z axis is directed along the line of sight. In order to obtain a manageable analytical expression for its total duration δt_D, defined as

$$\delta t_D := t_{\rm IV} - t_{\rm I} \tag{2.97}$$

and dubbed T_{14} by Eastman et al. (2019), some reasonable assumptions are to be made. (a) The distance between the planet and its parent star is assumed to be large enough so that the orbital period $T_{\rm K}$ is much longer than δt_D (Carter et al., 2008; Ford et al., 2008). Thus, the planetary disk moves across the stellar one along an approximately rectilinear segment at an essentially constant speed which can be assumed equal to that at midtransit; in general, the speed along an elliptical orbit is variable, as per Equation (2.8). (b) The star–planet separation r, generally variable according to Equation (2.11), remains substantially unchanged and equal to its value at midtransit. (c) The assumed rectilinear chord of the stellar disk traversed during the transit is parallel to the line of the nodes. Thus, let the reference x axis be aligned just along it, so that $\Omega = 0$. (d) Since it is assumed that, during the transit, the planet moves rectilinearly in front of the star without changing their mutual separation, as per (a), $y_{\rm p}$ does not change during δt_D, and it can be posed equal to its value at midtransit occurring when $x_{\rm p} = 0$; from Equation (2.60), written for $\Omega = 0$, the latter condition occurs for $u_{\rm mid} = 90°$, so that

$$f_{\rm mid} = 90° - \omega. \tag{2.98}$$

With such assumptions, it is possible to finally obtain (Iorio, 2023c)

$$\delta t_D = \frac{2R_\star\sqrt{1-e^2}}{n_{\rm K}a\sqrt{1+2e\sin\omega+e^2}}\sqrt{(1+\vartheta)^2-b^2}, \tag{2.99}$$

where

$$\vartheta := \frac{R_{\rm p}}{R_\star}, \tag{2.100}$$

$$b := \frac{a\left(1-e^2\right)\cos I}{R_\star\left(1+e\sin\omega\right)}. \tag{2.101}$$

Equation (2.101) is usually dubbed impact parameter; in it, $R_\star$ and $R_{\rm p}$ are the star's and planet's radii, respectively. Its nonvanishing partial derivatives are

$$\begin{aligned}\frac{\partial\delta t_D}{\partial a} = & \frac{\sqrt{1-e^2}}{n_{\rm K}R_\star a^2\sqrt{1+2e\sin\omega+e^2}\left(1+e\sin\omega\right)^2}\left[-3a^2\left(1-e^2\right)^2\cos^2 I\right.\\ & \left.+R_\star^2\left(1+\vartheta\right)^2\left(1+e\sin\omega\right)^2\right]\frac{1}{\sqrt{(1+\vartheta)^2-b^2}},\end{aligned} \tag{2.102}$$

$$\frac{\partial \delta t_D}{\partial e} = \frac{2\left[2e + \left(1 + e^2\right)\sin\omega\right]}{n_{\rm K} a R_\star \sqrt{1 - e^2}\,(1 + e\sin\omega)^3\left(1 + e^2 + 2e\sin\omega\right)^{3/2}}$$

$$\left[-R_\star^2 (1 + \vartheta)^2 (1 + e\sin\omega)^3\right.$$

$$\left. + a^2\left(1 - e^2\right)^2\cos^2 I\left(2 + e^2 + 3e\sin\omega\right)\right]\frac{1}{\sqrt{(1 + \vartheta)^2 - b^2}}, \quad (2.103)$$

$$\frac{\partial \delta t_D}{\partial I} = \frac{a\left(1 - e^2\right)^{5/2}\sin 2I}{n_{\rm K} R_\star \sqrt{1 + 2e\sin\omega + e^2}\,(1 + e\sin\omega)^2\sqrt{(1 + \vartheta)^2 - b^2}}, \quad (2.104)$$

$$\frac{\partial \delta t_D}{\partial \omega} = \frac{2e\sqrt{1 - e^2}\cos\omega}{n_{\rm K} a R_\star (1 + e\sin\omega)^3\left(1 + e^2 + 2e\sin\omega\right)^{3/2}}$$

$$\left[-R_\star^2 (1 + \vartheta)^2 (1 + e\sin\omega)^3\right.$$

$$\left. + a^2\left(1 - e^2\right)^2\cos^2 I\left(2 + e^2 + 3e\sin\omega\right)\right]\frac{1}{\sqrt{(1 + \vartheta)^2 - b^2}}. \quad (2.105)$$

The Ingress/Egress Transit Duration $\delta t_{\rm i/e}$

The total ingress duration $\delta t_{\rm i}$ is the time interval required for the planetary disk to pass from external to internal tangency to the stellar one in moving towards the star; thus, it can be defined as

$$\delta t_{\rm ing} := t_{\rm II} - t_{\rm I}. \quad (2.106)$$

By relying upon the same assumptions made in the previous section for the total transit duration δt_D, one finally obtains (Iorio, 2023c)

$$\delta t_{\rm ing} = \frac{R_\star \sqrt{1 - e^2}}{n_{\rm K} a \sqrt{1 + 2e\sin\omega + e^2}}\left[\sqrt{(1 + \vartheta)^2 - b^2} - \sqrt{(1 - \vartheta)^2 - b^2}\right]. \quad (2.107)$$

Its nonvanishing partial derivatives are

$$\frac{\partial \delta t_{\rm ing}}{\partial a} = \frac{R_\star \sqrt{1 - e^2}}{2 n_{\rm K} a^2 \sqrt{1 + e^2 + 2e\sin\omega}}\left\{\sqrt{(1 + \vartheta)^2 - b^2} - \sqrt{(1 - \vartheta)^2 - b^2}\right.$$

$$\left. + 2b^2\left[\frac{1}{\sqrt{(1 - \vartheta)^2 - b^2}} - \frac{1}{\sqrt{(1 + \vartheta)^2 - b^2}}\right]\right\}, \quad (2.108)$$

$$\frac{\partial \delta t_{\text{ing}}}{\partial e} = \frac{\left[2e + \left(1+e^2\right)\sin\omega\right]}{n_{\text{K}} a\sqrt{1-e^2} R_\star \left(1+e\sin\omega\right)^3 \left(1+e^2+2e\sin\omega\right)^{3/2}}$$

$$\left\{\frac{R_\star^2 \left(1-\vartheta\right)^2 \left(1+e\sin\omega\right)^3 - a^2\left(1-e^2\right)^2 \cos^2 I \left(2+e^2+3e\sin\omega\right)}{\sqrt{\left(1-\vartheta\right)^2 - b^2}}\right.$$

$$\left.+\frac{-R_\star^2 \left(1+\vartheta\right)^2 \left(1+e\sin\omega\right)^3 + a^2\left(1-e^2\right)^2 \cos^2 I \left(2+e^2+3e\sin\omega\right)}{\sqrt{\left(1+\vartheta\right)^2 - b^2}}\right\}, \tag{2.109}$$

$$\frac{\partial \delta t_{\text{ing}}}{\partial I} = \frac{a\left(1-e^2\right)^{5/2}\sin 2I}{2n_{\text{K}} R_\star \sqrt{1+2e\sin\omega + e^2}\left(1+e\sin\omega\right)^2}\left[\frac{1}{\sqrt{\left(1+\vartheta\right)^2 - b^2}}\right.$$

$$\left.-\frac{1}{\sqrt{\left(1-\vartheta\right)^2 - b^2}}\right], \tag{2.110}$$

$$\frac{\partial \delta t_{\text{ing}}}{\partial \omega} = \frac{eR_\star\sqrt{1-e^2}\cos\omega}{n_{\text{K}} a\left(1+2e\sin\omega+e^2\right)^{3/2}}\left\{\sqrt{\left(1-\vartheta\right)^2 - b^2} - \sqrt{\left(1+\vartheta\right)^2 - b^2}\right.$$

$$+\frac{a^2\left(1-e^2\right)^2\cos^2 I\left(1+2e\sin\omega+e^2\right)}{R_\star^2\left(1+e\sin\omega\right)^3}\left[\frac{1}{\sqrt{\left(1+\vartheta\right)^2 - b^2}}\right.$$

$$\left.\left.-\frac{1}{\sqrt{\left(1-\vartheta\right)^2 - b^2}}\right]\right\}. \tag{2.111}$$

It turns out (Iorio, 2023c) that the total egress duration δt_{egr}, that is, the time interval required to the planetary disk to pass from internal to external tangency to the stellar one in moving away from the star and defined as

$$\delta t_{\text{egr}} := t_{\text{IV}} - t_{\text{III}}, \tag{2.112}$$

is equal to Equation (2.107) for δt_{ing}. Thus, they will be commonly denoted as $\delta t_{\text{i/e}}$, dubbed as τ by Eastman et al. (2019).

The Full Width at Half Maximum Primary Transit Duration δt_H

Another measurable characteristic timescale of the primary transit is the full width at half maximum primary transit duration (Eastman et al., 2019), which can be defined as

$$t_H := \left(\frac{t_{\text{III}} + t_{\text{IV}}}{2}\right) - \left(\frac{t_{\text{I}} + t_{\text{II}}}{2}\right); \tag{2.113}$$

Eastman et al. (2019) dub it T_{FWHM}. It can be calculated with the same approximations adopted for the previous timescales, obtaining (Iorio, 2023c)

$$\delta t_H = \frac{R_\star\sqrt{1-e^2}}{n_{\rm K} a\sqrt{1+2e\sin\omega+e^2}}\left[\sqrt{(1-\vartheta)^2-b^2}+\sqrt{(1+\vartheta)^2-b^2}\right]. \tag{2.114}$$

Its nonvanishing partial derivatives are

$$\begin{aligned}\frac{\partial\delta t_H}{\partial a} = &\frac{\sqrt{1-e^2}}{n_{\rm K} a^2 R_\star\sqrt{1+e^2+2e\sin\omega}\,(1+e\sin\omega)^2}\left[3a^2\left(1-e^2\right)^2\cos^2 I\right.\\ &\left.- R_\star^2(1-\vartheta)^2(1+e\sin\omega)^2\right]\frac{1}{\sqrt{(1-\vartheta)^2-b^2}},\end{aligned} \tag{2.115}$$

$$\begin{aligned}\frac{\partial\delta t_H}{\partial e} = &\frac{\left[2e+\left(1+e^2\right)\sin\omega\right]}{n_{\rm K} a\sqrt{1-e^2}R_\star(1+e\sin\omega)^3\left(1+e^2+2e\sin\omega\right)^{3/2}}\\ &\left\{\frac{-R_\star^2(1-\vartheta)^2(1+e\sin\omega)^3+a^2\left(1-e^2\right)^2\cos^2 I\left(2+e^2+3e\sin\omega\right)}{\sqrt{(1-\vartheta)^2-b^2}}\right.\\ &\left.+\frac{-R_\star^2(1+\vartheta)^2(1+e\sin\omega)^3+a^2\left(1-e^2\right)^2\cos^2 I\left(2+e^2+3e\sin\omega\right)}{\sqrt{(1+\vartheta)^2-b^2}}\right\},\end{aligned} \tag{2.116}$$

$$\begin{aligned}\frac{\partial\delta t_H}{\partial I} = &\frac{a\left(1-e^2\right)^{5/2}\sin 2I}{2n_{\rm K}R_\star\sqrt{1+2e\sin\omega+e^2}\,(1+e\sin\omega)^2}\left[\frac{1}{\sqrt{(1+\vartheta)^2-b^2}}\right.\\ &\left.+\frac{1}{\sqrt{(1-\vartheta)^2-b^2}}\right],\end{aligned} \tag{2.117}$$

$$\begin{aligned}\frac{\partial\delta t_H}{\partial\omega} = &\frac{R_\star\sqrt{1-e^2}e\cos\omega}{n_{\rm K} a\left(1+2e\sin\omega+e^2\right)^{3/2}}\left\{-\sqrt{(1-\vartheta)^2-b^2}-\sqrt{(1+\vartheta)^2-b^2}\right.\\ &+\frac{a^2\left(1-e^2\right)^2\cos^2 I\left(1+2e\sin\omega+e^2\right)}{R_\star^2(1+e\sin\omega)^3}\left[\frac{1}{\sqrt{(1+\vartheta)^2-b^2}}\right.\\ &\left.\left.+\frac{1}{\sqrt{(1-\vartheta)^2-b^2}}\right]\right\}.\end{aligned} \tag{2.118}$$

The Time of Inferior Conjunction t_{cj}

A further measurable quantity in transiting exoplanets is the time of inferior conjunction[28] t_{cj} (Eastman et al., 2019), named T_C by Eastman et al. (2019). Its explicit expression can be obtained as follows.

By integrating Equation (2.19) with respect to the true anomaly from 0, which occurs at the pericentre, to an arbitrary value f, one gets (Capderou, 2005)

$$t = t_{\mathrm{p}} + \frac{1}{n_{\mathrm{K}}}\left\{2\arctan\left[\sqrt{\frac{1-e}{1+e}}\tan\left(\frac{f}{2}\right)\right] - \frac{e\sqrt{1-e^2}\sin f}{1+e\cos f}\right\}. \tag{2.119}$$

By calculating Equation (2.119) with the value of the true anomaly at midtransit, given by Equation (2.98), yields t_{cj}. Thus, one finally has (Iorio, 2023c)

$$t_{\mathrm{cj}} = t_{\mathrm{p}} + \frac{1}{n_{\mathrm{K}}}\left\{2\arctan\left[\sqrt{\frac{1-e}{1+e}}\tan\left(\frac{\pi}{4}-\frac{\omega}{2}\right)\right] - \frac{e\sqrt{1-e^2}\cos\omega}{1+e\sin\omega}\right\}. \tag{2.120}$$

Its nonvanishing partial derivatives are[29]

$$\begin{aligned}\frac{\partial t_{\mathrm{cj}}}{\partial a} = &-\frac{3}{2an_{\mathrm{K}}\left(1+e\sin\omega\right)}\left\{\eta\left(1+e\sin\omega\right) + e\cos\omega\sqrt{1-e^2}\right.\\ &\left.-2\left(1+e\sin\omega\right)\arctan\left[\sqrt{\frac{1-e}{1+e}}\cot\left(\frac{\pi}{4}-\frac{\omega}{2}\right)\right]\right\},\end{aligned} \tag{2.121}$$

$$\frac{\partial t_{\mathrm{cj}}}{\partial e} = -\frac{\sqrt{1-e^2}\cos\omega\left(2+e\sin\omega\right)}{n_{\mathrm{K}}\left(1+e\sin\omega\right)^2}, \tag{2.122}$$

$$\frac{\partial t_{\mathrm{cj}}}{\partial \omega} = -\frac{\left(1-e^2\right)^{3/2}}{n_{\mathrm{K}}\left(1+e\sin\omega\right)^2}, \tag{2.123}$$

$$\frac{\partial t_{\mathrm{cj}}}{\partial \eta} = -\frac{1}{n_{\mathrm{K}}}. \tag{2.124}$$

Note that both Equation (2.120) and Equations (2.121)–(2.124) depend neither on the radius of the star nor that of the planet.

2.4.3 The Sky-Projected Spin-Orbit Angle of Transiting Exoplanets

Another directly measurable quantity in transiting exoplanets is the angle λ between the projections of the system's orbital angular momentum and the star's

[28] It occurs when the star, the planet and the Earth are aligned with the planet in between the former and the latter.

[29] In calculating Equation (2.121), $t_{\mathrm{p}} = t_0 - \eta/n_{\mathrm{K}}$ is used.

angular momentum on the plane of the sky, known also as sky-projected spin-orbit angle, counted clockwise from the former to the latter (Albrecht et al., 2022). It is usually measured with the Rossiter–McLaughlin (McLaughlin, 1924; Rossiter, 1924) effect, widely used in exoplanetary research (Triaud, 2018).

By relying upon some of the same assumptions put forth in Section 2.4.2, it can be inferred that

$$\lambda = \Omega - \zeta_\star + 90^\circ, \tag{2.125}$$

where $\zeta_\star$ is the azimuthal angle of the projection of the stellar angular momentum onto the plane of the sky reckoned from some fixed direction chosen as the reference x axis; also the node Ω is counted from it.

By assuming that the star's angular momentum does not undergo any secular precession, so that $\zeta_\star$ can be considered constant, from Equation (2.125) it turns out

$$\frac{d\lambda}{dt} = \frac{d\Omega}{dt}. \tag{2.126}$$

Thus, measuring the rate of change of the sky-projected spin-orbit angle allows, in principle, to detect the combined effect of any pK acceleration displacing the node as well.

To date, it seems that there are not very many measurements of $d\lambda/dt$ in the literature. The case of XO-3b (Hébrard et al., 2008) seems to be spurious, being the reported discrepancy in the measurements of λ probably due to systematic errors in one or both of them (Worku et al., 2022). The current best level of accuracy in measuring λ is of the order of $\sigma_\lambda \simeq 2^\circ$. With repeated measurements over 10 yr, it may be pushed to the $\sigma_\lambda \simeq 1^\circ$ level; then, an accuracy of $\sigma_{\dot\lambda} \simeq 0.1^\circ/\mathrm{yr}$ over 10 yr may be hypothesized.[30] Nonetheless, there are a few systems for which the accuracy in measuring λ is currently at the sub-degree level; suffice it to say that, according to Table A.1 of Albrecht et al. (2022), $\lambda = 59.2 \pm 0.1^\circ$ for Kepler-13b, $\lambda = 112.9 \pm 0.2^\circ$ for WASP-33b, and $\lambda = 85.1 \pm 0.2^\circ$ for Kelt-9b, corresponding to a 0.2% relative accuracy. For Kelt-9b, a measurement of $d\lambda/dt$ accurate to $\sigma_{\dot\lambda} = 0.08^\circ/\mathrm{yr}$ is reported in the literature (Stephan et al., 2022), while for WASP-33b, the upper bound $|d\lambda/dt| \leq 0.07^\circ/\mathrm{yr}$ is quoted (Stephan et al., 2022).

2.4.4 The Variation of the Times of Arrival of Binary Pulsars

In a binary hosting at least one pulsar psr emitting electromagnetic radiation[31] (Lorimer, 2008), the TOAs $\widetilde{t}_{\mathrm{psr}}$ of the emitted pulses change primarily because of

30 Winn, J. W., personal communication, February 2024.

31 Although the pulsars' emission is usually in the radio portion of the electromagnetic spectrum (Beskin et al., 1988; Lyutikov et al., 1999), neutron stars emitting also optical, X and gamma radiation are known (Torres and Rea, 2011; Giraud and Pétri, 2021).

the orbital motion about the common centre of mass caused by the gravitational tug of the companion c which can be, in principle, either a main sequence star (Wex, 1998) or an astrophysical compact object like, for example, another neutron star which does not emit (Hulse and Taylor, 1975) or whose pulses, for some reason, are no longer[32] detectable, a white dwarf (Antoniadis et al., 2013; Ransom et al., 2014; Venkatraman Krishnan et al., 2020) or, perhaps, even a BH (Wex and Kopeikin, 1999). Such a periodic variation $\widetilde{\delta t}_{\rm psr}\left(f\right)$ can be modeled as the ratio of the projection of the barycentric orbit of the pulsar onto the line of sight to the speed of light c (Damour and Schäfer, 1991; Konacki et al., 2000). By assuming a coordinate system centred in the binary's centre of mass whose reference z axis is along the line of sight in such a way that the reference plane $\{x,\ y\}$ coincides with the plane of the sky, we have

$$\widetilde{\delta t}_{\rm psr}\left(f\right) = \frac{z_{\rm psr}}{c} = \frac{r_{\rm psr}\sin I\sin u}{c} = \frac{a_{\rm psr}\left(1-e^2\right)\sin I\sin u}{c\left(1+e\cos f\right)} \tag{2.127}$$

$$= \frac{m_{\rm c}}{M_{\rm b}}\frac{a\left(1-e^2\right)\sin I\sin u}{c\left(1+e\cos f\right)}. \tag{2.128}$$

In obtaining Equation (2.128), which is somewhat analogous to the range in Earth–Moon or Earth–planets studies (Damour and Schäfer, 1991), we used the fact that, to the Keplerian level, the barycentric semimajor axis of the pulsar psr is

$$a_{\rm psr} \simeq \frac{m_{\rm c}}{M_{\rm b}}a, \tag{2.129}$$

where a is referred to the relative orbit of psr with respect to the companion c. In a purely Keplerian scenario, there is no net variation $\overline{\widetilde{\delta t}}_{\rm psr}$ over a full orbital cycle.

The nonvanishing partial derivatives of Equation (2.128) are

$$\frac{\partial\widetilde{\delta t}_{\rm psr}}{\partial a} = \frac{m_{\rm c}}{M_{\rm b}}\frac{a\left(1-e^2\right)\sin I\sin u}{c\left(1+e\cos f\right)}, \tag{2.130}$$

$$\frac{\partial\widetilde{\delta t}_{\rm psr}}{\partial e} = -\frac{m_{\rm c}}{M_{\rm b}}\frac{a\left[2e+\left(1+e^2\right)\cos f\right]\sin I\sin u}{c\left(1+e\cos f\right)^2}, \tag{2.131}$$

$$\frac{\partial\widetilde{\delta t}_{\rm psr}}{\partial I} = \frac{m_{\rm c}}{M_{\rm b}}\frac{a\left(1-e^2\right)\cos I\sin u}{c\left(1+e\cos f\right)}, \tag{2.132}$$

[32] At least for some years (2003–2008), both the members of the double pulsar PSR J0737–3039 (Burgay et al., 2003; Lyne et al., 2004) were simultaneously detectable as emitting radio pulsars. Later, the beam by PSR J0737–3039B was displaced away from our line of sight by the geodetic precession (Damour and Ruffini, 1974; Barker and O'Connell, 1975) experienced by its spin, measured by Breton et al. (2008) to a $\simeq$ 13% accuracy, due to the 1pN gravitoelectric field of PSR J0737–3039A. PSR J0737–3039B should become visible again around 2024 at the latest (Noutsos et al., 2020).

$$\frac{\partial \delta \widetilde{t}_{\rm psr}}{\partial \omega} = \frac{m_{\rm c}}{M_{\rm b}} \frac{a\left(1-e^2\right)\sin I \cos u}{c\left(1+e\cos f\right)}, \tag{2.133}$$

$$\frac{\partial \delta \widetilde{t}_{\rm psr}}{\partial f} = \frac{m_{\rm c}}{M_{\rm b}} \frac{a\left(1-e^2\right)\sin I \left(e\cos\omega + \cos u\right)}{c\left(1+e\cos f\right)^2}. \tag{2.134}$$

The calculation of the pK instantaneous shifts $\Delta\delta\widetilde{t}_{\rm psr}$ is made according to Equation (2.77) along with Equations (2.130)–(2.134).

2.4.5 The Astrometric Angles RA and dec.

Given a test particle revolving about a massive primary, the body-centric RA α and dec. δ are directly connected with the astrometric angular variables which are actually measured in real astronomical observational campaigns. An example is provided, in the case of an Earth's satellite, by their relation with the spacecraft's topocentric or local tangent coordinates (Montenbruck and Gill, 2000; Moyer, 2003; Xu, 2008). Another important scenario characterized by the use, among other things, of the RA and the dec. is the GC and the S stars. Suffice it to say that the recent discovery of the star S4716 (Peißker et al., 2022) relied upon extensive collections of measurements of RA and dec. accurate to $\simeq 5-0.5$ milliarcseconds (mas).

The RA can be calculated from Equations (2.60)–(2.61) and Equation (2.62) in the case of a reference frame having the mean Earth's equator at the reference epoch J2000.0 as fundamental plane, so that $\phi \rightarrow \alpha$. The dec. is obtained from

$$z\left(t\right) = r\left(t\right)\sin I \sin u\left(t\right), \tag{2.135}$$

as

$$\delta\left(t\right) = \arcsin\left[\frac{z\left(t\right)}{r\left(t\right)}\right]. \tag{2.136}$$

Thus, the nonvanishing partial derivatives of α, δ turn out to be

$$\frac{\partial \alpha}{\partial I} = -\frac{2\sin I \sin 2u}{3+\cos 2I + 2\sin^2 I \cos 2u}, \tag{2.137}$$

$$\frac{\partial \alpha}{\partial \Omega} = 1, \tag{2.138}$$

$$\frac{\partial \alpha}{\partial \omega} = \frac{4\cos I}{3+\cos 2I + 2\sin^2 I \cos 2u}, \tag{2.139}$$

$$\frac{\partial \alpha}{\partial f} = \frac{4\cos I}{3+\cos 2I + 2\sin^2 I \cos 2u}, \tag{2.140}$$

$$\frac{\partial \delta}{\partial I} = \frac{\cos I \sin u}{\sqrt{1 - \sin^2 I \sin^2 u}}, \tag{2.141}$$

$$\frac{\partial \delta}{\partial \omega} = \frac{\sin I \cos u}{\sqrt{1 - \sin^2 I \sin^2 u}}, \tag{2.142}$$

$$\frac{\partial \delta}{\partial f} = \frac{\sin I \cos u}{\sqrt{1 - \sin^2 I \sin^2 u}}. \tag{2.143}$$

The calculation of the pK instantaneous shifts $\Delta\alpha$, $\Delta\delta$ is made according to Equation (2.77) along with Equations (2.137)–(2.143).

2.5 The pK Variations of the Two-Body Range and Range Rate

Let A and B be two test particles orbiting the same primary whose position and velocity vectors are $\boldsymbol{r}_{\mathrm{A}}(t)$, $\boldsymbol{r}_{\mathrm{B}}(t)$ and $\boldsymbol{v}_{\mathrm{A}}(t)$, $\boldsymbol{v}_{\mathrm{B}}(t)$, respectively. Typically, A and B may be a major body of our solar system orbited by a probe and the Earth to which electromagnetic signals are transmitted by the latter, or a pair of twin spacecraft orbiting the same primary along identical orbits. Examples of the latter scenario are the Gravity Recovery and Climate Experiment (GRACE) (Tapley et al., 2004b), GRACE Follow-On (GRACE-FO) (Kornfeld et al., 2019) and Gravity Recovery and Interior Laboratory (GRAIL) (Zuber et al., 2013) missions, the first two of which orbit around the Earth and the third around the Moon. In such a tandem configuration, whose idea dates back to Wolff (1969), one spacecraft follows the other along the same orbit as both continually measure their reciprocal distance by means of microwave (or, possibly, laser as well) ranging instruments. The first mission concepts were proposed by Fischell and Pisacane (1978) (GRAVSAT) and Reigber (1978) (SLALOM).

The two-body mutual range $\rho(t)$ and range rate $\dot{\rho}(t)$ are defined as (Cheng, 2002)

$$\rho := \sqrt{(\boldsymbol{r}_{\mathrm{A}} - \boldsymbol{r}_{\mathrm{B}}) \cdot (\boldsymbol{r}_{\mathrm{A}} - \boldsymbol{r}_{\mathrm{B}})}, \tag{2.144}$$

$$\dot{\rho} := (\boldsymbol{v}_{\mathrm{A}} - \boldsymbol{v}_{\mathrm{B}}) \cdot \hat{\boldsymbol{\rho}}, \tag{2.145}$$

where

$$\hat{\boldsymbol{\rho}} := \frac{\boldsymbol{r}_{\mathrm{A}} - \boldsymbol{r}_{\mathrm{B}}}{\rho}; \tag{2.146}$$

the range is the instantaneous distance between A and B, while the range rate is the projection of their relative velocity $\boldsymbol{v}_{\mathrm{A}} - \boldsymbol{v}_{\mathrm{B}}$ onto the line joining them whose unit vector is $\hat{\boldsymbol{\rho}}$. Both the range and the range rate vary over time as A and B go along their orbits about the common primary.

2.5.1 The Two-Body Range Shift

According to (Cheng, 2002), the time-dependent perturbation $\Delta\rho\left(f\right)$ of Equation (2.144) induced by a pK disturbing acceleration is given by

$$\Delta\rho\left(f\right)=\left(\boldsymbol{\Delta r}_{\mathrm{A}}-\boldsymbol{\Delta r}_{\mathrm{B}}\right)\cdot\hat{\boldsymbol{\rho}}, \tag{2.147}$$

where

$$\boldsymbol{\Delta r}=\mathcal{R}_r\left(f\right)\hat{\boldsymbol{r}}+\mathcal{R}_\tau\left(f\right)\hat{\boldsymbol{\tau}}+\mathcal{R}_h\left(f\right)\hat{\boldsymbol{h}} \tag{2.148}$$

denotes the perturbation experienced by the position vector $\boldsymbol{r}$ of any of the two bodies A, B. In turn, $\mathcal{R}_r\left(f\right), \mathcal{R}_\tau\left(f\right), \mathcal{R}_h\left(f\right)$ in Equation (2.148) are the radial, transverse and normal components of the instantaneous shift $\boldsymbol{\Delta r}\left(f\right)$ of the position vector given by the following (Casotto, 1993):

$$\mathcal{R}_r\left(f\right):=\boldsymbol{\Delta r}\cdot\hat{\boldsymbol{r}}=\frac{r}{a}\Delta a\left(f\right)-a\cos f\,\Delta e\left(f\right)+\frac{ae\sin f}{\sqrt{1-e^2}}\Delta\mathcal{M}\left(f\right), \tag{2.149}$$

$$\begin{aligned}\mathcal{R}_\tau\left(f\right):=\boldsymbol{\Delta r}\cdot\hat{\boldsymbol{\tau}}&=a\sin f\left(1+\frac{r}{p}\right)\Delta e\left(f\right)\\&+r\left[\cos I\,\Delta\Omega\left(f\right)+\Delta\omega\left(f\right)\right]+\frac{a^2}{r}\sqrt{1-e^2}\,\Delta\mathcal{M}\left(f\right),\end{aligned} \tag{2.150}$$

$$\mathcal{R}_h\left(f\right):=\boldsymbol{\Delta r}\cdot\hat{\boldsymbol{h}}=r\left[\sin u\,\Delta I\left(f\right)-\sin I\cos u\,\Delta\Omega\left(f\right)\right]. \tag{2.151}$$

The target two-way accuracy for the probe BepiColombo (Balogh et al., 2007), currently en route to Mercury, is $\simeq 20-30$ cm for range at 1,000 s integration time (Iess and Boscagli, 2001).

2.5.2 The Two-Body Range Rate Shift

The perturbation $\Delta\dot\rho\left(f\right)$ of the range rate can be written as (Cheng, 2002)

$$\Delta\dot\rho\left(f\right)=\left(\boldsymbol{\Delta v}_{\mathrm{A}}-\boldsymbol{\Delta v}_{\mathrm{B}}\right)\cdot\hat{\boldsymbol{\rho}}+\left(\boldsymbol{\Delta r}_{\mathrm{A}}-\boldsymbol{\Delta r}_{\mathrm{B}}\right)\cdot\hat{\boldsymbol{\rho}}_{\mathrm{v}}, \tag{2.152}$$

where

$$\hat{\boldsymbol{\rho}}_{\mathrm{v}}:=\frac{\left(\boldsymbol{v}_{\mathrm{A}}-\boldsymbol{v}_{\mathrm{B}}\right)-\dot\rho\,\hat{\boldsymbol{\rho}}}{\rho}, \tag{2.153}$$

and

$$\boldsymbol{\Delta v}=\mathcal{V}_r\left(f\right)\hat{\boldsymbol{r}}+\mathcal{V}_\tau\left(f\right)\hat{\boldsymbol{\tau}}+\mathcal{V}_h\left(f\right)\hat{\boldsymbol{h}} \tag{2.154}$$

is the perturbation of the velocity vector $\boldsymbol{v}$ of either of the two bodies A, B. Furthermore, $\mathcal{V}_r\left(f\right), \mathcal{V}_\tau\left(f\right), \mathcal{V}_h\left(f\right)$ in Equation (2.154) are the radial, transverse and normal components of the instantaneous shift $\boldsymbol{\Delta v}\left(f\right)$ of the velocity vector which are (Casotto, 1993)

$$\mathcal{V}_r(f) := \boldsymbol{\Delta v} \cdot \hat{\boldsymbol{r}} = -\frac{n_{\rm K} a \sin f}{\sqrt{1-e^2}}\left[\frac{e}{2a}\Delta a(f) + \frac{a}{r}\Delta e(f)\right]$$

$$- \frac{n_{\rm K} a^2}{r}\left[\cos I \Delta\Omega(f) + \Delta\omega(f)\right] - \frac{n_{\rm K} a^3}{r^2}\Delta\mathcal{M}(f), \tag{2.155}$$

$$\mathcal{V}_\tau(f) := \boldsymbol{\Delta v} \cdot \hat{\boldsymbol{\tau}} = -\frac{n_{\rm K} a\sqrt{1-e^2}}{2r}\Delta a(f)$$

$$+ \frac{n_{\rm K} a(e+\cos f)}{\left(1-e^2\right)^{3/2}}\Delta e(f) + \frac{n_{\rm K} a e \sin f}{\sqrt{1-e^2}}\left[\cos I \Delta\Omega(f) + \Delta\omega(f)\right], \tag{2.156}$$

$$\mathcal{V}_h(f) := \boldsymbol{\Delta v} \cdot \hat{\boldsymbol{h}} = \frac{n_{\rm K} a}{\sqrt{1-e^2}}\left[(\cos u + e\cos\omega)\Delta I(f)\right.$$

$$\left. + (\sin u + e \sin\omega)\sin I \Delta\Omega(f)\right]. \tag{2.157}$$

The shifts of a, e, I, Ω, ω in Equations (2.149)–(2.151) and in Equations (2.155)–(2.157) are computed, to the first order in A, according to Equation (2.12) calculated with Equations (2.13)–(2.17) and Equation (2.19), while the variation of the order of $\mathcal{O}(A)$ of the mean anomaly $\Delta\mathcal{M}(f)$ is calculated with Equation (2.79). About the latter, the calculation of $\Phi(f)$ as per Equation (2.80) is often quite cumbersome.

The target two-way accuracy for the probe BepiColombo (Balogh et al., 2007), currently en route to Mercury, is $\simeq 3 \times 10^{-4}$ cm/s for range rate at 10,000 s integration time (Iess and Boscagli, 2001).

2.5.3 How to Produce Time-Dependent Time Series

Equation (2.147) and Equation (2.152) allow one to obtain analytically calculated time series of the two-body range and range rate by means of Equation (2.4) or Equation (2.6); to this aim, the following expressions may be computationally useful as well (Murray and Dermott, 1999):

$$\cos f = -e + \frac{2(1-e^2)}{e}\sum_{s=1}^{\infty}\mathcal{J}_s(se)\cos s\mathcal{M}, \tag{2.158}$$

$$\sin f = 2\sqrt{1-e^2}\sum_{s=1}^{\infty}\frac{1}{s}\frac{d\mathcal{J}_s(se)}{de}\sin s\mathcal{M}, \tag{2.159}$$

and

$$\cos E = -\frac{e}{2} + 2\sum_{s=1}^{\infty} \frac{1}{s^2} \frac{d\mathcal{J}_s(se)}{de} \cos s\mathcal{M}, \tag{2.160}$$

$$\sin E = \frac{2}{e}\sum_{s=1}^{\infty} \frac{1}{s}\mathcal{J}_s(se) \sin s\mathcal{M}. \tag{2.161}$$

From a practical point of view, $s \leq s_{\max}$ where $s_{\max}$ is set by the desired accuracy level.

3

1pN Gravitoelectric Effects: Mass Monopole(s)

3.1 The 1pN Gravitoelectric Effects for a Test Particle

The 1pN gravitoelectric acceleration due to the mass monopole moment of the source, namely, its mass M, is (see, e.g., Soffel, 1989, Equation (4.2.1), p. 89; Brumberg, 1991, Equation (3.1.46), p. 82; Soffel and Han, 2019, p. 332)

$$\boldsymbol{A}^{\mathrm{1pN}} = \frac{\mu}{c^2 r^2}\left[\left(\frac{4\mu}{r} - v^2\right)\hat{\boldsymbol{r}} + 4 v_r \boldsymbol{v}\right], \tag{3.1}$$

where

$$v_r := \boldsymbol{v} \cdot \hat{\boldsymbol{r}} \tag{3.2}$$

is the projection of the velocity vector $\boldsymbol{v}$ onto the direction of the position vector $\boldsymbol{r}$. It induces the largest out of all the pN orbital effects, and historically played a landmark role in establishing the empirical basis of GTR since it allowed Einstein (1915b) to correctly find the cause of the then anomalous perihelion precession of Mercury (Roseveare, 1982) of 42.98 arcseconds per century $\left(''/\mathrm{cty}\right)$[1] (Nobili and Will, 1986; Pireaux and Rozelot, 2003) which puzzled astronomers since it was discovered in the second half of the nineteenth century by Le Verrier (1859a,b).[2] According to Brush (1989) and Weinberg (1992), the explanation by Einstein was particularly important since it was a successful *retrodiction* of an effect which was known for decades. The 1pN perihelion precession was later repeatedly measured with radar measurements of Mercury itself (Shapiro et al., 1972; Shapiro, 1990), of other inner planets (Anderson et al., 1978, 1993), and of the asteroid Icarus (Shapiro et al., 1968, 1971) as well. Also Earth's geodetic satellites (Lucchesi and Peron, 2010, 2014) and the star S2 around the SMBH in Sgr A*

[1] The value predicted by Einstein (1915b) amounted to 43 $''$/cty, later corrected to 43.03 $''$/cty by Clemence (1947).

[2] The value reported by Le Verrier (1859a,b) was 38.3 $''$/cty, later corrected to 41.2 ± 2.1 $''$/cty by Newcomb (1895). Modern determinations based on radar ranging led to 43.2 ± 0.9 $''$/cty (Shapiro et al., 1972). The latest determination, based on optical data, yields 42.8 $''$/cty (Constantin, 2010).

at the GC (GRAVITY Collaboration et al., 2020) were (or, possibly, laser as well) used so far.

By projecting Equation (3.1) onto the unit vectors $\hat{\boldsymbol{r}}, \hat{\boldsymbol{\tau}}, \hat{\boldsymbol{h}}$ defined in Equations (2.23)–(2.25) and by using Equations (2.7)–(2.8) along with Equation (2.11), its radial, transverse, and normal components are obtained; they turn out to be

$$A_r^{1\text{pN}} = \frac{4\mu^2\left(1+e\cos f\right)^2\left[e^2+3+2e\cos f-2\left(2\cos^2 f-1\right)\right]}{c^2a^3\left(1-e^2\right)^3}, \tag{3.3}$$

$$A_\tau^{1\text{pN}} = \frac{4e\mu^2\left(1+e\cos f\right)^3\sin f}{c^2a^3\left(1-e^2\right)^3}, \tag{3.4}$$

$$A_h^{1\text{pN}} = 0. \tag{3.5}$$

Since $A_h^{1\text{pN}}$ vanishes, as per Equation (3.5), the orbital plane, whose orientation is determined by I and Ω, remains fixed in space. Indeed, according to Equations (2.15)–(2.16), their rates of change are caused just by the normal component of a pK acceleration. For eccentric orbits, Equation (3.3) is always positive, that is, it is directed radially outward, while Equation (3.4) is positive from the pericentre to the apocentre and negative from the apocentre back to the pericentre, vanishing just at the apsides. Thus, the cosine of the angle between $\boldsymbol{A}_\text{N}$ and $\boldsymbol{A}^{1\text{pN}}$ is always negative; as a result, Equation (3.1) weakens the Newtonian attraction overall. Such a feature is particularly evident for circular orbits, in which case Equation (3.4) vanishes, leaving just Equation (3.3) directed outward.

3.1.1 The Orbital Shifts of the Keplerian Orbital Elements

The 1pN gravitoelectric instantaneous shifts $\Delta\kappa\left(f\right)$ of the Keplerian orbital elements $\kappa = a, e, I, \Omega, \omega, \eta$ due to Equation (3.1) can be calculated as per Equation (2.12) by using Equations (3.3)–(3.5) in Equations (2.13)–(2.18). They turn out to be

$$\Delta a\left(f\right)^{1\text{pN}} = -\frac{2e\mu}{c^2\left(1-e^2\right)^2}\left\{\left(\cos f-\cos f_0\right)\left[7+3e^2+5e\left(\cos f+\cos f_0\right)\right]\right\}, \tag{3.6}$$

$$\Delta e\left(f\right)^{1\text{pN}} = -\frac{\mu}{c^2a\left(1-e^2\right)}\left\{\left(\cos f-\cos f_0\right)\left[3+7e^2+5e\left(\cos f+\cos f_0\right)\right]\right\}, \tag{3.7}$$

$$\Delta I\left(f\right)^{1\text{pN}} = 0, \tag{3.8}$$

$$\Delta\Omega\left(f\right)^{1\text{pN}} = 0, \tag{3.9}$$

$$\Delta\omega\,(f)^{\rm 1pN} = -\frac{\mu}{c^2 a\left(1-e^2\right)}\{3e\,(-f+f_0) + \left(3-e^2+5e\cos f\right)\sin f + \left(-3+e^2-5e\cos f_0\right)\sin f_0\}, \quad (3.10)$$

$$\Delta\eta\,(f)^{\rm 1pN} = \frac{\mu}{c^2 a}\left\{-12\Delta\Theta\,(f) + \frac{1}{e\sqrt{1-e^2}}\left[15e\,(-f+f_0) + \left(3+7e^2+5e\cos f\right)\sin f - \left(3+7e^2+5e\cos f_0\right)\sin f_0\right]\right\}. \quad (3.11)$$

In Equation (3.11),

$$\Delta\Theta\,(f) := \arctan\left[\sqrt{\frac{1-e}{1+e}}\tan\left(\frac{f_0}{2}\right)\right] - \arctan\left[\sqrt{\frac{1-e}{1+e}}\tan\left(\frac{f}{2}\right)\right]. \quad (3.12)$$

By calculating Equations (3.6)–(3.11) with the replacement $f \rightarrow f_0 + 2\pi$ and dividing the result by $T_{\rm K}$, one obtains the averaged precessions

$$\left\langle\frac{da}{dt}\right\rangle^{\rm 1pN} = 0, \quad (3.13)$$

$$\left\langle\frac{de}{dt}\right\rangle^{\rm 1pN} = 0, \quad (3.14)$$

$$\left\langle\frac{dI}{dt}\right\rangle^{\rm 1pN} = 0, \quad (3.15)$$

$$\left\langle\frac{d\Omega}{dt}\right\rangle^{\rm 1pN} = 0, \quad (3.16)$$

$$\left\langle\frac{d\omega}{dt}\right\rangle^{\rm 1pN} = \frac{3n_{\rm K}\mu}{c^2 a\left(1-e^2\right)}, \quad (3.17)$$

$$\left\langle\frac{d\eta}{dt}\right\rangle^{\rm 1pN} = \frac{3n_{\rm K}\mu}{c^2 a}\left(2-\frac{5}{\sqrt{1-e^2}}\right). \quad (3.18)$$

Equations (3.13)–(3.18) coincide[3] with the definite integrals of $(1/T_{\rm K})\,d\kappa/df$, $\kappa = a, e, I, \Omega, \omega, \eta$ from f_0 to $f_0 + 2\pi$.

[3] It may be worthwhile to further investigate the fact that Equation (3.17) was recently obtained by Stepanov (2018) and Kubo (2022) with STR only.

Using Equations (2.27)–(2.32) with[4]

$$\langle\mathfrak{R}\rangle^{1\mathrm{pN}} = -\frac{3\mu^2}{8c^2a^2}\left(5-\frac{8}{\sqrt{1-e^2}}\right), \tag{3.19}$$

$$\langle\boldsymbol{v}\cdot\nabla_{v}\mathfrak{R}\rangle^{1\mathrm{pN}} = -\frac{\mu^2}{2c^2a^2}\left(9-\frac{16}{\sqrt{1-e^2}}\right), \tag{3.20}$$

obtained from Equation (C.12) and Equation (C.14), respectively, yields just Equations (3.13)–(3.18).

The Contribution of Φ to the Mean Anomaly

When Equation (3.1) enters the equations of motion, the analytical expression of the term Φ entering the shift of the mean anomaly $\mathcal{M}$, calculated with Equation (2.80), turns out to be

$$\begin{aligned}\Phi\left(f\right)^{1\mathrm{pN}} &= \frac{3\mu}{c^2a\left(1-e^2\right)^2}\left(\Delta\Theta\left(f\right)\left[10-e^2+6e^4+2e\left(7+3e^2\right)\cos f_0\right.\right.\\&\left.+5e^2\cos 2f_0\right]+\frac{\sqrt{1-e^2}}{1+e\cos f}\left\{-5\left(-1+e^2\right)\left(f-f_0\right)\right.\\&+e\left(2+3e^2+5e\cos f_0\right)\left(\sin f+e\cos f_0\sin f-\sin f_0\right)\\&\left.\left.+e\cos f\left[-5\left(-1+e^2\right)\left(f-f_0\right)-e\left(2+3e^2+5e\cos f_0\right)\sin f_0\right]\right\}\right).\end{aligned} \tag{3.21}$$

The net change per orbit of Φ can be obtained with the replacement $f\rightarrow f_0+2\pi$ in Equation (3.21), getting

$$\begin{aligned}\overline{\Phi}^{1\mathrm{pN}} &= -\frac{3\pi\mu}{c^2a\left(1-e^2\right)^2}\left\{10\left[1-\left(1-e^2\right)^{3/2}\right]+6e^4\right.\\&\left.+\;2e\left(7+3e^2\right)\cos f_0+e^2\left(-1+5\cos 2f_0\right)\right\}.\end{aligned} \tag{3.22}$$

3.1.2 The Anomalistic Period

When the 1pN gravitoelectric acceleration of Equation (3.1) is taken into account, the anomalistic period can be calculated by means of Equations (3.3)–(3.5), as explained in Section 2.3.1.

It turns out to be

$$T_{\mathrm{ano}}^{1\mathrm{pN}} = T_{\mathrm{K}}+\Delta T_{\mathrm{ano}}^{1\mathrm{pN}}, \tag{3.23}$$

[4] The scheme outlined in Appendix C is followed.

with

$$\Delta T_{\rm ano}^{\rm 1pN} = \frac{3\,\pi\,\sqrt{\mu a}}{c^2\,\left(1-e^2\right)^2}\,\left[6+7\,e^2+2\,e^4+2\,e\,\left(7+3\,e^2\right)\,\cos f_0+5\,e^2\,\cos 2f_0\right]. \tag{3.24}$$

Figure 3.1, obtained for generic values of the Keplerian orbital parameters, confirms the analytical result of Equation (3.24); over, say, three orbital revolutions, the test particle always reaches the precessing line of apsides *after* a time interval equal to $T_{\rm ano}^{\rm 1pN}$. It is *longer* than $T_{\rm K}$, in agreement with Equation (3.24), which is always *positive*.

Furthermore, Figure 3.2 plots the final part of the time series of the cosine $\hat{\boldsymbol{r}}\cdot\hat{\boldsymbol{C}}$ of the angle between the position vector $\boldsymbol{r}$ and the Laplace–Runge–Lenz unit vector $\hat{\boldsymbol{C}}$ versus time t, in units of $T_{\rm K}$, for a numerically integrated fictitious test particle with and without Equation (3.1) starting in both cases from, say, the moving pericentre, that is, for $\hat{\boldsymbol{r}}_0\cdot\hat{\boldsymbol{C}}_0=+1$. It can be seen that the orbiter comes back to the same position on the precessing line of apsides, namely it is $\hat{\boldsymbol{r}}\cdot\hat{\boldsymbol{C}}=+1$ again, just after $T_{\rm ano}^{\rm 1pN}=T_{\rm K}+\Delta T_{\rm ano}^{\rm 1pN}$ differing from $T_{\rm K}$ by a (*positive*) amount, in agreement with Equation (3.24).

3.1.3 The Draconitic Period

The draconitic period, calculated when the 1pN gravitoelectric acceleration of Equation (3.1) is taken into account as explained in Section 2.3.2, turns out to be

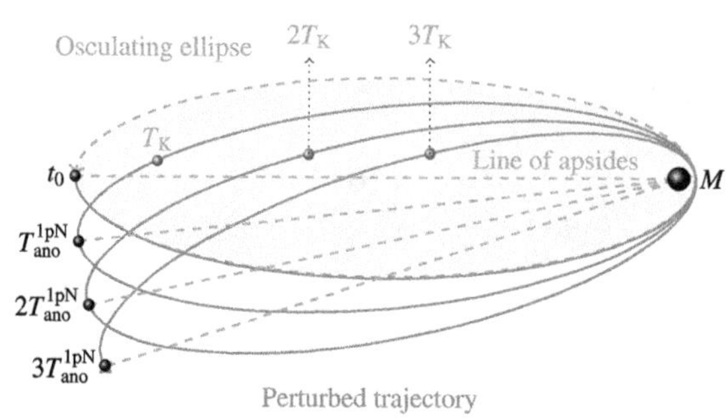

Figure 3.1 Perturbed 1pN trajectory (continuous curve) and its osculating Keplerian ellipse (dashed curve) at the initial instant of time t_0 characterized by $e=0.95$, $I=0$, $\Omega=0$, $\omega=90^\circ$, $f_0=180^\circ$ as seen from above the fixed orbital plane. In this example, it is assumed that both ω and η undergo the 1pN gravitoelectric secular precessions of Equations (3.17)–(3.18) due to the mass M of the primary. For a better visualization of their effect, their sizes are suitably rescaled. The positions on the perturbed trajectory after one, two and three Keplerian periods $T_{\rm K}$ are marked as well. At each orbit, the passages at the drifting line of apsides occur always *later* than in the Keplerian case by the amount given by Equation (3.24), which is always *positive*.

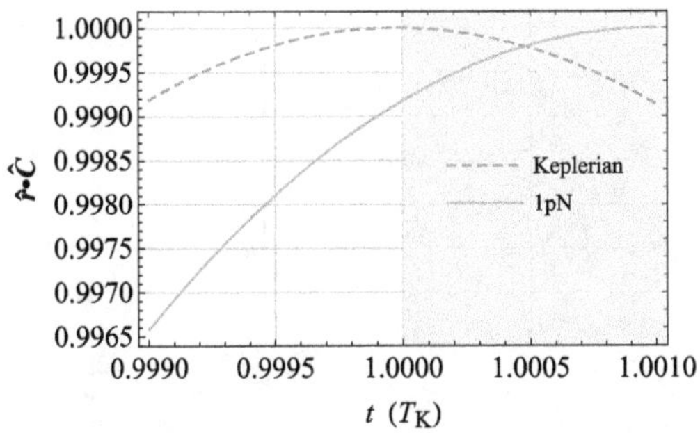

Figure 3.2 Plot of the numerically produced time series of the cosine $\hat{\boldsymbol{r}} \cdot \hat{\boldsymbol{C}}$ of the angle between the position vector $\boldsymbol{r}$ and the Laplace–Runge–Lenz vector $\boldsymbol{C}$ versus time t, in units of T_K, obtained by integrating the equations of motion of a fictitious test particle with and without the 1pN gravitoelectric acceleration of Equation (3.1) for an elliptical ($e = 0.665$) orbit arbitrarily oriented in space ($I = 40°$, $\Omega = 45°$, $\omega = 50°$) starting from the periapsis ($f_0 = 0$), that is, $\hat{\boldsymbol{r}}_0 \cdot \hat{\boldsymbol{C}}_0 = +1$; the semimajor axis is $a = 6R_e$. The physical parameters of the Earth are adopted. The 1pN acceleration is suitably rescaled in such a way that $\Delta T_{\text{ano}}^{\text{1pN}}/T_K = 0.001$. The time needed to come back to the initial position on the (moving) line of apsides, so that $\hat{\boldsymbol{r}} \cdot \hat{\boldsymbol{C}} = +1$ again, is *longer* than in the Keplerian case by the amount $\Delta T_{\text{ano}}^{\text{1pN}} = +0.001\,T_K$, shown by the shaded area, in agreement with Equation (3.24).

$$T_{\text{dra}}^{\text{1pN}} = T_K + \Delta T_{\text{dra}}^{\text{1pN}}, \tag{3.25}$$

with

$$\Delta T_{\text{dra}}^{\text{1pN}} = \frac{3\pi\sqrt{\mu a}}{c^2}\left[\frac{6 + 7e^2 + 2e^4 + 2e\left(7 + 3e^2\right)\cos f_0 + 5e^2\cos 2f_0}{\left(1 - e^2\right)^2}\right.$$
$$\left. - \frac{2\sqrt{1 - e^2}}{(1 + e\cos\omega)^2}\right] \simeq \frac{12\pi\sqrt{\mu a}}{c^2}\left[1 + e\frac{(7\cos f_0 + 2\cos\omega)}{2}\right.$$
$$\left. + e^2\frac{(17 + 5\cos 2f_0 - 3\cos 2\omega)}{4} + \mathcal{O}\left(e^3\right)\right]. \tag{3.26}$$

It can be noted that Equation (3.26) is always *positive* for all values of e, f_0 and ω; this means that the node is reached *later* than in the Keplerian case.

Figure 3.3, obtained for generic values of the Keplerian orbital parameters, confirms the analytical result of Equation (3.26); over, say, three orbital revolutions, the test particle reaches always the fixed line of nodes after a time interval equal to $T_{\text{dra}}^{\text{1pN}}$. It is *longer* than T_K, in agreement with Equation (3.26).

Furthermore, Figure 3.4 plots the final part of the time series of the cosine $\hat{\boldsymbol{r}} \cdot \hat{\boldsymbol{l}}$ of the angle between the position vector $\boldsymbol{r}$ and the node unit vector $\hat{\boldsymbol{l}}$ versus time t, in units of T_K, for a numerically integrated fictitious test particle with and without

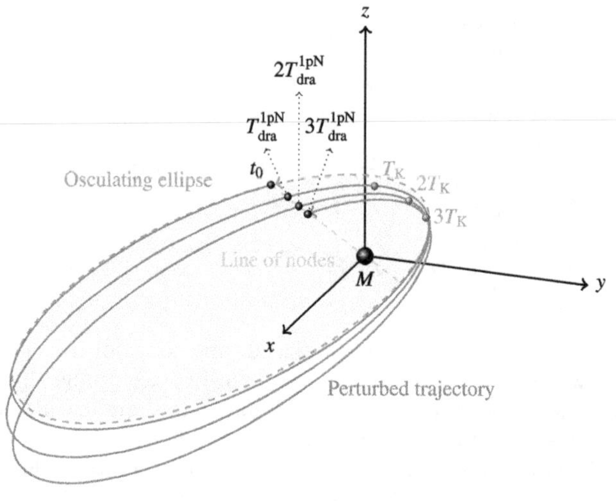

Figure 3.3 Perturbed 1pN trajectory (continuous curve) and its osculating Keplerian ellipse (dashed curve) at the initial instant of time t_0 characterized by $e = 0.7, I = 30°, \Omega = 45°, \omega = 50°, f_0 = 180° - \omega$. In this example, it is assumed that both ω and η undergo the 1pN gravitoelectric secular precessions of Equations (3.17)–(3.18) due to the mass M of the primary. For a better visualization of their effect, their sizes are suitably rescaled. The positions on the perturbed trajectory after one, two and three Keplerian periods $T_{\rm K}$ are marked as well. At each orbit, the passages at the fixed line of nodes occurs always *later* than in the Keplerian case by the amount given by Equation (3.26), which is always positive.

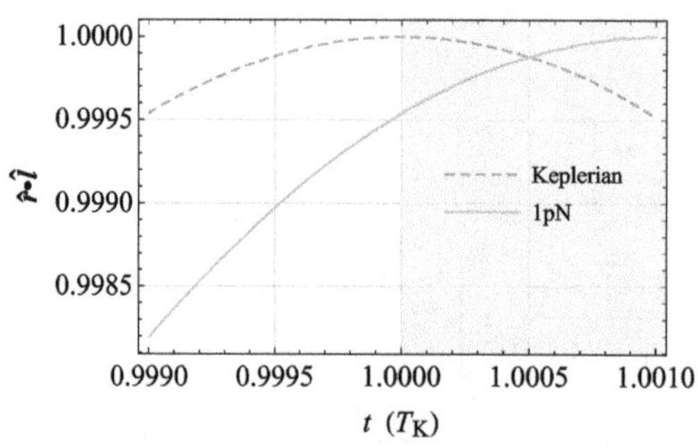

Figure 3.4 Plot of the numerically produced time series of the cosine $\hat{\boldsymbol{r}} \cdot \hat{\boldsymbol{l}}$ of the angle between the position vector $\boldsymbol{r}$ and the node unit vector $\hat{\boldsymbol{l}}$ versus time t, in units of $T_{\rm K}$, obtained by integrating the equations of motion of a fictitious test particle with and without the 1pN gravitoelectric acceleration of Equation (3.1) for an elliptical ($e = 0.665$) orbit arbitrarily oriented in space ($I = 40°, \Omega = 45°, \omega = 50°$) starting from the ascending node ☊ ($f_0 = -\omega + 360°$), that is, $\hat{\boldsymbol{r}}_0 \cdot \hat{\boldsymbol{l}}_0 = +1$; the semimajor axis is $a = 6R_{\rm e}$. The physical parameters of the Earth are adopted. The 1pN acceleration is suitably rescaled in such a way that $\Delta T_{\rm dra}^{\rm 1pN}/T_{\rm K} = 0.001$. The time needed to come back to the initial position on the (fixed) line of nodes, so that $\hat{\boldsymbol{r}} \cdot \hat{\boldsymbol{l}} = +1$ again, is *longer* than in the Keplerian case by the amount $\Delta T_{\rm dra}^{\rm 1pN} = +0.001T_{\rm K}$, shown by the shaded area, in agreement with Equation (3.26).

Equation (3.1) starting in both cases from, say, the fixed ascending node, namely, for $\hat{\boldsymbol{r}}_0 \cdot \hat{\boldsymbol{l}}_0 = +1$. It can be seen that it comes back to the same position on the constant line of nodes, that is, it is $\hat{\boldsymbol{r}} \cdot \hat{\boldsymbol{l}} = +1$ again, just after $T_{\mathrm{dra}}^{\mathrm{1pN}} = T_{\mathrm{K}} + \Delta T_{\mathrm{dra}}^{\mathrm{1pN}}$ differing from T_{K} by a (*positive*) amount, in agreement with Equation (3.26).

3.1.4 The Sidereal Period

As shown in Section 2.3.3, the sidereal period for a generic perturbed orbit is the sum of the draconitic period, calculated as explained in Section 2.3.2, and the term given by Equation (2.66). For Equation (3.1), Equation (2.66) turns out to be

$$\Delta T_{\mathrm{sid\,II}}^{\mathrm{1pN}} = 0. \tag{3.27}$$

Thus, in this case, the sidereal period coincides with the draconitic one.

This is shown in Figure 3.5. It plots the final part of the time series of the cosine of the angle ϕ, normalized to its initial value $\cos\phi_0$, versus time t, in units of T_{K}, for a numerically integrated fictitious test particle with and without Equation (3.1) starting from the same generic initial position. It can be seen that it comes back to the same position on the fixed direction chosen in the reference plane, namely it is $\cos\phi/\cos\phi_0 = +1$ again, just after $T_{\mathrm{sid}}^{\mathrm{1pN}} = T_{\mathrm{dra}}^{\mathrm{1pN}}$ differing from T_{K} by a *positive* amount, in agreement with Equation (3.26).

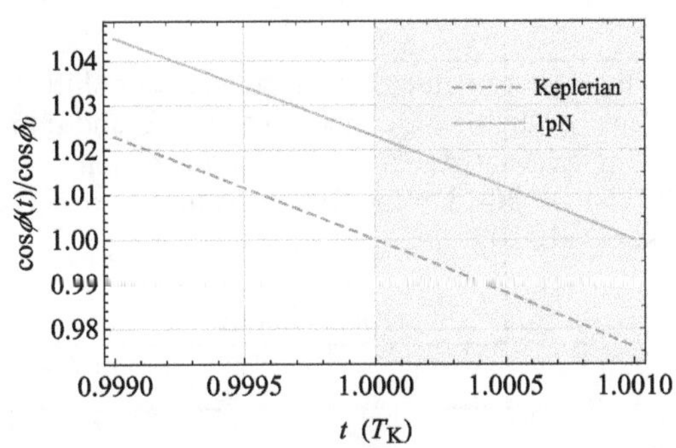

Figure 3.5 Plot of the numerically produced time series of the cosine $\cos\phi\,(t)$ of the azimuthal angle $\phi\,(t)$ normalized to its initial value $\cos\phi_0$ versus time t, in units of T_{K}, obtained by integrating the equations of motion of a fictitious test particle with and without the 1pN gravitoelectric acceleration of Equation (3.1) for an elliptical ($e = 0.665$) orbit arbitrarily oriented in space ($I = 40^\circ$, $\Omega = 45^\circ$, $\omega = 50^\circ$) starting from, say, the ascending node ☊ ($f_0 = -\omega + 360^\circ$); the semimajor axis is $a = 6R_{\mathrm{e}}$. The physical parameters of the Earth are adopted. The 1pN acceleration is suitably rescaled in such a way that $\Delta T_{\mathrm{sid}}^{\mathrm{1pN}}/T_{\mathrm{K}} = 0.001$. The time needed to $\cos\phi\,(t)$ to assume again its initial value $\cos\phi_0$ is *longer* than in the Keplerian case by the amount $\Delta T_{\mathrm{sid}}^{\mathrm{1pN}} = +0.001T_{\mathrm{K}}$, shown by the shaded area, in agreement with the sum of Equation (3.26).

3.1.5 The Astrometric Angles RA and dec.

For a test particle and a massive primary, as in the case of the S stars and Sgr A*, the instantaneous shifts of the RA and dec. can be obtained from Equation (2.77) calculated with Equations (2.137)–(2.143), Equations (3.6)–(3.10), and Equation (3.11) and Equation (3.21). By replacing f with $f_0 + 2\pi$ in the resulting expressions, the net variations per orbit are inferred.

Figures 3.6 and 3.7 display the instantaneous shifts $\Delta\alpha\,(t)\,,\Delta\delta\,(t)$ of the RA and the dec. of a fictitious S star obtained by varying T_{K} and e in such a way that the stellar perinigricon distance is kept fixed to $r_{\mathrm{min}} = 1250\,\mathcal{R}_\bullet$ where $R_\bullet := 2\mu_\bullet/c^2$ is the BH's Schwarzschild radius.

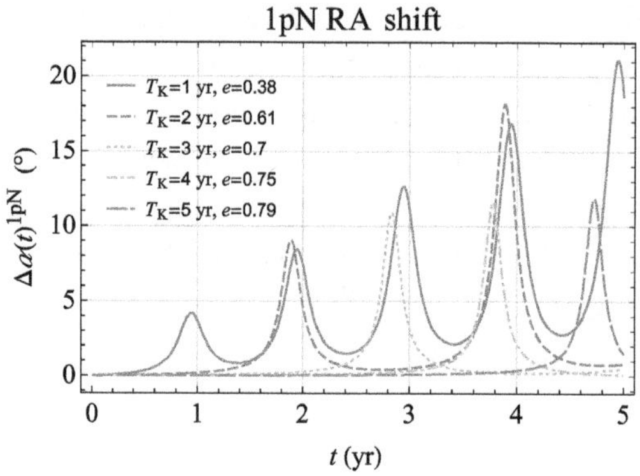

Figure 3.6 Plot of the 1pN gravitoelectric instantaneous shift $\Delta\alpha\,(t)^{1\mathrm{pN}}$, in °, of the RA of a fictitious S star for different values of T_{K} and e in such a way that $r_{\mathrm{min}} = 1250\,\mathcal{R}_\bullet$. The relevant stellar orbital parameters are $I = 161.24°$, $\Omega = 151.54°$, $\omega = 0.073°$, $\eta = 20°$. For the BH, the value $M_\bullet = 4.1 \times 10^6\,M_\odot$ is used for its mass (Peißker et al., 2022).

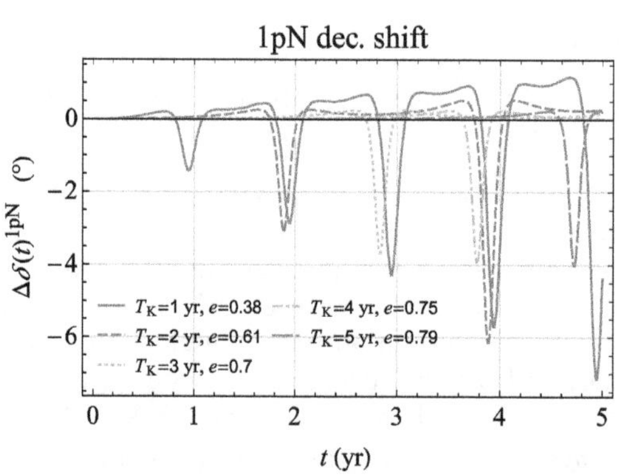

Figure 3.7 Plot of the 1pN gravitoelectric instantaneous shift $\Delta\delta\,(t)^{1\mathrm{pN}}$, in °, of the dec. of a fictitious S star for different values of T_{K} and e in such a way that $r_{\mathrm{min}} = 1250\,\mathcal{R}_\bullet$. The relevant stellar orbital parameters are $I = 161.24°$, $\Omega = 151.54°$, $\omega = 0.073°$, $\eta = 20°$. For the BH, the value $M_\bullet = 4.1 \times 10^6\,M_\odot$ is used (Peißker et al., 2022).

It turns out that the 1pN gravitoelectric astrometric signatures can be as large as $\lesssim 5 - 20^\circ$, depending on the star's orbital period and eccentricity.

3.1.6 The Two-Body Range and Range Rate

Here, Equations (3.6)–(3.10), along with Equation (3.11) and Equation (3.21) for the shift of the mean anomaly, are used in Equation (2.147) and Equation (2.152) to analytically calculate the time series of the range and range rate shifts for A = Mercury and B = Earth, both moving in the 1pN gravitoelectric field of the Sun induced by its mass $M_\odot$.

Figures 3.8–3.9 plot the resulting signals, obtained introducing the dependence on time t through the mean anomaly by means of Equation (2.4) and Equations (2.158)–(2.159), over 2 yr, which is the expected duration of the extended phase of the BepiColombo mission (Benkhoff et al., 2010, 2021) from[5] 2026 to 2028. It turns out that the peak-to-peak nominal amplitudes of the range and range rate shifts can reach the $\simeq 2000$ km and $\simeq 1.5$ m/s level, respectively.

An integration of the equations of motion with and without Equation (3.1) generated corresponding numerical time series differing from the aforementioned analytical ones by $\lesssim 0.02$ km and $\lesssim 8$ mm/s level over 2 years, respectively.

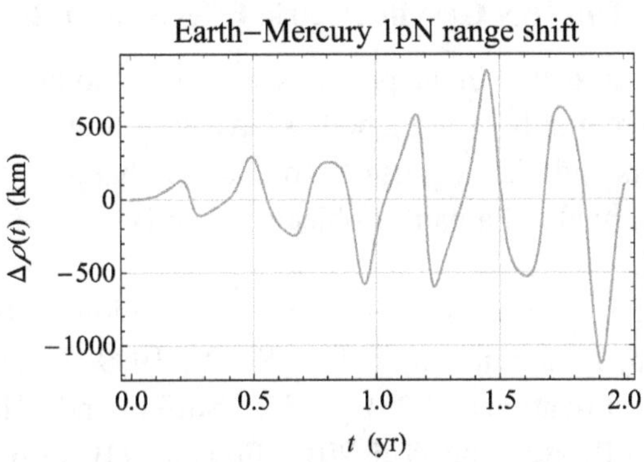

Figure 3.8 Analytically produced time series, in km, of the two-body 1pN range shift $\Delta\rho\,(t)$ due to the Sun's mass $M_\odot$ for A=Mercury and B=Earth plotted over 2 yr. It was worked out by calculating Equations (2.149)–(2.151) entering Equations (2.147)–(2.148) with Equations (3.6)–(3.11), Equation (3.21) and Equation (2.4). The initial values of the Keplerian orbital elements of both planets, referred to the International Celestial Reference Frame (ICRF), were retrieved from the WEB interface Horizons System at https://ssd.jpl.nasa.gov/horizons/, maintained by the Jet Propulsion Laboratory (JPL) of the National Aeronautics and Space Administration (NASA), and accessed 12 February 2024. For the Sun's standard gravitational parameter, see Table J.1.

[5] See www.esa.int/Science_Exploration/Space_Science/BepiColombo/BepiColombo_factsheet. Accessed 19th January 2024.

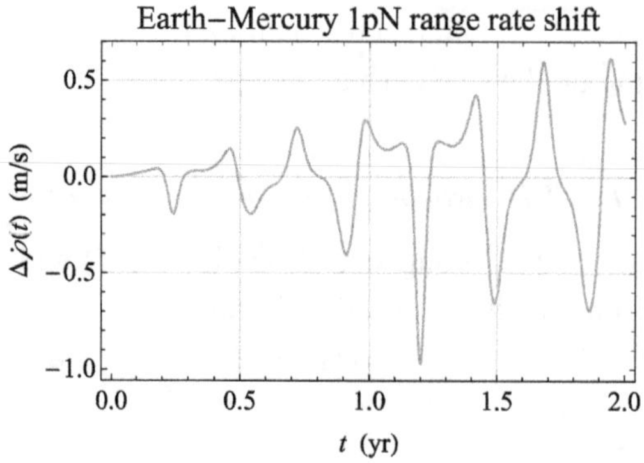

Figure 3.9 Analytically produced time series, in m/s, of the two-body 1pN range rate shift $\Delta\dot{\rho}\,(t)$ due to the Sun's mass $M_\odot$ for A=Mercury and B=Earth plotted over 2 yr. It was worked out by calculating Equations (2.155)–(2.157) entering Equations (2.152)–(2.154) with Equations (3.6)–(3.11), Equation (3.21) and Equation (2.4). The initial values of the Keplerian orbital elements of both planets, referred to the International Celestial Reference Frame (ICRF), were retrieved from the WEB interface Horizons System at https://ssd.jpl.nasa.gov/horizons/, maintained by the Jet Propulsion Laboratory (JPL) of the National Aeronautics and Space Administration (NASA), and accessed 12 February 2024. For the Sun's standard gravitational parameter, see Table J.1.

3.2 The 1pN Gravitoelectric Effects for a Binary

The test particle limit treated in the previous sections is no longer applicable to the case where the two mutually orbiting bodies have both comparable masses like, for example, exoplanets and binary pulsars. To this aim, here the 1pN gravitoelectric effects for a system made of a pair of objects A and B both with finite masses M_A and M_B are treated.

In such a case, the 1pN gravitoelectric acceleration is (see, e.g., Damour and Deruelle, 1985, Equation (2.5), p. 111; Soffel, 1989, Equation (A2.6), p. 166; Brumberg, 1991, Equation (4.4.28), p. 154; Soffel and Han, 2019, Equation (10.3.7), p. 381; Poisson and Will, 2014, Equation (10.1), p. 482)

$$\boldsymbol{A}^{\mathrm{1pN}} = \frac{\mu_\mathrm{b}}{c^2 r^2}\left\{\left[(4+2\nu)\frac{\mu_\mathrm{b}}{r} + \frac{3}{2}\nu v_r^2 - (1-3\nu)\,v^2\right]\hat{\boldsymbol{r}} + (4-2\nu)\,v_r\boldsymbol{v}\right\}. \tag{3.28}$$

In Equation (3.28),

$$\mu_\mathrm{b} := GM_\mathrm{b} \tag{3.29}$$

is the standard gravitational parameter of the binary whose total mass is

$$M_\mathrm{b} := M_\mathrm{A} + M_\mathrm{B}. \tag{3.30}$$

Furthermore, the symmetric mass ratio ν is defined as

$$\nu := \frac{M_{\mathrm{A}} M_{\mathrm{B}}}{M_{\mathrm{b}}^2}, \tag{3.31}$$

with $0 \leq \nu \leq 1/4 = 0.25$; the value 0 corresponds to the test particle limit, while $\nu = 1/4$ if both bodies have the same mass.

The radial, transverse, and normal components of Equation (3.28) turn out to be

$$A_r^{\mathrm{1pN}} = \frac{\mu_{\mathrm{b}}^2 (1 + e\cos f)^2}{4c^2 a^3 \left(1 - e^2\right)^3} \left[e^2 (4 - 13\nu) - 4(-3 + \nu) + 8e(1 - 2\nu)\cos f + e^2 (-8 + \nu)\cos 2f\right], \tag{3.32}$$

$$A_\tau^{\mathrm{1pN}} = \frac{2e\mu_{\mathrm{b}}^2 (1 + e\cos f)^3 (2 - \nu)\sin f}{c^2 a^3 \left(1 - e^2\right)^3}, \tag{3.33}$$

$$A_h^{\mathrm{1pN}} = 0. \tag{3.34}$$

They agree with Equations (A2.77a)–(A2.77c), calculated with GTR, by Soffel (1989, p. 178). In the limit $\nu \to 0$, Equations (3.32)–(3.34) reduce just to Equations (3.3)–(3.5).

3.2.1 The Orbital Shifts of the Keplerian Orbital Elements

The instantaneous shifts of the Keplerian orbital elements, calculated according to Equation (2.12) with Equations (3.32)–(3.34) in Equations (2.13)–(2.18), are

$$\Delta a(f)^{\mathrm{1pN}} = \frac{e\mu_{\mathrm{b}} (\cos f - \cos f_0)}{2c^2 \left(1 - e^2\right)^2} \{4\left[-7 + 3\nu + e^2(-3 + 4\nu)\right] + e\left[e\nu\cos 2f + 4(-5 + 4\nu)\cos f_0 + 2\cos f(-10 + 8\nu + e\nu\cos f_0) + e\nu\cos 2f_0\right]\}, \tag{3.35}$$

$$\Delta e(f)^{\mathrm{1pN}} = \frac{\mu_{\mathrm{b}} (\cos f - \cos f_0)}{4c^2 a \left(1 - e^2\right)} \{4\left[-3 + \nu + e^2(-7 + 6\nu)\right] + e\left[e\nu\cos 2f + 4(-5 + 4\nu)\cos f_0 + 2\cos f(-10 + 8\nu + e\nu\cos f_0) + e\nu\cos 2f_0\right]\}, \tag{3.36}$$

$$\Delta I(f)^{\mathrm{1pN}} = 0, \tag{3.37}$$

$$\Delta\Omega(f)^{\mathrm{1pN}} = 0, \tag{3.38}$$

$$\Delta\omega(f)^{\mathrm{1pN}} = -\frac{\mu_{\mathrm{b}}}{8c^2 a e \left(1 - e^2\right)} \{-2\left[4(-3 + \nu) + e^2(4 + 11\nu) + 4e(-5 + 4\nu)\cos f + e^2\nu\cos 2f\right]\sin f + \left[8(-3 + \nu) + e^2(8 + 21\nu)\right]\sin f_0 + e\left[24(-f + f_0) + 4(-5 + 4\nu)\sin 2f_0 + e\nu\sin 3f_0\right]\}, \tag{3.39}$$

$$\begin{aligned}\Delta\eta\left(f\right)^{1\text{pN}} = & \frac{\mu_b}{8c^2ae\sqrt{1-e^2}}\Big(16e\sqrt{1-e^2}\left(-6+7\nu\right)\Delta\Theta\left(f\right) \\ & -2\left[4\left(-3+\nu\right)+e^2\left(-28+15\nu\right)\right. \\ & \left.+4e\left(-5+4\nu\right)\cos f+e^2\nu\cos 2f\right]\sin f \\ & +8\left\{e\left[\left(f-f_0\right)\left(-15+9\nu\right)\right]\right. \\ & \left.-3\sin f_0\right\}+\left[8\nu+e^2\left(-56+29\nu\right)\right]\sin f_0 \\ & +4e\left(-5+4\nu\right)\sin 2f_0+e^2\nu\sin 3f_0\Big). \end{aligned} \tag{3.40}$$

The net shifts per orbit of the Keplerian osculating elements, calculated by replacing f with $f_0 + 2\pi$ in Equations (3.35)–(3.40), are

$$\overline{\Delta a}^{1\text{pN}} = 0, \tag{3.41}$$

$$\overline{\Delta e}^{1\text{pN}} = 0, \tag{3.42}$$

$$\overline{\Delta I}^{1\text{pN}} = 0, \tag{3.43}$$

$$\overline{\Delta \Omega}^{1\text{pN}} = 0, \tag{3.44}$$

$$\overline{\Delta \omega}^{1\text{pN}} = \frac{6\pi\mu_b}{c^2a\left(1-e^2\right)}, \tag{3.45}$$

$$\overline{\Delta \eta}^{1\text{pN}} = \frac{2\pi\mu_b\left[-15+6\sqrt{1-e^2}+\nu\left(9-7\sqrt{1-e^2}\right)\right]}{c^2a\sqrt{1-e^2}}. \tag{3.46}$$

The pericentre advance of Equation (3.45) has been measured several times so far by monitoring the motion of the periastron of various binary pulsars (Weisberg and Taylor, 1984; Stairs, 2003; Champion et al., 2004; Weisberg and Taylor, 2005; Kramer et al., 2006). In the limit $\nu \to 0$, Equations (3.41)–(3.46) reduce to the shifts obtainable from Equations (3.13)–(3.18) by rescaling the latter ones by $2\pi/n_K$.

The Contribution of Φ to the Mean Anomaly

The analytical expression of the term Φ entering the shift of the mean anomaly $\mathcal{M}$, calculated with Equation (2.80), turns out to be

$$\begin{aligned}\Phi\left(f\right)^{1\text{pN}} = & \frac{3\mu_b}{8c^2a\left(1-e^2\right)^2\left(1+e\cos f\right)}\left[-2\Delta\Theta\left(f\right)\left(1+e\cos f\right)\left\{-40+24\nu\right.\right. \\ & +4e^2\left[1-3\nu+e^2\left(-6+7\nu\right)\right]+e\left[8\left(-7+3\nu\right)\right. \\ & \left.+e^2\left(-24+31\nu\right)\right]\cos f_0+4e^2\left(-5+4\nu\right)\cos 2f_0 \\ & \left.+e^3\nu\cos 3f_0\right\}+\sqrt{1-e^2}\left(8\left(-1+e^2\right)\left(f-f_0\right)\left(-5+3\nu\right)\right. \end{aligned}$$

$$+e\left\{4\left[4+e^2(11-10\nu)\right]\sin f+e\left[56+e^2(24-31\nu)\right.\right.$$

$$\left.-24\nu\right]\cos f_0\sin f+4e^2(5-4\nu)\cos 2f_0\sin f$$

$$-e^3\nu\cos 3f_0\sin f+4\left(-1+e^2\right)\cos f\left[2\left(f-f_0\right)(-5+3\nu)\right.$$

$$\left.+e\nu\sin f\right]+\left[-16+e^2(-24+25\nu)\right](1+e\cos f)\sin f_0$$

$$\left.\left.+2e(-10+7\nu)(1+e\cos f)\sin 2f_0+e^2\nu(1+e\cos f)\sin 3f_0\right\}\right)\Big]. \tag{3.47}$$

By replacing f with $f_0+2\pi$ in Equation (3.47) yields

$$\overline{\Phi}^{1\text{pN}}=\frac{3\pi\mu_b}{2c^2a\left(1-e^2\right)^2}\left\{-4\left(-1+\sqrt{1-e^2}\right)(-5+3\nu)+2e^4(-6+7\nu)\right.$$

$$+2e^2\left[1-10\sqrt{1-e^2}+\left(-3+6\sqrt{1-e^2}\right)\nu\right]+e\left[4(-7+3\nu)\right.$$

$$\left.\left.+3e^2(-4+5\nu)\right]\cos f_0+e^2\left(-10+8\nu+e\nu\cos f_0\right)\cos 2f_0\right\}. \tag{3.48}$$

In the limit $\nu\rightarrow 0$, Equations (3.47)–(3.48) reduce to Equations (3.21)–(3.22).

3.2.2 The Anomalistic Period

The anomalistic period calculated with Equations (3.32)–(3.34) as explained in Section 2.3.1, turns out to be

$$T_{\text{ano}}^{1\text{pN}}=T_{\text{K}}+\Delta T_{\text{ano}}^{1\text{pN}}, \tag{3.49}$$

with

$$\begin{aligned}\Delta T_{\text{ano}}^{1\text{pN}}=&\frac{\pi\sqrt{\mu_b a}}{2c^2\left(1-e^2\right)^2}\left(36+e^2(42-38\nu)+2e^4(6-7\nu)-8\nu\right.\\&+3e\left\{\left[28+3e^2(4-5\nu)-12\nu\right]\cos f_0\right.\\&\left.\left.-e\left(-10+8\nu+e\nu\cos f_0\right)\cos 2f_0\right\}\right).\end{aligned} \tag{3.50}$$

In the limit $\nu\rightarrow 0$, Equation (3.50) reduces to Equation (3.24).

3.2.3 The Draconitic Period

The draconitic period, calculated with Equations (3.32)–(3.34) as explained in Section 2.3.2, turns out to be

$$T_{\text{dra}}^{1\text{pN}}=T_{\text{K}}+\Delta T_{\text{dra}}^{1\text{pN}}, \tag{3.51}$$

with

$$\Delta T_{\rm dra}^{\rm 1pN} = \frac{\pi\sqrt{\mu_{\rm b}a}}{4c^2}\left(72 + e^2(84 - 76\nu) + 4e^4(6 - 7\nu) - 16\nu \right.$$
$$- 3e\left\{\left[8(-7 + 3\nu) + e^2(-24 + 31\nu)\right]\cos f_0\right.$$
$$\left.\left. + e\left[4(-5 + 4\nu)\cos 2f_0 + e\nu\cos 3f_0\right]\right\} - \frac{24\sqrt{1 - e^2}}{(1 + e\cos\omega)^2}\right). \quad (3.52)$$

In the limit $\nu \rightarrow 0$, Equation (3.52) reduces to Equation (3.26).

3.2.4 The Sidereal Period

The considerations presented in Section 3.1.4 hold also in this case: the sidereal period coincides with the draconitic one.

3.2.5 The Radial Velocity of a Spectroscopic Binary

From Equation (2.88) and Equations (3.13)–(3.15), it straightforwardly turns out that the semiamplitude of the radial velocity curve does not experience any 1pN gravitoelectric net shift per orbit.

Nonetheless, the radial velocity curve exhibits a generally nonvanishing 1pN gravitoelectric instantaneous shift ΔV which can be analytically worked out according to Equations (2.77)–(2.78) by using Equations (2.83)–(2.87) along with Equations (3.35)–(3.40) and Equation (3.47). By replacing f with $f_0 + 2\pi$ in the resulting expression allows one to obtain the net shift per orbit.

Figure 3.10 shows the plot of the analytically computed times series for a fictitious tight exoplanetary system made of a main sequence star and a gaseous giant planet with the same masses of the Sun and Jupiter, respectively. By varying the orbital period $T_{\rm K}$ and the eccentricity e in such a way that the star-planet minimum distance, in units of $R_\star + R_{\rm p}$, is 2, it turns out that the peak-to-peak amplitude of the 1pN gravitolectric shift can reach the level of about $\simeq 0.02 - 0.05\,{\rm m/s}$ over 1 day.

Figure 3.11 shows the plot of the analytically computed times series for a fictitious S star. By varying the orbital period $T_{\rm K}$ and the eccentricity e in such a way that the perinigricon distance is $r_{\rm min} = 1250\,\mathcal{R}_\bullet$, it turns out that the peak-to-peak amplitude of the 1pN gravitolectric shift can reach the level of about $\simeq 800\,{\rm km/s}$ over 5 yr.

3.2.6 The Characteristic Timescales of Transiting Exoplanets

Here, the 1pN gravitoelectric net shifts per orbit of the characteristic timescales of transiting exoplanets are calculated in their full generality.

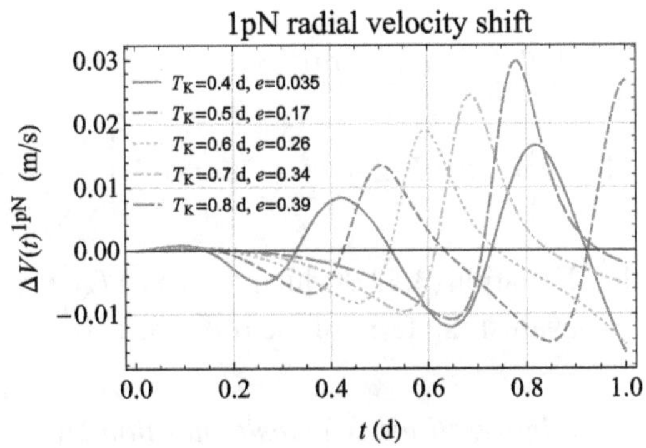

Figure 3.10 Analytically produced time series, in m/s, of the 1pN gravitoelectric shift $\Delta V(t)^{1\mathrm{pN}}$ of the radial velocity curve of a fictitious detached exoplanetary system made of a Sunlike star and a gaseous giant planet p with the same mass and radius of Jupiter for different values of the e and the orbital period T_{K} in such a way that $r_{\min} = 2\left(R_\star + R_{\mathrm{p}}\right)$. The other relevant orbital parameters, chosen arbitrarily, are $I = 45°, \omega = 50°, \eta = 20°$. Since p cannot be considered a test particle, Equations (3.35)–(3.36) and Equations (3.39)–(3.47) are used along with Equations (2.83)–(2.87) in Equation (2.77) and Equation (2.78).

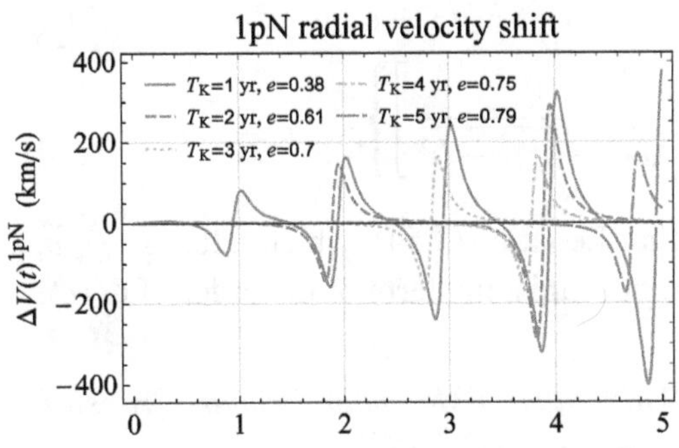

Figure 3.11 Analytically produced time series, in km/s, of the 1pN gravitoelectric shift $\Delta V(t)^{1\mathrm{pN}}$ of the radial velocity curve of a fictitious S star for different values of the e and the orbital period T_{K} in such a way that $r_{\min} = 1250\,\mathcal{R}_\bullet$. The other relevant orbital parameters, chosen arbitrarily, are $I = 161.24°, \omega = 0.073°, \eta = 20°$. Since the star can be considered a test particle with respect to such a BH, Equations (3.6)–(3.7) and Equations (3.10)–(3.21) are used along with Equations (2.83)–(2.87) in Equation (2.77) and Equation (2.78).

The Total Transit Duration δt_D

From Equations (2.102)–(2.105) and Equations (3.41)–(3.46), it turns out

$$\overline{\Delta\delta t_D}^{1\mathrm{pN}} = \frac{12\pi\mu_{\mathrm{b}} e\cos\omega}{c^2 n_{\mathrm{K}} a^2 R_\star \sqrt{1-e^2}\,(1+e\sin\omega)^3\left(1+e^2+2e\sin\omega\right)^{3/2}}$$

$$\left[-R_\star^2(1+\vartheta)^2(1+e\sin\omega)^3\right.$$
$$\left.+a^2\left(1-e^2\right)^2\cos^2 I\left(2+e^2+3e\sin\omega\right)\right]\frac{1}{\sqrt{(1+\vartheta)^2-b^2}}. \quad (3.53)$$

It should be noted that Equation (3.53), which falls as $1/\sqrt{a}$, vanishes for circular orbits, being its first nonvanishing term of the order of $\mathcal{O}(e)$.

The Ingress/Egress Transit Duration $\delta t_{\mathrm{i/e}}$

From Equations (2.108)–(2.111) and Equations (3.41)–(3.46), it turns out

$$\begin{aligned}\overline{\Delta\delta t_{\mathrm{i/e}}}^{\,\mathrm{1pN}} &= \frac{6\pi\mu_{\mathrm{b}}R_\star e\cos\omega}{c^2 n_{\mathrm{K}}a^2\sqrt{1-e^2}\left(1+2e\sin\omega+e^2\right)^{3/2}}\left\{\sqrt{(1-\vartheta)^2-b^2}\right.\\ &-\sqrt{(1+\vartheta)^2-b^2}\\ &+\frac{a^2\left(1-e^2\right)^2\cos^2 I\left(1+2e\sin\omega+e^2\right)}{R_\star^2(1+e\sin\omega)^3}\left[\frac{1}{\sqrt{(1+\vartheta)^2-b^2}}\right.\\ &\left.\left.-\frac{1}{\sqrt{(1-\vartheta)^2-b^2}}\right]\right\}. \quad (3.54)\end{aligned}$$

It should be noted that Equation (3.54), which falls as $1/\sqrt{a}$, vanishes for circular orbits, being its first nonvanishing term of the order of $\mathcal{O}(e)$.

The Full Width at Half Maximum Primary Transit Duration δt_H

From Equations (2.115)–(2.118) and Equations (3.41)–(3.46), it turns out

$$\begin{aligned}\overline{\Delta\delta t_H}^{\,\mathrm{1pN}} &= \frac{6\pi\mu_{\mathrm{b}}R_\star e\cos\omega}{c^2 n_{\mathrm{K}}a^2\sqrt{1-e^2}\left(1+2e\sin\omega+e^2\right)^{3/2}}\left\{-\sqrt{(1-\vartheta)^2-b^2}\right.\\ &-\sqrt{(1+\vartheta)^2-b^2}\\ &+\frac{a^2\left(1-e^2\right)^2\cos^2 I\left(1+2e\sin\omega+e^2\right)}{R_\star^2(1+e\sin\omega)^3}\left[\frac{1}{\sqrt{(1+\vartheta)^2-b^2}}\right.\\ &\left.\left.+\frac{1}{\sqrt{(1-\vartheta)^2-b^2}}\right]\right\}. \quad (3.55)\end{aligned}$$

It should be noted that Equation (3.55), which falls as $1/\sqrt{a}$, vanishes for circular orbits, being its first nonvanishing term of the order of $\mathcal{O}(e)$.

The Time of Inferior Conjunction t_{cj}

From Equations (2.121)–(2.124) and Equations (3.41)–(3.46), it turns out that

$$\overline{\Delta t_{\mathrm{cj}}}^{\,\mathrm{1pN}} = \frac{6\pi\mu_{\mathrm{b}}\left[4 + e^2 - 3\nu - e\left(-5 + 3\nu\right)\sin\omega\left(2 + e\sin\omega\right)\right]}{c^2 n_{\mathrm{K}} a\sqrt{1 - e^2}\left(1 + e\sin\omega\right)^2}. \tag{3.56}$$

To the zeroth order in e, Equation (3.56), which increases with the distance from the star as $\sqrt{a}$, reduces to

$$\overline{\Delta t_{\mathrm{cj}}}^{\,\mathrm{1pN}} \simeq \frac{\pi\mu_{\mathrm{b}}\left(4 - 3\nu\right)}{c^2 n_{\mathrm{K}} a} \tag{3.57}$$

A Numerical Evaluation

Figure 3.12 displays the plots of Equations (3.53)–(3.56) for a Jovian-type exoplanet transiting in front of its Sunlike host star, seen edge-on, as functions of the

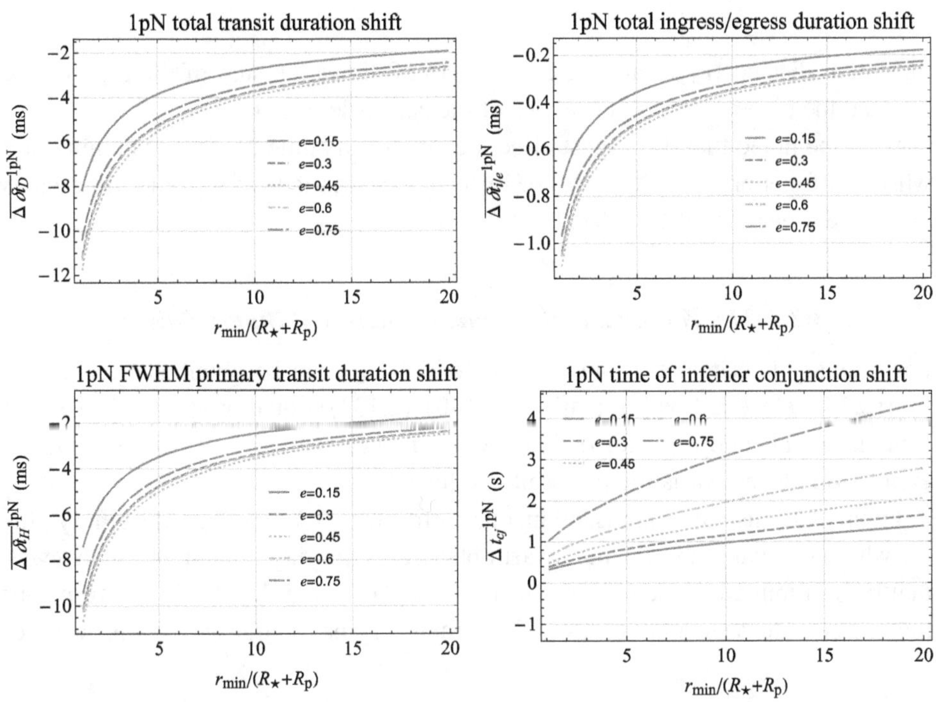

Figure 3.12 Plots of the net shifts per orbit $\overline{\Delta\delta t_D}^{\,\mathrm{1pN}}, \overline{\Delta\delta t_{\mathrm{i/e}}}^{\,\mathrm{1pN}}, \overline{\Delta\delta t_H}^{\,\mathrm{1pN}}, \overline{\Delta t_{\mathrm{cj}}}^{\,\mathrm{1pN}}$ of a fictitious Sun-Jupiter exoplanetary system, seen edge-on, as functions of the minimum star-planet distance r_{min}, in units of $R_\star + R_{\mathrm{p}}$, for different values of the eccentricity e according to Equations (3.53)–(3.56). The values $I = 90^\circ, \omega = 50^\circ$ are used for the relevant orbital parameters. The units are s for the time of inferior conjunction and ms for the other timescales.

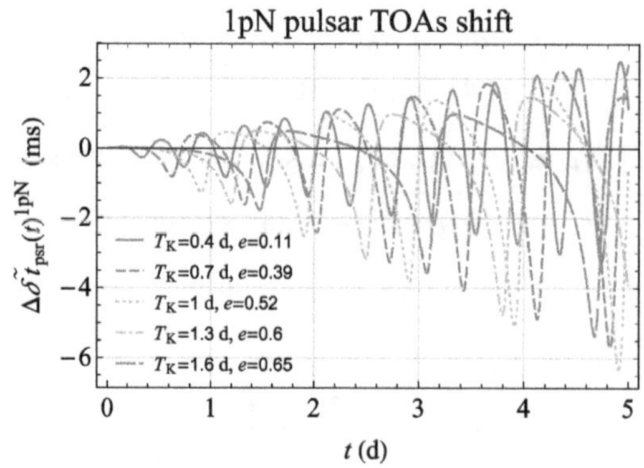

Figure 3.13 Plot of the 1pN gravitoelectric instantaneous shift $\Delta\widetilde{\delta t}_{\mathrm{psr}}(t)^{1\mathrm{pN}}$, in ms, of a binary pulsar with a white dwarf as companion for different values of T_K and e in such a way that $r_{\min} = 1.8 \times 10^6$ km. The relevant orbital parameters are $I = 75^\circ$, $\omega = 42.457^\circ$, $\eta = 20^\circ$. For the stellar corpses, the values $M_{\mathrm{psr}} = 1.27\,M_\odot$, $M_{\mathrm{wd}} = 1.02\,M_\odot$, taken from those of PSR J1141–6545 (Antoniadis et al., 2011), are used.

minimum distance $r_{\min}$, ranging from 1.1 to 20 times the sum of the radii of the star and the planet, for various values of the eccentricity e.

It can be noted that the largest effect occurs for the time of inferior conjunction whose shift per orbit is at the s level. Instead, the variations of the other timescales are of the order of $\simeq 0.1 - 10$ ms.

3.2.7 The Variation of the Times of Arrival of Binary Pulsars

For a binary pulsar, the instantaneous shift of $\widetilde{\delta t}_{\mathrm{psr}}$ can be obtained from Equation (2.77) calculated with Equations (2.130)–(2.134), Equations (3.6)–(3.10), and Equation (3.11) and Equation (3.21). By replacing f with $f_0 + 2\pi$ in the resulting expression, the net variation per orbit is inferred.

Figure 3.13 displays the instantaneous shifts $\Delta\widetilde{\delta t}_{\mathrm{psr}}(t)$ of a fictitious binary pulsar whose companion is a white dwarf obtained by varying T_K and e in such a way that the minimum distance is kept fixed to,[6] say, $r_{\min} = 1.8 \times 10^6$ km. It turns out that the peak-to-peak amplitudes of the signals may be as large as $\simeq 8$ ms over 5 d.

[6] It corresponds to actual minimum relative distance between the pulsar and the white dwarf of PSR J1141–6545 (Antoniadis et al., 2011).

4

2pN Gravitoelectric Effects: Mass Monopoles

To the second post-Newtonian (2pN) order, the gravitoelectric net shifts of the osculating Keplerian orbital elements occurring in a binary system made of two (non-rotating) spherical bodies A and B of comparable masses $M_{\rm A}$ and $M_{\rm B}$ can be calculated as explained in Section 2.2.2; see also Iorio (2020b, 2021b,a) where some mistakes, corrected from time to time, occurred. Among the several calculations dedicated solely to the pericentre advance existing in the literature that rely upon different computational schemes and parameterizations like, for example, Schäfer and Wex (1993), Kopeikin and Potapov (1994), Wex (1995), and Tucker and Will (2019), see, in particular, the one by Damour and Schäfer (1988), based on the Damour–Deruelle parametrization (Damour and Deruelle, 1985) and the Hamilton–Jacobi equation, which became widely adopted in binary analyses of pulsars.

The *total* 2pN net orbital shifts arise from two contributions.

The first one is due to the *direct* action of the 2pN gravitoelectric acceleration (see, for example, Brumberg, 1991, Equation (4.4.29), p. 154; Kidder, 1995, Equation (2.2d), p. 825; Gergely, 2010, Equation (B11), p. 10)

$$\begin{aligned} \boldsymbol{A}^{\rm 2pN} = \frac{\mu_{\rm b}}{c^4 r^2}\Bigg\{\Bigg[&\nu\left(-3+4\nu\right)v^4 + \frac{15}{8}\nu\left(-1+3\nu\right)v_r^4 \\ &+\nu\left(\frac{9}{2}-6\nu\right)v^2 v_r^2 + \nu\left(\frac{13}{2}-2\nu\right)\frac{\mu_{\rm b}}{r}v^2 \\ &+\left(2+25\nu+2\nu^2\right)\frac{\mu_{\rm b}}{r}v_r^2 - \left(9+\frac{87}{4}\nu\right)\frac{\mu_{\rm b}^2}{r^2}\Bigg]\hat{\boldsymbol{r}} + \Bigg[\nu\left(\frac{15}{2}+2\nu\right)v^2 \\ &-\nu\left(\frac{9}{2}+3\nu\right)v_r^2 - \left(2+\frac{41}{2}\nu+4\nu^2\right)\frac{\mu_{\rm b}}{r}\Bigg]v_r\boldsymbol{v}\Bigg\}. \end{aligned} \tag{4.1}$$

The second one arises from the *indirect* effect of the 1pN acceleration of Equation (3.28) for a two-body system when the average is calculated by accounting for the 1pN instantaneous changes of either the line of apsides (Equation (2.36)) and of the orbital elements themselves (Equation (2.37)) during the orbital revolution.

The 2pN contribution to the pericentre's evolution has been investigated so far especially in connection with binary pulsars, when the fractional[1] periastron shift per orbit $w_\omega := \overline{\Delta\omega}/2\pi$ (Damour and Schäfer, 1988) is estimated as one of the solve-for parameters (Damour and Deruelle, 1986; Damour and Taylor, 1992), S stars (Capuzzo-Dolcetta and Sadun-Bordoni, 2023), and BH binaries as that in the blazar OJ 287 (Dey et al., 2018, 2019).

4.1 The Direct Net Orbital Shifts

By projecting Equation (4.1) on the unit vectors $\hat{\boldsymbol{r}}, \hat{\boldsymbol{\tau}}, \hat{\boldsymbol{h}}$ defined in Equations (2.23)–(2.25) and by using Equations (2.7)–(2.8) along with Equation (2.11), its radial, transverse, and normal components, evaluated onto the Keplerian ellipse, are obtained; they turn out to be

$$\begin{aligned}
A_r^{2\text{pN}} &= \frac{\mu_\text{b}^3 (1+e\cos f)^2}{64c^4a^4\left(1-e^2\right)^4}\left(e^4\nu\,(39+191\nu)+16\,[-36+\nu\,(-73+8\nu)]\right.\\
&\quad +8e^2\,[-36+\nu\,(-13+72\nu)]+8e\,\{-144\\
&\quad +\nu\left[-288+80\nu+e^2\,(13+92\nu)\right]\}\cos f\\
&\quad +e^2\left(4\left\{-72+\nu\left[-298+144\nu+e^2\,(-45+11\nu)\right]\right\}\cos 2f\right.\\
&\quad \left.\left.+e\nu\,[8\,(-57+20\nu)\cos 3f+3e\,(-17+7\nu)\cos 4f]\right)\right), \qquad (4.2)\\
A_\tau^{2\text{pN}} &= -\frac{e\mu_\text{b}^3 (1+e\cos f)^3\sin f}{2c^4a^4\left(1-e^2\right)^4}\left\{4+\nu\left[26+4\nu-e^2\,(15+4\nu)\right]\right.\\
&\quad \left.+e\,(4+11\nu)\cos f+3e^2\nu\,(3+2\nu)\sin^2 f\right\}, \qquad (4.3)\\
A_h^{2\text{pN}} &= 0. \qquad (4.4)
\end{aligned}$$

By inserting Equations (4.2)–(4.4) in Equations (2.13)–(2.18) and integrating their right-hand sides evaluated onto the unperturbed Keplerian ellipse from f_0 to $f_0+2\pi$ by means of Equation (2.11) and Equation (2.19), the *direct* net orbital shifts per revolution of the order of $\mathcal{O}\left(1/c^4\right)$ are obtained; they are

$$\overline{\Delta a}_\text{dir}^{2\text{pN}} = 0, \qquad (4.5)$$

$$\overline{\Delta e}_\text{dir}^{2\text{pN}} = 0, \qquad (4.6)$$

$$\overline{\Delta I}_\text{dir}^{2\text{pN}} = 0, \qquad (4.7)$$

[1] Here, the symbol w_ω is used instead of k, adopted by Damour and Schäfer (1988), Damour and Deruelle (1986), and Damour and Taylor (1992), in order to avoid confusion with the nonsingular orbital element $k := e\sin\omega$.

$$\overline{\Delta\Omega}^{\text{2pN}}_{\text{dir}} = 0, \tag{4.8}$$

$$\overline{\Delta\omega}^{\text{2pN}}_{\text{dir}} = \frac{\pi\mu_{\text{b}}^2}{4c^4a^2\left(1-e^2\right)^2}\left\{e^2\left[-2+3\left(7-16\nu\right)\nu\right]+8\left[7+\left(5-7\nu\right)\nu\right]\right\}, \tag{4.9}$$

$$\begin{aligned}\overline{\Delta\eta}^{\text{2pN}}_{\text{dir}} = & \frac{\pi\mu_{\text{b}}^2}{4c^4a^2\left(1-e^2\right)^2}\left(88\sqrt{1-e^2}-42e^4\nu\right.\\ & +6\nu\left[-7+49\sqrt{1-e^2}-7\left(1-e^2\right)\nu+11\sqrt{1-e^2}\nu\right]\\ & +e^2\left\{2\sqrt{1-e^2}+\nu\left[84-200\sqrt{1-e^2}+42\left(1-e^2\right)\nu\right.\right.\\ & \left.\left.\left.-95\sqrt{1-e^2}\nu\right]\right\}\right).\end{aligned} \tag{4.10}$$

4.2 The Mixed Net Orbital Shifts

4.2.1 I: The 1pN Instantaneous Change of the Apsidal Line

Here, the mixed net orbital shifts arising from Equation (2.36), marked conventionally with the subscript I, are calculated by means of the radial, transverse, and normal components of Equation (3.28) given by Equations (3.32)–(3.34).

One finally obtains

$$\overline{\Delta a}^{\text{2pN}}_{\text{I}} = 0, \tag{4.11}$$

$$\overline{\Delta e}^{\text{2pN}}_{\text{I}} = 0, \tag{4.12}$$

$$\overline{\Delta I}^{\text{2pN}}_{\text{I}} = 0, \tag{4.13}$$

$$\overline{\Delta\Omega}^{\text{2pN}}_{\text{I}} = 0, \tag{4.14}$$

$$\begin{aligned}\overline{\Delta\omega}^{\text{2pN}}_{\text{I}} = & \frac{\pi\mu_{\text{b}}^2}{32c^4a^2e^2\left(1-e^2\right)^2}\left\{32\left(-3+\nu\right)^2\right.\\ & \left.+8e^2\left[148+5\nu\left(-43+17\nu\right)\right]+e^4\left[32+3\nu\left(56+75\nu\right)\right]\right\},\end{aligned} \tag{4.15}$$

$$\begin{aligned}\overline{\Delta\eta}^{\text{2pN}}_{\text{I}} = & -\frac{\pi\mu_{\text{b}}^2}{32c^4a^2e^2\left(1-e^2\right)^{3/2}}\left(e^4\left[160-\nu\left(56+699\nu\right)\right]\right.\\ & +32\left\{-75+84\sqrt{1-e^2}+\nu\left[158-164\sqrt{1-e^2}\right.\right.\\ & \left.\left.+\left(-76+77\sqrt{1-e^2}\right)\nu\right]\right\}-8e^2\left\{-820+336\sqrt{1-e^2}\right.\\ & \left.\left.+\nu\left[1183-656\sqrt{1-e^2}+\left(-523+308\sqrt{1-e^2}\right)\nu\right]\right\}\right).\end{aligned} \tag{4.16}$$

4.2.2 II: The 1pN Instantaneous Orbital Shifts

Here, the mixed net orbital shifts arising from Equation (2.37) are calculated; they are marked conventionally with the subscript II. To compute them, the expressions for the 1pN instantaneous shifts of a and e for a two-body system, given by Equations (3.35)–(3.36), are needed.

As a result, they turn out to be

$$\overline{\Delta a}_{\mathrm{II}}^{\mathrm{2pN}} = 0, \tag{4.17}$$

$$\overline{\Delta e}_{\mathrm{II}}^{\mathrm{2pN}} = 0, \tag{4.18}$$

$$\overline{\Delta I}_{\mathrm{II}}^{\mathrm{2pN}} = 0, \tag{4.19}$$

$$\overline{\Delta \Omega}_{\mathrm{II}}^{\mathrm{2pN}} = 0, \tag{4.20}$$

$$\begin{aligned}\overline{\Delta \omega}_{\mathrm{II}}^{\mathrm{2pN}} &= \frac{\pi \mu_{\mathrm{b}}^2}{32 c^4 a^2 e^2 \left(1-e^2\right)^2} \left\{-32\left(-3+\nu\right)^2\right. \\ &\quad - 8e^2\left(-3+\nu\right)\left(-64+29\nu\right) + e^4\left[32+3\nu\left(48+53\nu\right)\right] \\ &\quad \left. + 768 e^3 \left(-2+\nu\right)\cos f_0\right\},\end{aligned} \tag{4.21}$$

$$\begin{aligned}\overline{\Delta \eta}_{\mathrm{II}}^{\mathrm{2pN}} &= -\frac{\pi \mu_{\mathrm{b}}^2}{32 c^4 a^2 e^2 \left(1-e^2\right)^{5/2}} \left(e^6\left(960-3136\nu+2458\nu^2\right)\right. \\ &\quad + e^4\left[48\left(55+64\sqrt{1-e^2}\right) - 8\left(467+976\sqrt{1-e^2}\right)\nu\right. \\ &\quad \left. + \left(-29+4928\sqrt{1-e^2}\right)\nu^2\right] - 32\left\{-75+168\sqrt{1-e^2}\right. \\ &\quad \left. + 2\nu\left[79-164\sqrt{1-e^2}+\left(-38+77\sqrt{1-e^2}\right)\nu\right]\right\} \\ &\quad + 8e^2\left\{48\left(-25+36\sqrt{1-e^2}\right)+\nu\left[2031-2976\sqrt{1-e^2}\right.\right. \\ &\quad \left.\left. + \left(-767+1120\sqrt{1-e^2}\right)\nu\right]\right\} + 16e^3\left[\cos f_0\left(e^2\left\{60\right.\right.\right. \\ &\quad + 288\sqrt{1-e^2}+\nu\left[69-708\sqrt{1-e^2}\right. \\ &\quad \left.\left. + 7\left(-9+62\sqrt{1-e^2}\right)\nu\right]\right\} + 4\left\{3\left(-55+56\sqrt{1-e^2}\right)\right. \\ &\quad \left. + \nu\left[174-268\sqrt{1-e^2}+\left(-45+84\sqrt{1-e^2}\right)\nu\right]\right\} \\ &\quad + 3e^2\left(5-3\nu\right)\nu\cos 2f_0\Big) + 2e\left[\left(-5+4\nu\right)\left(15-24\sqrt{1-e^2}\right.\right. \\ &\quad \left.\left.\left.\left. - 9\nu+28\sqrt{1-e^2}\,\nu\right)\cos 2f_0 + e\sqrt{1-e^2}\,\nu\left(-6+7\nu\right)\cos 3f_0\right]\right]\right).\end{aligned} \tag{4.22}$$

4.2.3 The Total Mixed Net Orbital Shifts

From the sum of Equations (4.11)–(4.16) and Equations (4.17)–(4.22), the total 2pN mixed net orbital shifts turn out to be

$$\overline{\Delta a}^{\rm 2pN}_{\rm mix} = 0, \tag{4.23}$$

$$\overline{\Delta e}^{\rm 2pN}_{\rm mix} = 0, \tag{4.24}$$

$$\overline{\Delta I}^{\rm 2pN}_{\rm mix} = 0, \tag{4.25}$$

$$\overline{\Delta \Omega}^{\rm 2pN}_{\rm mix} = 0, \tag{4.26}$$

$$\begin{aligned}\overline{\Delta \omega}^{\rm 2pN}_{\rm mix} = & \frac{\pi\mu_b^2}{4c^4a^2\left(1-e^2\right)^2}\left[-44+8\nu\left(-8+7\nu\right)\right.\\ & \left.+e^2\left(8+39\nu+48\nu^2\right)+96e\left(-2+\nu\right)\cos f_0\right],\end{aligned} \tag{4.27}$$

$$\begin{aligned}\overline{\Delta \eta}^{\rm 2pN}_{\rm mix} = & -\frac{\pi\mu_b^2}{32c^4a^2e^2\left(1-e^2\right)^{5/2}}\left(-32\sqrt{1-e^2}\left(-6+7\nu\right)\left(-14+11\nu\right)\right.\\ & +e^6\left[800+77\nu\left(-40+41\nu\right)\right]+32e^2\left\{4\left(-5+66\sqrt{1-e^2}\right)\right.\\ & \left.+\nu\left[54-416\sqrt{1-e^2}+3\left(5+42\sqrt{1-e^2}\right)\nu\right]\right\}\\ & +8e^4\left\{-470+720\sqrt{1-e^2}+\nu\left[709-1632\sqrt{1-e^2}\right.\right.\\ & \left.\left.+\left(-614+924\sqrt{1-e^2}\right)\nu\right]\right\}+16e^3\left[\cos f_0\left(e^2\left\{60\right.\right.\right.\\ & +288\sqrt{1-e^2}+\nu\left[69-708\sqrt{1-e^2}\right.\\ & \left.\left.+7\left(-9+62\sqrt{1-e^2}\right)\nu\right]\right\}+4\left\{3\left(-55+56\sqrt{1-e^2}\right)\right.\\ & \left.+\nu\left[174-268\sqrt{1-e^2}+\left(-45+84\sqrt{1-e^2}\right)\nu\right]\right\}\\ & \left.+3e^2\left(5-3\nu\right)\nu\cos 2f_0\right)+2e\left[\left(-5+4\nu\right)\left(15\right.\right.\\ & \left.-24\sqrt{1-e^2}-9\nu+28\sqrt{1-e^2}\nu\right)\cos 2f_0\\ & \left.\left.\left.+e\sqrt{1-e^2}\nu\left(-6+7\nu\right)\cos 3f_0\right]\right]\right).\end{aligned} \tag{4.28}$$

4.3 The Total Net Orbital Shifts

The total 2pN net shifts per orbit can be obtained by summing Equations (4.5)–(4.10) and Equations (4.23)–(4.28). One finally gets

$$\overline{\Delta a}^{\rm 2pN} = 0, \tag{4.29}$$

$$\overline{\Delta e}^{\rm 2pN} = 0, \tag{4.30}$$

$$\overline{\Delta I}^{\rm 2pN} = 0, \tag{4.31}$$

$$\overline{\Delta \Omega}^{\rm 2pN} = 0, \tag{4.32}$$

$$\overline{\Delta \omega}^{\rm 2pN} = \frac{3\pi\mu_{\rm b}^2\left[2-4\nu+e^2\left(1+10\nu\right)+16e\left(-2+\nu\right)\cos f_0\right]}{2c^4a^2\left(1-e^2\right)^2}, \tag{4.33}$$

$$\begin{aligned}\overline{\Delta \eta}^{\rm 2pN} = & -\frac{\pi\mu_{\rm b}^2}{32c^4a^2e^2\left(1-e^2\right)^2}\left\{-32\sqrt{1-e^2}\left(-6+7\nu\right)\left(-14+11\nu\right)\right.\\ & +3e^6\left[272+\nu\left(-1560+112\sqrt{1-e^2}+799\nu\right.\right.\\ & \left.\left.+112\sqrt{1-e^2}\nu\right)\right]+16e^2\left[-84+528\sqrt{1-e^2}\right.\\ & \left.+\nu\left(-39-811\sqrt{1-e^2}+273\sqrt{1-e^2}\nu\right)\right]\\ & +24e^4\left\{16\left(-8+15\sqrt{1-e^2}\right)+\nu\left[401-572\sqrt{1-e^2}\right.\right.\\ & \left.\left.+\left(-151+280\sqrt{1-e^2}\right)\nu\right]\right\}+8e^3\left(\{8\left[3\left(-55\right.\right.\right.\\ & \left.+56\sqrt{1-e^2}\right)+\nu\left(174-268\sqrt{1-e^2}-45\nu\right.\\ & \left.\left.+84\sqrt{1-e^2}\nu\right)\right]+e^2\left[24\left(5+24\sqrt{1-e^2}\right)\right.\\ & \left.\left.+\nu\left(153-1416\sqrt{1-e^2}-135\nu+868\sqrt{1-e^2}\nu\right)\right]\right\}\cos f_0\\ & +e\left[15-24\sqrt{1-e^2}+\left(-9+28\sqrt{1-e^2}\right)\nu\right]\left[4\left(-5+4\nu\right)\cos 2f_0\right.\\ & \left.\left.\left.+e\nu\cos 3f_0\right]\right)\right\}. \end{aligned} \tag{4.34}$$

4.4 The Total Net Orbital Precessions

As remarked in Section 2.2.2, care is needed in correctly calculating the total 2pN orbital *precessions* since the mere division of Equations (4.29)–(4.34) by the Keplerian orbital period $T_{\rm K}$ is *not* enough to obtain them.

The correct procedure consists in expanding the ratios of the 1pN net shifts per orbit to the 1pN apsidal period in powers of $1c$, and adding the resulting terms of the order $\mathcal{O}\left(1/c^4\right)$ to the ratios of Equations (4.29)–(4.34) to $T_{\rm K}$.

By taking the ratios of the sum of Equations (4.29)–(4.34) and Equations (3.41)–(3.46) to the anomalistic period of Equation (3.49), calculated with Equation (3.50), and expanding the resulting expressions in powers of $1/c$ to the order $\mathcal{O}\left(1/c^4\right)$, one finally gets the *total* 2pN *precessions* of the Keplerian orbital elements:

$$\left\langle\frac{da}{dt}\right\rangle^{\rm 2pN} = 0, \tag{4.35}$$

$$\left\langle \frac{de}{dt} \right\rangle^{\text{2pN}} = 0, \tag{4.36}$$

$$\left\langle \frac{dI}{dt} \right\rangle^{\text{2pN}} = 0, \tag{4.37}$$

$$\left\langle \frac{d\Omega}{dt} \right\rangle^{\text{2pN}} = 0, \tag{4.38}$$

$$\begin{aligned}\left\langle \frac{d\omega}{dt} \right\rangle^{\text{2pN}} &= -\frac{3 n_{\text{K}} \mu_{\text{b}}^2}{4c^4 a^2 \left(1-e^2\right)^3} \left(34 + e^2 \left(43 - 52\nu\right) + e^4 \left(13 - 4\nu\right) - 4\nu \right. \\ &\quad + e\left\{\left[116 + e^2 \left(4 - 29\nu\right) - 52\nu\right] \cos f_0 \right. \\ &\quad \left.\left. - 3e\left[-10 + \nu \left(8 + e\cos f_0\right)\right] \cos 2f_0\right\}\right),\end{aligned} \tag{4.39}$$

$$\begin{aligned}\left\langle \frac{d\eta}{dt} \right\rangle^{\text{2pN}} &= \frac{n_{\text{K}} \mu_{\text{b}}^2}{64c^4 a^2 e^2 \left(1-e^2\right)^{5/2}} \left[32\sqrt{1-e^2} \left(-6 + 7\nu\right) \left(-14 + 11\nu\right) \right. \\ &\quad - 16e^2 \left\{-624 + 744\sqrt{1-e^2} + \nu \left[405 - 1111\sqrt{1-e^2} \right.\right. \\ &\quad \left.\left. + \left(-75 + 329\sqrt{1-e^2}\right) \nu\right]\right\} - 8e^4 \left\{-1644 + 1224\sqrt{1-e^2} \right. \\ &\quad \left. + \nu \left[3099 - 2760\sqrt{1-e^2} + \left(-1137 + 1372\sqrt{1-e^2}\right) \nu\right]\right\} \\ &\quad - e^6 \left\{48\left(-43 + 24\sqrt{1-e^2}\right) + \nu \left[408 - 2352\sqrt{1-e^2} \right.\right. \\ &\quad \left.\left. + \left(381 + 1904\sqrt{1-e^2}\right) \nu\right]\right\} + 8e^3 \left(\cos f_0 \left\{-7\sqrt{1-e^2} \left(-6 \right.\right.\right. \\ &\quad \left. + 7\nu\right) \left[8\left(-7 + 3\nu\right) + e^2 \left(-24 + 31\nu\right)\right] + 24 \left(-5 + 3\nu\right) \left[-32 \right. \\ &\quad \left.\left. + \nu \left(14 + e^2 \cos 2f_0\right) + e^2 \left(-8 + 13\nu\right)\right]\right\} \\ &\quad + e\left[-4\left(-5 + 4\nu\right) \left(60 - 42\sqrt{1-e^2} - 36\nu \right.\right. \\ &\quad \left.\left.\left.\left. + 49\sqrt{1-e^2}\nu\right) \cos 2f_0 + 7e\sqrt{1-e^2} \left(6 - 7\nu\right) \nu \cos 3f_0\right]\right)\right].\end{aligned} \tag{4.40}$$

In the limit of small eccentricities, Equations (4.39)–(4.40) reduce to

$$\left\langle \frac{d\omega}{dt} \right\rangle^{\text{2pN}} \simeq \frac{3 n_{\text{K}} \mu_{\text{b}}^2 \left[-17 + 2\nu + 2e\left(-29 + 13\nu\right) \cos f_0\right]}{2c^4 a^2} + \mathcal{O}\left(e^2\right), \tag{4.41}$$

$$\begin{aligned}\left\langle \frac{d\eta}{dt} \right\rangle^{\text{2pN}} &\simeq \frac{n_{\text{K}} \mu_{\text{b}}^2}{2c^4 a^2 e^2} \left\{84 + \nu \left(-164 + 77\nu\right) \right. \\ &\quad \left. + e^2 \left[108 + \nu \left(25 + 27\nu\right)\right] - 2e^3 \left[-186 + \nu \left(29 + 21\nu\right)\right] \cos f_0\right\} \\ &\quad + \mathcal{O}\left(e^2\right).\end{aligned} \tag{4.42}$$

It can be noted that, while Equation (4.41) is regular for $e \to 0$, Equation (4.42) is divergent.

4.5 Confrontation with Other Approaches in the Literature

Here, it is shown that, as far as the pericentre is concerned, Equation (4.33) is in agreement with other results which can be found in the literature: the derivation by Damour and Schäfer (1988), based on the Hamilton–Jacobi method and the Damour–Deruelle parametrization (Damour and Deruelle, 1985), and the calculation by Kopeikin and Potapov (1994) making use of the osculating Keplerian orbital elements and the Gauss equations, although with a different computational approach with respect to the present one.

4.5.1 The Damour–Deruelle Parameterization

The net shift per orbit of the pericentre of Equation (4.33) is in agreement with the corresponding expression, written in terms of the osculating Keplerian orbital elements, which can be extracted from the sum of the fractional 1pN and 2pN pericentre advances of Damour and Schäfer (1988, Equation (5.18)):

$$w_{\omega}^{1\mathrm{pN}} + w_{\omega}^{2\mathrm{pN}} = \frac{3\left(\mu_{\mathrm{b}} n_{\mathrm{DD}}\right)^{2/3}}{c^2\left(1 - e_{\mathrm{T}}^2\right)}\left[1 + \frac{\left(\mu_{\mathrm{b}} n_{\mathrm{DD}}\right)^{2/3}}{c^2\left(1 - e_{\mathrm{T}}^2\right)}\left(\frac{39}{4}x_{\mathrm{A}}^2 + \frac{27}{4}x_{\mathrm{B}}^2 + 15 x_{\mathrm{A}} x_{\mathrm{B}}\right)\right.$$
$$\left. - \frac{\left(\mu_{\mathrm{b}} n_{\mathrm{DD}}\right)^{2/3}}{c^2}\left(\frac{13}{4}x_{\mathrm{A}}^2 + \frac{1}{4}x_{\mathrm{B}}^2 + \frac{13}{3}x_{\mathrm{A}} x_{\mathrm{B}}\right)\right]. \tag{4.43}$$

In Equation (4.43),

$$x_{\mathrm{A}} := \frac{M_{\mathrm{A}}}{M_{\mathrm{b}}}, \tag{4.44}$$

$$x_{\mathrm{B}} := \frac{M_{\mathrm{B}}}{M_{\mathrm{b}}} = 1 - x_{\mathrm{A}} \tag{4.45}$$

are the normalized masses of the bodies A and B, respectively, to the system's total mass, while e_{T} and n_{DD} are members of the Damour–Deruelle formalism (Damour and Deruelle, 1985) which, in the limit $c \to \infty$, reduce to the Keplerian eccentricity e and mean motion n_{K}, as will be shown in what follows.

The so-called 'proper time' eccentricity e_{T} reads (Damour and Deruelle, 1986, pag. 272)

$$e_{\mathrm{T}} = e_t\left(1 + \tilde{\delta}\right) + e_\theta - e_r, \tag{4.46}$$

where (Damour and Deruelle, 1985, Equation (3.8b))

$$e_t = \frac{e_R}{1 + \frac{\mu_{\mathrm{b}}}{c^2 a_R}\left(4 - \frac{3}{2}\nu\right)}, \tag{4.47}$$

(Damour and Deruelle, 1985, Equation (4.13))

$$e_\theta = e_R \left(1 + \frac{\mu_b}{2c^2 a_R}\right), \tag{4.48}$$

(Damour and Deruelle, 1985, Equation (6.3b))

$$e_r = e_R \left[1 - \frac{\mu_b}{2c^2 a_R}\left(x_A^2 - \nu\right)\right], \tag{4.49}$$

and (Damour and Deruelle, 1986, Equation (20))[2]

$$\tilde{\delta} = \frac{\mu_b}{c^2 a_R}\left(x_A x_B + 2x_B^2\right). \tag{4.50}$$

In Equations (4.47)–(4.50), a_R is another member of the Damour–Deruelle parametrization. According to Equations (4.47)–(4.50), Equation (4.46) can be expressed in terms of only a_R, e_R as

$$\frac{e_T}{e_R} = \frac{1 + \frac{\mu_b}{2c^2 a_R}\left[4 + 3\left(x_A - 2\right)x_A\right] + \frac{\mu_b^2}{4c^4 a_R^2}\left(8 - 3\nu\right)x_A^2}{1 + \frac{\mu_b}{2c^2 a_R}\left(8 - 3\nu\right)}. \tag{4.51}$$

The Damour–Deruelle mean motion is (Damour and Deruelle, 1985, Equation (3.7))

$$n_{DD} := \sqrt{\frac{\mu_b}{a_R^3}}\left[1 + \frac{\mu_b}{2c^2 a_R}\left(-9 + \nu\right)\right]. \tag{4.52}$$

Equations (4.51)–(4.52) are both functions of a_R, e_R which, in turn, can be expressed in terms of the osculating Keplerian a and e by means of (Klioner and Kopeikin, 1994, Equations (28)–(29))

$$a_R = a - da_0 - \frac{\mu_b}{c^2\left(1 - e^2\right)^2}\left[-3 + \nu + e^2\left(-13 + e^2 + 7\nu + 2e^2\nu\right)\right], \tag{4.53}$$

$$e_R = e - de_0 - \frac{e\mu_b}{2c^2 a\left(1 - e^2\right)}\left[-17 + 6\nu + e^2\left(2 + 4\nu\right)\right], \tag{4.54}$$

with (Klioner and Kopeikin, 1994, Equation (14))

$$\begin{aligned} da_0 = {} & \frac{e\mu_b}{4c^2\left(1 - e^2\right)^2}\left\{\left[8\left(-7 + 3\nu\right) + e^2\left(-24 + 31\nu\right)\right]\cos f_0\right. \\ & \left. + e\left[4\left(-5 + 4\nu\right)\cos 2f_0 + e\nu\cos 3f_0\right]\right\}, \end{aligned} \tag{4.55}$$

and (Klioner and Kopeikin, 1994, Equation (16))

[2] Here, the symbol $\tilde{\delta}$ is used instead of δ as in Damour and Deruelle (1986) in order to avoid confusion with the astrometric angle dec.

$$de_0 = \frac{\mu_b}{8c^2 a\left(1-e^2\right)}\left\{\left[8\left(-3+\nu\right)+e^2\left(-56+47\nu\right)\right]\cos f_0\right.$$
$$\left.+e\left[4\left(-5+4\nu\right)\cos 2f_0+e\nu\cos 3f_0\right]\right\}. \tag{4.56}$$

Note that Equations (4.53)–(4.56) are written for GTR; their general expressions for a given class of alternative theories of gravitation can be found in Klioner and Kopeikin (1994). The final expressions for a_R, e_R are

$$\frac{a_R}{a} = 1-\frac{\mu_b}{c^2 a\left(1-e^2\right)^2}\left[-3+\nu+e^4\left(1+2\nu\right)+e^2\left(-13+7\nu\right)\right]$$
$$+\,e\frac{\mu_b}{4c^2 a\left(1-e^2\right)^2}\left\{\left[56+e^2\left(24-31\nu\right)-24\nu\right]\cos f_0\right.$$
$$\left.+e\left[4\left(5-4\nu\right)\cos 2f_0-e\nu\cos 3f_0\right]\right\}, \tag{4.57}$$
$$\frac{e_R}{e} = 1-\frac{\mu_b}{2c^2 a\left(1-e^2\right)}\left[-17+6\nu+e^2\left(2+4\nu\right)\right]$$
$$-\,\frac{\mu_b}{8c^2 ae\left(1-e^2\right)}\left\{\left[8\left(-3+\nu\right)+e^2\left(-56+47\nu\right)\right]\cos f_0\right.$$
$$\left.+\,e\left[4\left(-5+4\nu\right)\cos 2f_0+e\nu\cos 3f_0\right]\right\}. \tag{4.58}$$

By using Equations (4.57)–(4.58), Equations (4.51)–(4.52) can be finally expressed, to the order of $\mathcal{O}\left(c^{-2}\right)$, as

$$\frac{8c^2 a\left(e-e_{\mathrm{T}}\right)\left(1-e^2\right)}{\mu_b} = \left[8\left(-3+\nu\right)+e^2\left(-56+47\nu\right)\right]\cos f_0$$
$$+\,e\left(4\left\{-13+3\nu-3\left(-2+x_{\mathrm{A}}\right)x_{\mathrm{A}}\right.\right.$$
$$\left.+\,e^2\left[-2+7\nu+3\left(-2+x_{\mathrm{A}}\right)x_{\mathrm{A}}\right]\right\}$$
$$\left.+\,4\left(-5+4\nu\right)\cos 2f_0+e\nu\cos 3f_0\right), \tag{4.59}$$
$$\left(\frac{n_{\mathrm{DD}}}{n_{\mathrm{K}}}-1\right)\frac{8c^2 a\left(1-e^2\right)^2}{\mu_b} = 8\left(-9+2\nu\right)+4e^4\left(-6+7\nu\right)+e^2\left(-84+76\nu\right)$$
$$+\,3e\left\{\left[8\left(-7+3\nu\right)+e^2\left(-24+31\nu\right)\right]\cos f_0\right.$$
$$\left.+\,e\left[4\left(-5+4\nu\right)\cos 2f_0+e\nu\cos 3f_0\right]\right\}. \tag{4.60}$$

A power expansion to the order of $\mathcal{O}\left(1/c^4\right)$ of Equation (4.43), calculated with Equations (4.59)–(4.60), yields just the ratio of Equation (4.33) to 2π.

4.5.2 The Calculation by Kopeikin and Potapov

The calculation by Kopeikin and Potapov (1994) is, perhaps, the most similar to the present one; it is based on the use of the osculating Keplerian orbital elements, although the calculational approach is different.

The sum of the fractional 1pN and 2pN pericentre shifts per orbit is[3] (Kopeikin and Potapov, 1994, Equation (5.2))

$$w_{\omega}^{1\mathrm{pN}} + w_{\omega}^{2\mathrm{pN}} = \frac{3\mu_{\mathrm{b}}}{c^2\xi_1\left(1-\xi_2^2\right)}\left[1+\frac{\mu_{\mathrm{b}}}{c^2\xi_1\left(1-\xi_2^2\right)}\left(\frac{3}{2}\nu+\frac{3}{4}\right)-\frac{\mu_{\mathrm{b}}}{\xi_1}\left(\frac{5}{2}\nu+\frac{1}{4}\right)\right]. \tag{4.61}$$

In Equation (4.61), ξ_1 and ξ_2 are the constants of integration of the solutions of the Gauss equations for the semimajor axis and the eccentricity to the 1pN level to be determined with the initial conditions at $t = t_0$. They can be obtained, for example, by evaluating Kopeikin and Potapov (1994, Equations (4.5)–(4.6)) at $t = t_0$ by replacing f, that is, V in the notation of Kopeikin and Potapov (1994), with f_0, and by recalling that, in the present notation, $a_0 \rightarrow \xi_1$, $e_0 \rightarrow \xi_2$. Moreover, a and e are, in the present book, the osculating numerical values of the semimajor axis and the eccentricity, respectively, at the same arbitrary instant t_0; thus, $a\left(t_0\right), e\left(t_0\right)$ in the left-hand side of Kopeikin and Potapov (1994, Equations (4.5)–(4.6)) are just a, e here. Then, one gets

$$\xi_1 \simeq a\left\{1-\frac{\mu_{\mathrm{b}}}{c^2a\left(1-e^2\right)^2}\left[\left(-14+6\nu-6e^2+\frac{31}{4}\nu e^2\right)e\cos f_0\right.\right.$$
$$\left.\left.+\left(-5+4\nu\right)e^2\cos 2f_0+\frac{1}{4}\nu e^3\cos 3f_0\right]\right\}+\mathcal{O}\left(1/c^4\right), \tag{4.62}$$

$$\xi_2 \simeq e\left[1-\frac{\mu_{\mathrm{b}}}{c^2a\left(1-e^2\right)e^2}\left\{\left(-3+\nu-7e^2+\frac{47}{8}\nu e^2\right)e\cos f_0\right.\right.$$
$$\left.\left.+\left(-\frac{5}{2}+2\nu\right)e^2\cos 2f_0+\frac{1}{8}\nu e^3\cos 3f_0\right\}\right]+\mathcal{O}\left(1/c^4\right). \tag{4.63}$$

By substituting Equations (4.62)–(4.63) in Equation (4.61) and expanding the resulting expression in powers of $1/c$ to the order of $\mathcal{O}\left(1/c^4\right)$, one gets just the ratio of Equation (4.33) to 2π.

[3] Here, the notation ξ_1, ξ_2 is adopted instead of a_0, e_0 used by Kopeikin and Potapov (1994).

5

1pN Gravitomagnetic Effects: Spin Dipole(s)

To the 1pN order, the gravitational field of an isolated, slowly rotating massive object is characterized by a stationary, magnetlike component, dubbed as 'gravitomagnetic', sourced by its angular momentum $\boldsymbol{J}$.

For an ordinary material body of mass M and equatorial radius $R_{\rm e}$, it is

$$J = \mathfrak{I}\omega, \tag{5.1}$$

where

$$\mathfrak{I} := \mathfrak{i}MR_{\rm e}^2 \tag{5.2}$$

is the moment of inertia, $\mathfrak{i}$ is the normalized moment of inertia (NMoI), and ω is the angular speed.

For a fluid object in hydrostatic equilibrium like, for example, a main sequence star or a gaseous giant planet, the Darwin–Radau approximate relation between the NMoI and the dimensionless quadrupole mass moment J_2 yields (Murray and Dermott, 1999, Equation (4.112), p. 153)

$$\mathfrak{i} = \frac{2}{3}\left(1 - \frac{2}{5}\sqrt{\frac{5}{2}\frac{q_{\rm c}}{\widetilde{f}} - 1}\right). \tag{5.3}$$

In Equation (5.3), it is[1]

$$q_{\rm c} := \frac{\omega^2 R_{\rm e}^3}{\mu}, \tag{5.4}$$

$$\widetilde{f} := \left(\frac{1+k_2}{2}\right) q_{\rm c}, \tag{5.5}$$

[1] Equation (5.4) comes from Murray and Dermott (1999, Equation (4.102), p. 150), while Equation (5.5) is Murray and Dermott (1999, Equation (4.110), p. 152) rewritten by means of $J_2 \simeq (k_2/3)\, q_{\rm c}$ (Ragozzine and Wolf, 2009; Correia and Rodríguez, 2013) for the body's first even zonal harmonic (see Chapter 7); in it, the possible tidal effects raised by the second object in the binary system are a priori neglected because the latter is assumed to be a test particle.

where k_2 is the Love number[2] (Love, 1911; Sterne, 1939; Kopal, 1959; Ragozzine and Wolf, 2009; Leconte et al., 2011). For a Jupiter-like planet, it lies in the range $0.1 \lesssim k_2^{\mathrm{p}} \lesssim 0.6$ (Ragozzine and Wolf, 2009), while for a main sequence star it is of the order of[3] $k_2^\star \simeq 0.015$ (Claret, 1995). In the stellar case, the angular speed $\omega_\star$ can be inferred from the measured values, when available, of the inclination $i_\star$ of $\hat{\boldsymbol{J}}_\star$ to the line of sight and the projected stellar rotational velocity[4]

$$\mathfrak{u}_\star := v_{\mathrm{e}}^\star \sin i_\star, \tag{5.6}$$

where $v_{\mathrm{e}}^\star$ is the star's equatorial rotational velocity, as

$$\omega_\star = \frac{2\pi R_{\mathrm{e}}^\star \sin i_\star}{\mathfrak{u}_\star}. \tag{5.7}$$

On the other hand, the stellar rotational period $P_\star := 2\pi/\omega_\star$ can also be estimated, for example, from photometric time series or spectroscopic time series of activity indices (Rainer et al., 2023); the star's equatorial radius R_{e} can be derived, for example, from spectral energy distribution (SED) fitting (Rainer et al., 2023). Helioseismology (Basu, 2016; Buldgen et al., 2019) yields for the Sun's angular momentum the value (Pijpers, 1998)

$$J_\odot = 1.90 \times 10^{41}\ \mathrm{kg\,m^2/s}; \tag{5.8}$$

for a collection of other values for it, see Iorio (2012a) and references therein.

As far as white dwarfs are concerned, the moment of inertia $\mathfrak{I}_{\mathrm{wd}}$ of the companion (Antoniadis et al., 2011) of the pulsar in the binary system PSR J1141–6545 (Kaspi et al., 2000) is of the order of (Boshkayev et al., 2017; Iorio, 2020a)

$$\mathfrak{I}_{\mathrm{wd}} \simeq 10^{43}\ \mathrm{kg\,m^2}. \tag{5.9}$$

By assuming for the white dwarf a rotational frequency (Boshkayev et al., 2017; Iorio, 2020a)

$$\omega_{\mathrm{wd}} = \sqrt{\frac{\mu_{\mathrm{wd}}}{R_{\mathrm{wd}}^3}}, \tag{5.10}$$

[2] It measures the mass concentration towards the centre of a fluid body assumed in hydrostatic equilibrium like, for example, a main sequence star. Its possible values range from 0 for the mass point approximation to $3/2$ for a fully homogeneous fluid body (Kellermann et al., 2018; Hellard et al., 2019). It should be noted that the Love number k_2 entering Equation (7.4) is the one used by geophysicists; astronomers often adopt a different definition of it, known as apsidal constant and equally denoted as k_2, being half the previous one (Poisson and Will, 2014, p. 115). Here, the apsidal constant is dubbed $k_2^{'}$, so that $k_2^{'} = k_2/2$; compare with Poisson and Will (2014, Equation (2.249), p. 118) where $k_2^{'\,\max} = 3/4$.

[3] For example, Claret et al. (2021) obtained for the apsidal constants of the members of the binary star DI Herculis (Hoffmeister, 1930) the values $\log k_2^{'\,\mathrm{A}} = -2.146$, $\log k_2^{'\,\mathrm{B}} = -2.171$, quoted also by Liang et al. (2022), corresponding to $k_2^{\mathrm{A}} = 0.0142$, $k_2^{\mathrm{B}} = 0.0135$.

[4] It can be measured from full-width at half-maximum (FWHM) of the cross-correlation function (CCF) of the reduced spectra with a stellar mask (Rainer et al., 2023).

where $M_{\rm wd} = 1.02\,M_\odot$, $R_{\rm wd} = 5400\,{\rm km}$ (Antoniadis et al., 2011), its angular momentum turns out to be of the order of

$$J_{\rm wd} \simeq 2.2 \times 10^{43}\,{\rm kg\,m^2/s}. \tag{5.11}$$

For a typical neutron star, the moment of inertia $\mathfrak{I}_{\rm ns}$ is of the order of (Ravenhall and Pethick, 1994; Lattimer and Prakash, 2001; Bejger and Haensel, 2002; Lattimer and Schutz, 2005; Worley et al., 2008; Breu and Rezzolla, 2016; Zhao, 2016; Greif et al., 2020)

$$\mathfrak{I}_{\rm ns} \simeq 1 \times 10^{38}\,{\rm kg\,m^2}; \tag{5.12}$$

for PSR J0737–3039A, recent estimates yield (Silva et al., 2021)

$$\mathfrak{I}_{\rm A} \simeq 1.6 \times 10^{38}\,{\rm kg\,m^2}. \tag{5.13}$$

Since the spin period of such a pulsar is 22.699 ms (Kramer et al., 2006), its angular momentum can be evaluated to be

$$J_{\rm A} \simeq 4.4 \times 10^{40}\,{\rm kg\,m^2/s}. \tag{5.14}$$

Instead, since PSR J0737–3039B rotates about 100 times slower, with its spin period amounting to just 2.77 s (Kramer et al., 2006), its angular momentum, calculated by assuming that its moment of inertia is equal to that of PSR J0737–3039A, is

$$J_{\rm B} \simeq 3.6 \times 10^{38}\,{\rm kg\,m^2/s}. \tag{5.15}$$

From a comparison of Equation (5.11) and Equations (5.14)–(5.15), it may be noted that white dwarfs have angular momenta which are orders of magnitude larger than neutron stars.

In the case of a rotating Kerr BH (Kerr, 1963; Teukolsky, 2015), its angular momentum is (Shapiro and Teukolsky, 1986)

$$J_\bullet = \chi_g \frac{M_\bullet^2 G}{c}, \quad \left|\chi_g\right| \leq 1. \tag{5.16}$$

If $\left|\chi_g\right| > 1$, a naked singularity without a horizon would occur, along with the possibility of causality violations because of closed timelike curves (Chandrasekhar, 1983). Incidentally, we remark that, although not yet proven, the cosmic censorship conjecture (Penrose, 2002) states that naked singularities cannot be formed via the gravitational collapse of a body. $J_\bullet$ is the BH's spin dipole moment since, according to the celebrated 'no-hair theorems' (Israel, 1967; Carter, 1971; Robinson, 1975), the mass and the spin moments $\mathbb{M}_\bullet^\ell$ and $\mathbb{J}_\bullet^\ell$ of a Kerr BH (Geroch, 1970; Hansen, 1974) are connected by the relation

$$\mathbb{M}_\bullet^\ell + i\mathbb{J}_\bullet^\ell = M_\bullet \left(i J_\bullet / c M_\bullet\right)^\ell, \tag{5.17}$$

where $i := \sqrt{-1}$ is the imaginary unit; the odd mass moments and even spin moments are identically zero.

As far as main sequence stars are concerned, it is known that their dimensionless spin parameter χ_g depends in a nonnegligible way on the stellar mass, and it can well be $\chi_g \gg 1$ (Kraft, 1969, 1970; Dicke, 1970; Gray, 1982; Grenier et al., 1999). For example, from the analysis in Iorio (2016a), it can be inferred that $\chi_g \simeq 36$ for the star HD15082 (WASP–33) (Grenier et al., 1999). Instead, from Equation (5.8), the value $\chi_g^{\odot} \simeq 0.216$ is obtained for the Sun.

The gravitomagnetic field is encoded in the off-diagonal components g_{0i}, $i = 1, 2, 3$ of the spacetime metric tensor accounting, in general, for the mass-energy currents of the source (Einstein, 1955; Thirring, 1918, 1921; Lense and Thirring, 1918; Mashhoon et al., 1984). Actually, the previous denomination has nothing to do with electric charges and currents; it is only due to the formal resemblance of the linearized equations of GTR, in its weak–field and slow-motion approximation, with the Maxwell equations of electromagnetism. In this general relativistic framework, the paradigm of 'gravitoelectromagnetism' arose (Cattaneo, 1958; White, 1958; Forward, 1961; Teyssandier, 1977, 1978; Thorne et al., 1986; Thorne, 1986, 1988; Harris, 1991; Jantzen et al., 1992b; Maartens and Bassett, 1998; Clark and Tucker, 2000; Mashhoon, 2001; Rindler, 2001; Mashhoon, 2007; Costa and Herdeiro, 2008; Costa and Natário, 2014; Costa et al., 2021; Costa and Natário, 2021; Ruggiero, 2021; Bini et al., 2022; Schmid, 2023). For an historical overview, see Section IV of Jantzen et al. (1992a), and references therein. See also the webpage[5] www.phy.olemiss.edu/ luca/Topics/grav/gravitomagnetism.html maintained by Luca Bombelli. Gravitoelectromagnetism encompasses a series of entirely gravitational phenomena affecting orbiting test particles, precessing gyroscopes, moving clocks and atoms, and propagating electromagnetic waves (Braginsky et al., 1977; Dymnikova, 1986; Tartaglia, 2002; Ruggiero and Tartaglia, 2002; Schäfer, 2004, 2009). Gravitomagnetism should play a major role in several complex processes which take place near spinning BHs and involve accretion disks and relativistic jets (Bardeen and Petterson, 1975; Rees, 1978; MacDonald and Thorne, 1982; Rees, 1984; Thorne, 1988; Armitage and Natarajan, 1999; Ingram et al., 2009; Stella and Possenti, 2009; Veledina et al., 2013; Franchini et al., 2016). Also various hypothesized effects like the Penrose process (Penrose, 2002; Penrose and Floyd, 1971; Stuchlík et al., 2021), the Blandford–Znajek effect (Blandford and Znajek, 1977), and superradiance (Zel'Dovich, 1971) are attributable to the gravitomagnetic field of a rotating BH; see Teukolsky (2015) and references therein. Thus, it is important to experimentally check such a prediction of GTR in as many different scenarios as possible in a reliable way in order to trustworthily extrapolate

[5] Accessed 17 January 2024.

its validity also to other realms in which testing it is much more difficult and uncertain. Here, the orbital effects are treated.

The 1pN gravitomagnetic LT acceleration due to the angular momentum $\boldsymbol{J}$ of the central body, is,[6] for an arbitrary orientation of the latter (Soffel, 1989; Huang et al., 1990; Damour et al., 1994; Petit and Luzum, 2010; Poisson and Will, 2014; Soffel and Han, 2019),

$$\boldsymbol{A}^{\mathrm{LT}} = \frac{2GJ}{c^2 r^3}\left(3 r_J \hat{\boldsymbol{r}} \times \boldsymbol{v} + \boldsymbol{v} \times \hat{\boldsymbol{J}}\right), \tag{5.18}$$

where

$$r_J := \hat{\boldsymbol{J}} \cdot \hat{\boldsymbol{r}} \tag{5.19}$$

is the cosine of the angle between the primary's spin axis $\hat{\boldsymbol{J}}$ and the satellite's position vector $\boldsymbol{r}$. Equation (5.18) is responsible for, among other things, perhaps the most famous gravitomagnetic feature of motion: the so–called LT effect (Lense and Thirring, 1918; Mashhoon et al., 1984), despite recent studies (Pfister, 2007, 2008, 2014) showing that it would be more appropriately renamed the Einstein–Thirring–Lense effect. It consists of small secular precessions of (some of) the Keplerian orbital elements, which are treated in detail in Section 5.1.

After initial proposals to use Earth's satellites to measure them dating back to the dawn of space era (Ginzburg, 1957; Bogorodskii, 1959; Ginzburg, 1959), they are nowadays under experimental scrutiny in the field of Earth (Lucchesi et al., 2019a, 2020; Ciufolini et al., 2023) since 1996 (Ciufolini et al., 1996) with some SLR geodetic satellites, as earlier suggested by Cugusi and Proverbio (1978). Some aspects of such tests are controversial (Renzetti, 2012a; Iorio et al., 2013; Renzetti, 2013a, 2014, 2015; Iorio, 2023b). For comprehensive overviews, see, for example, Iorio et al. (2011), Renzetti (2013b), and references therein. Proposals were made to measure the LT effect around Jupiter with its Galilean moons (Iorio, 2023d), following an earlier suggestion by Lense and Thirring (1918)[7] themselves, who, at that time, concluded that they were too small to be detected; the Juno spacecraft (Iorio, 2010b; Schärer et al., 2017), currently orbiting it (Bolton et al., 2017; Bolton, 2018); and another hypothetical mission (Iorio, 2019g). Among the outer planets of the solar system, also the gravitomagnetic fields of Saturn and Uranus were recently taken into consideration to be probed by spacecraft (Schärer et al., 2017; Iorio et al., 2023). As far as the rocky planets are concerned, it was suggested to measure the LT effect of a dedicated artificial satellite orbiting Mars (Iorio, 2009a), after a criticized attempt to spot it in the data of the Mars Global

[6] See, for example, Kidder (1995, Equation (2.2.c)) for a generalization of Equation (5.18) to a two-body system with comparable masses and spins. See also Soffel (1989).

[7] In fact, Lense and Thirring (1918) looked also at the then known moons of the other planets, finding that their gravitomagnetic precessions were all too small.

Surveyor (MGS) spacecraft (Iorio, 2006b; Krogh, 2007). The LT effect induced by the Sun's angular momentum $\boldsymbol{J}_\odot$ (Pijpers, 1998) on the Hermean[8] orbital motion has been long deemed too small to be detected since the pioneering work by[9] de Sitter (1916a); nowadays, the situation may become more favourable (Iorio, 2005, 2012a; Park et al., 2017; Iorio, 2018) in view of the expected improvement of the planetary ephemerides from the analysis of the data collected by the past mission MErcury Surface, Space ENvironment, GEochemistry, and Ranging (MESSENGER) and the ongoing one BepiColombo (Balogh et al., 2007). Going outside the solar system, some exoplanets around main sequence stars were considered in view of a possible detection of the LT effect (Iorio, 2011b, 2016a). Moving to astrophysical compact objects, a successful detection of the gravitomagnetic orbital precession of the inclination (see Section 5.1) of the binary system PSR J1141–6545 (Kaspi et al., 2000), made of a white dwarf and a pulsar, was recently claimed (Venkatraman Krishnan et al., 2020); later analysis by Iorio (2020a) raised concerns about such a test. Attempts to measure the gravitomagnetic periastron precession of the double pulsar PSR J0737-3039 (Burgay et al., 2003; Lyne et al., 2004) and of other binary pulsars in the near future are underway (Kehl et al., 2017; Hu et al., 2020; Iorio, 2021c; Hu and Freire, 2024). Also, some of the S stars were proposed as probes to test, among other things, also the LT effect induced by the spin of Sgr A*, or, conversely, to assume the validity of GTR and use the former as a tool to measure or constrain the latter (Jaroszynski, 1998; Kraniotis, 2007; Will, 2008; Preto and Saha, 2009; Angélil et al., 2010; Merritt et al., 2010; Iorio, 2011c; Han, 2014; Zhang et al., 2015; Psaltis et al., 2016; Yu et al., 2016; Zhang and Iorio, 2017; Waisberg et al., 2018; Fragione and Loeb, 2020; Iorio, 2020c; Peißker et al., 2020; Fragione and Loeb, 2022; Peißker et al., 2022; Iorio, 2023a). Evidence for manifestations of the gravitomagnetic field in the strong-field regime was often claimed over the years. Gravitomagnetism could be responsible for the[10] quasi-periodic oscillations in X-ray binaries (Cui et al., 1998; Marković and Lamb, 1998; Stella and Vietri, 1998). The LT precession induced by a slowly rotating compact object could be compatible with the daily variations of the ejecta angle observed in the microquasar LS I + 61°303 (Massi and Zimmermann, 2010). It was recently reported that the observed quasi-periodic modulation of the iron

[8] From Ἑρμῆς, -οῦ, ὁ, 'Hermē̃s', a Greek god identified with the Roman deity Mercury.

[9] Remarkably, de Sitter (1916a, p. 727) drew his pessimistic conclusion by obtaining a value of the Hermean gravitomagnetic perihelion precession, later quoted also by Lense and Thirring (1918) themselves and Soffel (1989, p. 111), which is about *one order of magnitude larger* than its currently accepted value, of the order of a milliarcsecond per century (mas/cty) (Iorio, 2012a), because he used an incorrect value for the angular momentum of the Sun, wrongly assumed homogeneous.

[10] From the adverb *quăsĭ*, 'somewhat like, about, nearly, almost, not far from'; univerbation of *quăm* 'how, as' + *sī* 'if'.

line centroid energy in the microquasar H1743–322 (Kaluzienski and Holt, 1977) may be produced by the LT effect (Ingram et al., 2016). Recently, the LT–driven precession of the misaligned accretion disk around the SMBH in M87 may have been observed (Cui et al., 2023). In general, the reliability and accuracy of such tests are difficult to properly assess because of the relatively poor knowledge of the astrophysical environments in which the phenomena of interest take place.

So far, the only unquestioned measurement of another consequence of the gravitomagnetic field of the Earth was performed with the quite expensive, dedicated spaceborne mission GP-B (Everitt, 1974). Its timeframe, ranging from its early conception to the release of its final results, lasted for about 40 years at a cost of about 750 million (Will, 2011a). It measured the Pugh–Schiff precessions (Pugh, 1959; Schiff, 1960) of the axes of four gyroscopes carried onboard to an accuracy of the order of 19% (Everitt et al., 2011, 2015), despite its originally expected level of about 1% (Everitt et al., 2001). Other proposals to measure the gravitomagnetic Pugh–Schiff spin precessions with dedicated spacecraft orbiting the Sun and Jupiter (Haas and Ross, 1975) were not followed up.

By projecting Equation (5.18) onto the unit vectors $\hat{\boldsymbol{r}}, \hat{\boldsymbol{\tau}}, \hat{\boldsymbol{h}}$ defined in Equations (2.23)–(2.25) and by using Equations (2.7)–(2.8) along with Equation (2.11), its radial, transverse, and normal components are obtained; they turn out to be

$$A_r^{\rm LT} = \frac{2n_{\rm K}GJ\,(1+e\cos f)^4\,\texttt{Jh}}{c^2a^2\left(1-e^2\right)^{7/2}}, \tag{5.20}$$

$$A_\tau^{\rm LT} = -\frac{2en_{\rm K}GJ\,(1+e\cos f)^3\,\sin f\,\texttt{Jh}}{c^2a^2\left(1-e^2\right)^{7/2}}, \tag{5.21}$$

$$A_h^{\rm LT} = -\frac{2n_{\rm K}GJ\,(1+e\cos f)^3}{c^2a^2\left(1-e^2\right)^{7/2}}\left\{[e\cos\omega - (2+3e\cos f)\cos u]\,\texttt{Jl}\right.$$
$$\left. -\frac{1}{2}\left[e\sin\omega + 4\sin u + 3e\sin(2f+\omega)\right]\texttt{Jm}\right\}, \tag{5.22}$$

where Jl, Jm and Jh are defined in Equations (D.1)–(D.3) of Appendix D. For an arbitrary orbital configuration and a generic direction of $\hat{\boldsymbol{J}}$, it is not possible to spot any particular feature of Equations (5.20)–(5.22). For polar orbits, set by Equations (D.26)–(D.28) of Appendix D, Equation (5.18) is entirely perpendicular to the orbital plane, as per Equations (D.26)–(D.28) of Appendix D. Instead, for equatorial orbits characterized by Equations (D.17)–(D.19) of Appendix D, the LT acceleration is entirely in-plane, becoming fully radial for circular orbits. For *prograde* orbits ($\texttt{Jh} = +1$), namely when the test particle moves along its orbit in the *same* direction as the rotation of the primary, the motion is *slowed down* because Equation (5.20) is directed *outward*, and *vice versa* for *retrograde* orbits (`Jh = -1`), namely when the direction of the orbital motion is *opposite* that of the rotation of the central body. This leads to a difference δT_{gvm} in the times required for two counter-revolving particles to come back to some fixed direction in space, usually known as gravitomagnetic clock effect (Mitskevich and Pulido Garcia, 1970; Vladimirov et al., 1987; Cohen and Mashhoon, 1993; Mashhoon, 1997; You, 1998; Mashhoon et al., 1999; Tartaglia, 2000b; Mashhoon and Santos, 2000; Mashhoon et al., 2001; Hackmann and Lämmerzahl, 2014; Scheumann et al., 2023; Iorio, 2024b; Jiale Li et al., 2024); for a recent review, see Iorio and Mashhoon (2024) and references therein. It turns out that δT_{gvm} is proportional to $J/\left(M c^2\right)$ through a numerical scaling coefficient that has been calculated in the literature to be equal to 4π; as will be shown later in this chapter, it is, in fact, larger by a factor of four (Iorio, 2024b; Jiale Li et al., 2024). Such an intriguing relativistic feature of motion was the subject of several papers investigating its possible detection as well; see Gronwald et al. (1997), Mashhoon et al. (1999), Lichtenegger et al. (2000), Tartaglia (2000a,c), Iorio (2001a,b), Iorio and Lichtenegger (2005), Lichtenegger et al. (2006), and Scheumann et al. (2023). It also has relevant consequences in astrophysical contexts such as Kerr BH spacetime (de Felice, 1995; Bonnor and Steadman, 1999; Bini and Jantzen, 2003; Faruque, 2004; Bini et al., 2005). For other versions of the gravitomagnetic clock effect involving spinning orbiters in the Kerr spacetime, see, for example, Bini et al. (2004) and Mashhoon and Singh (2006).

Equations (5.20)–(5.22) allow one to calculate in full generality several gravitomagnetic orbital effects, as shown in the next sections. Such a feature is particularly important in scenarios characterized by poorly known or even completely unconstrained orientation of $\hat{\boldsymbol{J}}$ in space, as in the case of the S stars. The spin axis of Sgr A* can be parameterized as

$$\hat{\boldsymbol{J}}_{\bullet} = \begin{cases} \sin i_{\bullet} \cos \zeta_{\bullet} \\ \sin i_{\bullet} \sin \zeta_{\bullet} \\ \cos i_{\bullet} \end{cases} . \tag{5.23}$$

In Equation (5.23), $i_{\bullet}$ is the tilt of $\hat{\boldsymbol{J}}_{\bullet}$ to the line of sight; if $i_{\bullet} = 90^{\circ}$, $\hat{\boldsymbol{J}}_{\bullet}$ lies entirely in the plane of the sky, while, if $i_{\bullet} = 0$, it is aligned with the line of sight. The angle $\zeta_{\bullet}$ reckons the position of the projection of the spin axis in the plane of the sky from the reference x axis within the latter. For performed attempts to somehow constrain $\hat{\boldsymbol{J}}_{\bullet}$ of Sgr A* with different non-dynamical approaches, see Falanga et al. (2007); Meyer et al. (2007); Broderick et al. (2009, 2011); Shcherbakov et al. (2012); Jørgensen et al. (2016); Yu et al. (2016), and references therein; it turns out that $i_{\bullet}$, $\zeta_{\bullet}$ are, in fact, still poorly constrained. On the one hand, $\hat{\boldsymbol{J}}_{\bullet}$ would be far from aligned with the line of sight. Indeed, according to, for example, Meyer et al. (2007), who used polarimetric observations of the near-infrared emission of Sgr A*, it is $i_\bullet \simeq 55^\circ$. Falanga et al. (2007) obtained $i_\bullet \simeq 77^\circ$ on the basis of their fit of a simulated Rossby wave-induced spiral pattern in the BH's accretion disk to the X-ray lightcurve detected with the mission X-ray Multi-Mirror–Newton (XMM–Newton). Shcherbakov et al. (2012) yielded the range $42^\circ \lesssim i_\bullet \lesssim 75^\circ$ by comparing polarized submillimetre infrared observations with spectra computed using three-dimensional general relativistic magnetohydrodynamical simulations. Methods based on gravitational lensing for determining the BH's spin direction independently of orbital dynamics were outlined, for example, in Saida (2017). Takahashi (2004) investigated the possibility of measuring, among other things, $i_\bullet$ from the shape and position of the BH's shadow under certain assumptions. On the other hand, the first EHT observations (Event Horizon Telescope Collaboration, 2022) disfavour, among other things, scenarios where the BH is viewed at high inclination ($i_\bullet > 50^\circ$).

Also the rotational axis of the main-sequence stars hosting exoplanets is, in general, poorly constrained; as an example, the determined values of $i_\star$ released in table 5 by Rainer et al. (2023) exhibit remarkably large uncertainties, while in several other cases, only lower bounds on such an important stellar parameter are available.[11] Also the spin axis's azimuthal angle in the plane of the sky is sometimes measured with a variety of techniques, but always with modest accuracy. As an example, Winn et al. (2005) by exploiting the Rossiter–McLaughlin effect (McLaughlin, 1924; Rossiter, 1924; Triaud, 2018), determined the angle λ between the projection of the spin axis of the transiting star HD 209458 (Charbonneau et al., 2000; Henry et al., 2000) onto the plane of the sky, assumed as the $\{x, y\}$ reference plane in a coordinate system whose z axis is directed away from the observer and the x axis is aligned with the line of nodes, and the y axis, which coincides with the sky-projected orbital angular momentum in view of the previously mentioned choice for the x axis, to a 32% accuracy level.

5.1 The Orbital Shifts of the Keplerian Orbital Elements

The LT instantaneous shifts $\Delta\kappa\,(f)$ of the Keplerian orbital elements $\kappa = a, e, I, \Omega, \omega, \eta$ due to Equation (5.18) can be calculated as per Equation (2.12) by using Equations (5.20)–(5.22) in Equations (2.13)–(2.18). They turn out to be

[11] As pointed out by Rainer et al. (2023), $i_\star$ heavily affects exoplanets' parameters (Hirano et al., 2014). Furthermore, it is also important to compute the spin-orbit angle of exoplanetary systems, which, among other things, is an important observational probe of the origin and evolution of the systems (Queloz et al., 2000; Winn et al., 2005).

$$\Delta a\,(f)^{\rm LT} = 0, \tag{5.24}$$

$$\Delta e\,(f)^{\rm LT} = \frac{2GJ\,\mathtt{Jh}\,(-\cos f + \cos f_0)}{c^2 n_{\rm K} a^3\sqrt{1-e^2}}, \tag{5.25}$$

$$\Delta I\,(f)^{\rm LT} = -\frac{GJ}{2c^2 n_{\rm K} a^3\left(1-e^2\right)^{3/2}}\left(\mathcal{I}_1^{\rm LT}\,\mathtt{Jl} + \mathcal{I}_2^{\rm LT}\,\mathtt{Jm}\right), \tag{5.26}$$

$$\Delta\Omega\,(f)^{\rm LT} = -\frac{GJ\csc I}{2c^2 n_{\rm K} a^3\left(1-e^2\right)^{3/2}}\left(\mathcal{N}_1^{\rm LT}\,\mathtt{Jl} + \mathcal{N}_2^{\rm LT}\,\mathtt{Jm}\right), \tag{5.27}$$

$$\Delta\omega\,(f)^{\rm LT} = -\frac{GJ}{2c^2 n_{\rm K} e a^3\left(1-e^2\right)^{3/2}}\left(\mathcal{G}_1^{\rm LT}\,\mathtt{Jl} + \mathcal{G}_2^{\rm LT}\,\mathtt{Jm} + \mathcal{G}_3^{\rm LT}\,\mathtt{Jh}\right), \tag{5.28}$$

$$\Delta\eta\,(f)^{\rm LT} = \frac{2GJ\,\mathtt{Jh}\,(\sin f - \sin f_0)}{c^2 n_{\rm K} a^3 e}, \tag{5.29}$$

where the coefficients $\mathcal{I}_1^{\rm LT}, \ldots \mathcal{G}_3^{\rm LT}$ entering Equations (5.26)–(5.28) are displayed in Appendix E.

By calculating Equations (5.24)–(5.29) with the replacement $f \rightarrow f_0 + 2\pi$ and dividing the result by $T_{\rm K}$, one obtains the LT averaged precessions

$$\left\langle\frac{da}{dt}\right\rangle^{\rm LT} = 0, \tag{5.30}$$

$$\left\langle\frac{de}{dt}\right\rangle^{\rm LT} = 0, \tag{5.31}$$

$$\left\langle\frac{dI}{dt}\right\rangle^{\rm LT} = \frac{2GJ\,\mathtt{Jl}}{c^2 a^3\left(1-e^2\right)^{3/2}}, \tag{5.32}$$

$$\left\langle\frac{d\Omega}{dt}\right\rangle^{\rm LT} = \frac{2GJ\,\mathtt{Jm}}{c^2 \sin I a^3\left(1-e^2\right)^{3/2}}, \tag{5.33}$$

$$\left\langle\frac{d\omega}{dt}\right\rangle^{\rm LT} = -\frac{2GJ\,(2\mathtt{Jh} + \cot I\,\mathtt{Jm})}{c^2 a^3\left(1-e^2\right)^{3/2}}, \tag{5.34}$$

$$\left\langle\frac{d\eta}{dt}\right\rangle^{\rm LT} = 0. \tag{5.35}$$

Equations (5.30)–(5.35) represent the LT effect in the most general case with respect to the earlier derivations,[12] valid only for the particular case in which the z axis is aligned with $\hat{\boldsymbol{J}}$ (Lense and Thirring, 1918; Bogorodskii, 1959); also, more

[12] Lense and Thirring (1918) calculated the LT precession of the longitude of pericentre ϖ.

recent calculation has made use of the same particular orbital configuration (Soffel, 1989; Ashby and Allison, 1993; Soffel and Han, 2019). For more general calculations based on different approaches and formalisms which take into account also the masses and the spins of both bodies,[13] see Kalitzin (1959), Michalska (1960), Barker and O'Connell (1975), Damour and Schäfer (1988), Soffel (1989), Brumberg (1991), Damour and Taylor (1992), Wex (1995), Wex and Kopeikin (1999), Königsdörffer and Gopakumar (2006), and Iorio (2017).

Using Equations (2.27)–(2.31) with

$$\langle \mathfrak{R} \rangle^{\mathrm{LT}} = \frac{n_{\mathrm{K}} G J}{c^2 a \left(1 - e^2\right)} \left[\left(\hat{\boldsymbol{J}} \times \hat{\boldsymbol{m}} \right) \cdot \hat{\boldsymbol{l}} - \left(\hat{\boldsymbol{J}} \times \hat{\boldsymbol{l}} \right) \cdot \hat{\boldsymbol{m}} \right], \tag{5.36}$$

obtained by averaging Equation (C.17) of Appendix C over one orbit, yields just Equations (5.30)–(5.34) because of Equation (D.16). About the mean anomaly at epoch, by noting that

$$\boldsymbol{v} \cdot \nabla_{\boldsymbol{v}} \mathfrak{R} = \mathfrak{R} \tag{5.37}$$

for Equation (C.15) since it is *linear* in $\boldsymbol{v}$, and by treating n_{K} as an *explicit* function of a when the partial derivative of Equation (5.36) with respect to the latter is performed, Equation (2.32) yields just Equation (5.35).

For a binary system made of two extended bodies with comparable masses M_{A} and M_{B} and angular momenta $\boldsymbol{J}_{\mathrm{A}}$ and $\boldsymbol{J}_{\mathrm{B}}$, the gravitomagnetic precessions of the relative orbit can be obtained from Equations (5.30)–(5.35) by replacing $\boldsymbol{J}$ with (Damour and Schäfer, 1988, Equation (4.14), p. 150; Soffel, 1989, Equation (4.7.70), p. 134; Brumberg, 1991, Equations (4.4.32)–(4.4.33), p. 155)

$$\boldsymbol{S} := \left(1 + \frac{3}{4}\frac{M_{\mathrm{B}}}{M_{\mathrm{A}}}\right) \boldsymbol{J}_{\mathrm{A}} + \left(1 + \frac{3}{4}\frac{M_{\mathrm{A}}}{M_{\mathrm{B}}}\right) \boldsymbol{J}_{\mathrm{B}}. \tag{5.38}$$

5.1.1 Some Special Orbital Configurations

Here, some peculiar orbital configurations, characterized in Appendix D, are examined.

Let, first, the satellite's orbital plane be assumed coincident with the body's equatorial one, irrespective of the orientation of the latter in the adopted reference frame, namely for generic values of α_J, δ_J. Then, according to Equations (D.17)–(D.25) of Appendix D, Equations (5.30)–(5.35) reduce to

$$\left\langle \frac{da}{dt} \right\rangle^{\mathrm{LT}} = 0, \tag{5.39}$$

[13] Kalitzin (1959) calculated the in-plane pericentre precession using the angle $\Omega \cos I + \omega$.

$$\left\langle \frac{de}{dt} \right\rangle^{\mathrm{LT}} = 0, \tag{5.40}$$

$$\left\langle \frac{dI}{dt} \right\rangle^{\mathrm{LT}} = 0, \tag{5.41}$$

$$\left\langle \frac{d\Omega}{dt} \right\rangle^{\mathrm{LT}} = 0, \tag{5.42}$$

$$\left\langle \frac{d\omega}{dt} \right\rangle^{\mathrm{LT}} = \mp \frac{4GJ}{c^2 a^3 \left(1 - e^2\right)^{3/2}}, \tag{5.43}$$

$$\left\langle \frac{d\eta}{dt} \right\rangle^{\mathrm{LT}} = 0. \tag{5.44}$$

Equation (5.43) is a genuine secular trend; the minus sign corresponds to a prograde motion, namely $\mathtt{Jh} = +1$, while the plus sign occurs if the motion is retrograde, namely $\mathtt{Jh} = -1$.

Let, now, the body's spin axis, irrespective of its orientation in the adopted coordinate system, namely for generic values of $\alpha_J,\ \delta_J$, lie somewhere in the satellite's orbital plane between $\hat{l}$ and $\hat{m}$. Then, according to Equations (D.26)–(D.34) of Appendix D, Equations (5.30)–(5.35) can be written as

$$\left\langle \frac{da}{dt} \right\rangle^{\mathrm{LT}} = 0, \tag{5.45}$$

$$\left\langle \frac{de}{dt} \right\rangle^{\mathrm{LT}} = 0, \tag{5.46}$$

$$\left\langle \frac{dI}{dt} \right\rangle^{\mathrm{LT}} = \frac{2GJ \cos \delta_J}{c^2 a^3 \left(1 - e^2\right)^{3/2}}, \tag{5.47}$$

$$\left\langle \frac{d\Omega}{dt} \right\rangle^{\mathrm{LT}} = \frac{2GJ \sin \delta_J}{c^2 a^3 \left(1 - e^2\right)^{3/2}}, \tag{5.48}$$

$$\left\langle \frac{d\omega}{dt} \right\rangle^{\mathrm{LT}} = 0, \tag{5.49}$$

$$\left\langle \frac{d\eta}{dt} \right\rangle^{\mathrm{LT}} = 0. \tag{5.50}$$

Equations (5.47)–(5.48), which do not vanish for a generic orientation of $\hat{\boldsymbol{J}}$, are genuine secular trends.

5.1.2 The Contribution of Φ to the Mean Anomaly

Since the semimajor axis is left unaffected by Equation (5.18), as per Equation (5.24), it is

$$\Phi\left(t\right)^{\mathrm{LT}} = \overline{\Phi}^{\mathrm{LT}} = 0. \tag{5.51}$$

5.1.3 Gravitomagnetic Ring Currents in Triple Systems

So far, the angular momentum $\boldsymbol{J}$ of an isolated rotating massive body was assumed as the source of the gravitomagnetic field inducing the LT effect encoded by the precessions of Equations (5.30)–(5.35).

In principle, the inner binary $\mathcal{S}$ of a hierarchical triple system $\mathcal{T}$ in which a distant companion orbits the former can also be thought as the source of a gravitomagnetic field through its *orbital* angular momentum

$$\boldsymbol{H}_{\mathrm{b}} = M_{\mathrm{red}}\sqrt{\mu_{\mathrm{b}} a \left(1 - e^2\right)}\,\hat{\boldsymbol{h}}. \tag{5.52}$$

In Equation (5.52),

$$M_{\mathrm{red}} := \frac{M_{\mathrm{A}} M_{\mathrm{B}}}{M_{\mathrm{b}}} \tag{5.53}$$

is the reduced mass, respectively. Stated differently, the third body would see the other two members of $\mathcal{T}$ closely orbiting one about each other as a rotating matter ring whose orbital angular momentum $\boldsymbol{H}_{\mathrm{b}}$ generates its own gravitomagnetic field (Ruggiero, 2016) which may be orders of magnitude larger than those due to the individual spin angular momenta $\boldsymbol{J}_{\mathrm{A}}$ and $\boldsymbol{J}_{\mathrm{B}}$ of each body of $\mathcal{S}$ (Iorio, 2022). Such a scenario corresponds, in principle, to the so-called circumbinary planets (CBPs) (Haghighipour, 2010; Thebault and Haghighipour, 2015), where an exoplanet revolves about a binary system typically made of two main sequence stars, and to triple pulsars like, for example, PSR J0337+1715 (Ransom et al., 2014) made of a millisecond pulsar–white dwarf pair in relative motion around each other, and another white dwarf as a distant companion. Several CBPs have already been discovered using different techniques: see Thorsett et al. (1993), Correia et al. (2005), Lee et al. (2009), Qian et al. (2010), Beuermann et al. (2010), Doyle et al. (2011), Orosz et al. (2012b,a), Qian et al. (2012b,a), Welsh et al. (2012), Schwamb et al. (2013), Kraus et al. (2014), Kostov et al. (2014), Welsh et al. (2015), Bennett et al. (2016), Kostov et al. (2016), Getley et al. (2017), Jain et al. (2017), Asensio–Torres et al. (2018), and Kostov et al. (2021). About triple pulsars, so far, PSR J0337+1715 is the sole relatively tight member so far discovered of the class of hierarchical triple systems hosting stellar corpses, apart from B1620–26 whose pulsar has a white dwarf as inner companion and a roughly Jupiter–mass at

35 au as outer orbiter (Thorsett et al., 1999). According to Ransom et al. (2014), $\lesssim 1\%$ of the millisecond pulsars' population resides in stellar triples, and $\lesssim 100$ such systems exist in the Galaxy. The possibility of measuring such 'annular' gravitomagnetic field is quantitatively investigated in Section K.5 of Appendix K.

5.2 The Anomalistic Period

The LT anomalistic period can be calculated by means of Equations (5.20)–(5.22) as explained in Section 2.3.1. It turns out to be

$$\Delta T_{\rm ano}^{\rm LT} = 0; \tag{5.54}$$

it is an exact result, valid to all orders in the eccentricity e.

Figure 5.1, obtained for generic values of the Keplerian orbital parameters, shows just that; over three orbital revolutions, the test particle reaches always the precessing line of apsides after a time interval equal to the Keplerian orbital period after each orbit.

Furthermore, Figure 5.2 plots the final part of the time series of the cosine $\hat{\boldsymbol{r}} \cdot \hat{\boldsymbol{C}}$ of the angle between the position vector $\boldsymbol{r}$ and the Laplace–Runge–Lenz vector $\boldsymbol{C}$

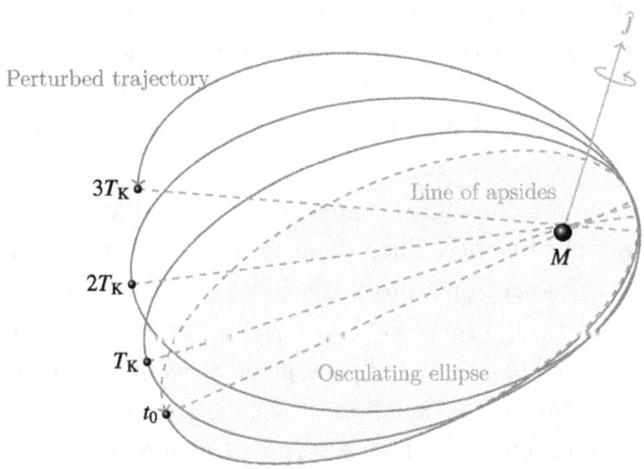

Figure 5.1 Perturbed LT trajectory (continuous curve) and its osculating Keplerian ellipse (dashed curve) at the initial instant of time t_0 characterized by $e = 0.7$, $I = 30°$, $\Omega = 72°$, $\omega = 50°$, $f_0 = 180°$. The orientation of the spin axis $\hat{\boldsymbol{J}}$ of the central body is set by $\alpha_J = 45°$, $\delta_J = 60°$. In this example, I, Ω, and ω undergo the LT precessions of Equations (5.32)–(5.34); their magnitudes are suitably rescaled by enhancing them for a better visualization. The initial position is chosen at the apocentre instead of the pericentre solely for the sake of better visualization. The positions on the perturbed trajectory after one, two, and three Keplerian periods are marked as well. At each orbit, the passage at the drifting line of apsides occurs always as in the Keplerian case because, according to Equation (5.54), $\Delta T_{\rm ano}^{\rm LT} = 0$.

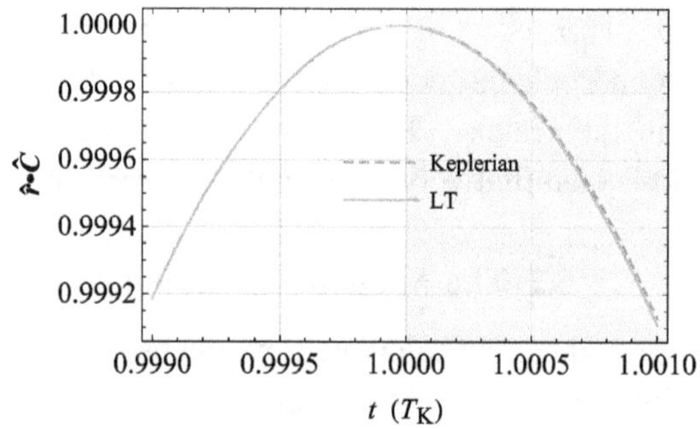

Figure 5.2 Plot of the numerically produced time series of the cosine $\hat{\boldsymbol{r}} \cdot \hat{\boldsymbol{C}}$ of the angle between the position vector $\boldsymbol{r}$ and the Laplace–Runge–Lenz vector $\boldsymbol{C}$ versus time t, in units of T_K, obtained by integrating the equations of motion of a fictitious test particle with and without the LT acceleration of Equation (5.18) for an elliptical ($e = 0.665$) orbit arbitrarily oriented in space ($I = 40°$, $\Omega = 45°$, $\omega = 50°$) starting from the periapsis ($f_0 = 0$), that is, $\hat{\boldsymbol{r}}_0 \cdot \hat{\boldsymbol{C}}_0 = +1$, and the semimajor axis is $a = 6R_e$. The physical parameters of the Earth are adopted, apart from the spin-axis position set by $\alpha_J = 45°$, $\delta_J = 60°$. The time needed to come back to the initial position on the (moving) line of apsides, so that $\hat{\boldsymbol{r}} \cdot \hat{\boldsymbol{C}} = +1$ again, is as in the Keplerian case.

versus time t, in units of T_K, for a numerically integrated fictitious test particle acted upon by Equation (5.18) starting from, say, the moving pericentre, namely for $\hat{\boldsymbol{r}}_0 \cdot \hat{\boldsymbol{C}}_0 = +1$. It can be seen that it comes back to the same position on the precessing line of apsides, that is, it is $\hat{\boldsymbol{r}} \cdot \hat{\boldsymbol{C}} = +1$ again, just after one Keplerian orbital period.

The fact that the gravitomagnetic apsidal period is identical to the Keplerian one can be intuitively justified since there is no net shift per orbit of the mean anomaly at epoch η, as per Equation (5.35). Indeed, from the definition of the mean anomaly of Equation (2.2), Equation (2.3) follows; it tells us that the mean anomaly at epoch is proportional to the time of passage at pericentre t_p. Thus, since n_K stays constant because, according to Equation (5.30), the semimajor axis is not secularly affected by the gravitomagnetic field, the rate of change of the mean anomaly at epoch is proportional to the opposite of the pace of variation of the time of passage at pericentre according to Equation (2.26). Should η *increase*, the crossing of the pericentre would be *anticipated* with respect to the Keplerian case since t_p would *decrease*, and vice versa. In this case, the variation of η would result in an orbit-by-orbit *advance* or *delay* of the passages at the pericentre, which does not occur in the present case because, in fact, $\langle d\eta/dt\rangle^{LT} = 0$.

5.3 The Draconitic Period

The gravitomagnetic LT draconitic period, calculated as explained in Section 2.3.2, turns out to be

$$T_{\text{dra}}^{\text{LT}} = T_{\text{K}} + \Delta T_{\text{dra}}^{\text{LT}}, \tag{5.55}$$

with

$$\Delta T_{\text{dra}}^{\text{LT}} = \frac{4\pi J\left(2\mathtt{Jh} + \mathtt{Jm}\cot I\right)}{c^2 M\left(1 + e\cos\omega\right)^2}. \tag{5.56}$$

The explicit form of the geometric coefficient in the numerator of Equation (5.56) depending on the orientation in space of both the orbital plane and the primary's spin axis is

$$2\mathtt{Jh} + \mathtt{Jm}\cot I = 3\cos I\sin\delta + \cos\delta\left(\csc I - 3\sin I\right)\sin\left(\alpha_J - \Omega\right). \tag{5.57}$$

In general, it can be either positive and negative. For a polar orbit, namely for $\Omega = \alpha_J$ and $I = 90°$, the gravitomagnetic correction to the draconitic period vanishes, as per Equation (5.57). For an equatorial orbit arbitrarily oriented in space, however, corresponding to Equations (D.17)–(D.19) of Appendix D, it does not vanish, amounting to

$$\Delta T_{\text{dra}}^{\text{LT}} = \pm\frac{8\pi J}{c^2 M\left(1 + e\cos\omega\right)^2}. \tag{5.58}$$

Furthermore, for circular orbits, Equation (5.58) reduces to

$$\Delta T_{\text{dra}}^{\text{LT}} = \pm\frac{8\pi J}{c^2 M}. \tag{5.59}$$

If the orbital plane lies in the reference plane, that is, for $I = 0$, Equation (5.56) loses its meaning, as is expected since, in this case, the line of nodes is no longer defined.

Figure 5.3, obtained for generic values of the Keplerian orbital parameters, confirms the analytical result of Equation (5.56); over three orbital revolutions, the test particle always reaches the precessing line of nodes after a time interval equal to $T_{\text{dra}}^{\text{LT}}$ after each orbit. For the particular choice of the values of the primary's spin and orbital parameters, it turns out to be *longer* than T_{K}, in agreement with Equation (5.56).

Furthermore, Figure 5.4 plots the final part of the time series of the cosine $\hat{\boldsymbol{r}}\cdot\hat{\boldsymbol{l}}$ of the angle between the position vector $\boldsymbol{r}$ and the node unit vector $\hat{\boldsymbol{l}}$ versus time t, in units of T_{K}, for a numerically integrated fictitious test particle with and without Equation (5.18), starting in both cases from, say, the moving ascending node, namely for $\hat{\boldsymbol{r}}_0\cdot\hat{\boldsymbol{l}}_0 = +1$. It can be seen that it comes back to the same position on the precessing line of nodes, that is, it is $\hat{\boldsymbol{r}}\cdot\hat{\boldsymbol{l}} = +1$ again, just after $T_{\text{dra}}^{\text{LT}} = T_{\text{K}} + \Delta T_{\text{dra}}^{\text{LT}}$, differing from T_{K} by a (*positive*) amount, in agreement with Equation (5.56) for the particular choice of the generic values of the spin and the orbital parameters adopted in the numerical integrations.

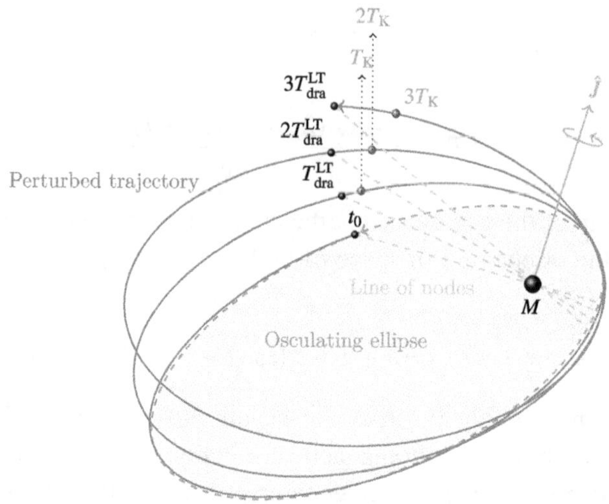

Figure 5.3 Perturbed LT trajectory (continuous curve) and its osculating Keplerian ellipse (dashed curve) at the initial instant of time t_0 characterized by $e = 0.7$, $I = 30°$, $\Omega = 72°$, $\omega = 50°$, $f_0 = 180° - \omega$. The orientation of the spin axis $\hat{\boldsymbol{J}}$ of the central body is set by $\alpha_J = 45°$, $\delta_J = 60°$. In this example, I, Ω, and ω undergo the LT precessions of Equations (5.32)–(5.34); their magnitudes are suitably rescaled by enhancing them for a better visualization. The positions on the perturbed trajectory after one, two, and three Keplerian periods $T_{\rm K}$ are marked as well. At each orbit, the passages at the precessing line of nodes occur always *later* than in the Keplerian case by the amount given by Equation (5.56), which is *positive* for the given values of the spin and orbital parameters.

5.4 The Sidereal Period

As shown in Section 2.3.3, the sidereal period for a generic perturbed orbit is the sum of the draconitic period, calculated as explained in Section 2.3.2, and the term given by Equation (2.66). For Equation (5.18), Equation (2.66) yields

$$\Delta T^{\rm LT}_{\rm sid\,II} = \frac{4\pi \boldsymbol{J}\cot I}{c^2 M e^2\sqrt{1-e^2}} \cdot \left\{\hat{\boldsymbol{m}}\left[-e^2 + 2\left(2 - e^2 - 2\sqrt{1-e^2}\right)\cos 2\omega\right] + 2\hat{\boldsymbol{l}}\left(-2 + e^2 + 2\sqrt{1-e^2}\right)\sin 2\omega\right\} \simeq -\frac{4\pi J\cot I}{c^2 M}\mathrm{Jm} + \mathcal{O}\left(e^2\right). \quad (5.60)$$

In the equatorial case set by Equations (D.17)–(D.19) of Appendix D, assuring that the orbital plane stays constant in space, Equation (5.60) vanishes, and the sidereal period coincides with the draconitic one, as is expected since neither the line of nodes nor the orbit's projection onto the reference plane change. By taking the sum of Equation (5.56) and Equation (5.60), the full expression of the gravitomagnetic correction of the sidereal period $\Delta T^{\rm LT}_{\rm sid}$ is obtained. It can be noted that, for a generic

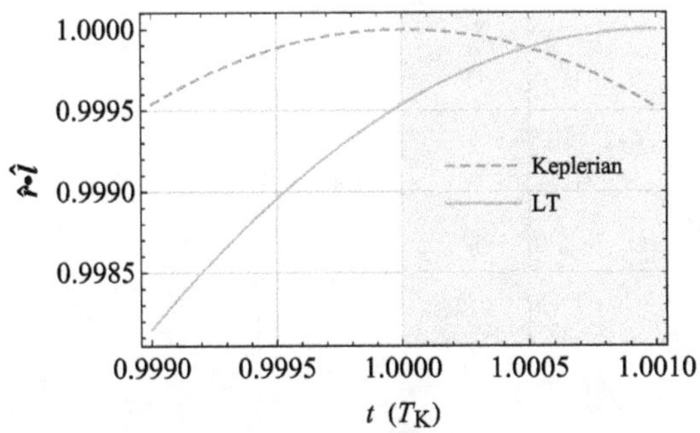

Figure 5.4 Plot of the numerically produced time series of the cosine $\hat{\boldsymbol{r}} \cdot \hat{\boldsymbol{l}}$ of the angle between the position vector $\boldsymbol{r}$ and the node unit vector $\hat{\boldsymbol{l}}$ versus time t, in units of T_K, obtained by integrating the equations of motion of a fictitious test particle with and without the LT acceleration of Equation (5.18) for an elliptical ($e = 0.665$) orbit arbitrarily oriented in space ($I = 40°$, $\Omega = 45°$, $\omega = 50°$) starting from the ascending node ☊ ($f_0 = -\omega + 360°$), namely $\hat{\boldsymbol{r}}_0 \cdot \hat{\boldsymbol{l}}_0 = +1$; the semimajor axis is $a = 6R_e$. The physical parameters of the Earth are adopted, apart from the spin axis position set by $\alpha_J = 45°$, $\delta_J = 60°$. The LT acceleration is suitably rescaled in such a way that $\left|\Delta T_{\rm dra}^{\rm LT}\right| / T_K = 0.001$. The time needed to come back to the initial position on the (moving) line of nodes, so that $\hat{\boldsymbol{r}} \cdot \hat{\boldsymbol{l}} = +1$ again, is *longer* than in the Keplerian case by the amount $\Delta T_{\rm dra}^{\rm LT} = +0.001 T_K$, shown by the shaded area, in agreement with Equation (5.56).

eccentric orbit, $\Delta T_{\rm sid}^{\rm LT}$ is not defined if the orbital plane lies in the fundamental one. Nonetheless, for $e = 0$, it reduces to

$$\Delta T_{\rm sid}^{\rm LT} = \frac{8\pi J}{c^2 M}\left[\cos I \sin\delta_J - \cos\delta_J \sin I \sin(\alpha_J - \Omega)\right], \tag{5.61}$$

which is not singular in $I = 0$. By using the true longitude l in the case $I = 0$, it turns out

$$\Delta T_{\rm sid}^{\rm LT} = \frac{8\pi J \sin\delta_J}{c^2 M (1 + e\cos\varpi)^2}. \tag{5.62}$$

In the limit $e \to 0$, it reduces to

$$\Delta T_{\rm sid}^{\rm LT} = \frac{8\pi J \sin\delta_J}{c^2 M}, \tag{5.63}$$

which agrees with Equation (5.61) calculated with $I = 0$. In turn, if $\delta_J = \pm 90°$, corresponding to the case of an equatorial orbit whose orbital plane coincides with the reference plane, Equation (5.63) becomes

$$\Delta T_{\rm sid}^{\rm LT} = \pm\frac{8\pi J}{c^2 M}, \tag{5.64}$$

in agreement with Equation (5.59).

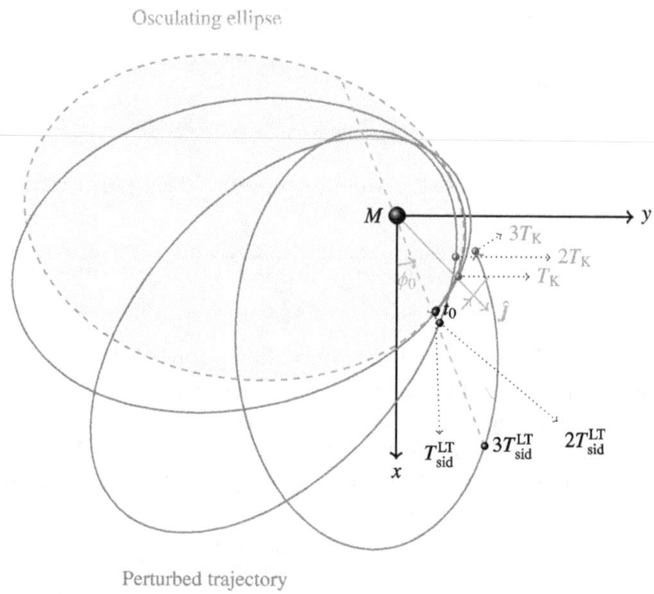

Figure 5.5 Projections of the perturbed LT trajectory (continuous curve) and of its osculating Keplerian ellipse (dashed curve) in the reference plane $\{x, y\}$ at the initial instant of time t_0 characterized by the generic initial conditions $e = 0.7$, $I = 30°$, $\Omega = 45°$, $\omega = 50°$, $f_0 = 285°$. The orientation of the spin axis $\hat{\boldsymbol{J}}$ of the central body, whose projection in the fundamental plane is depicted as well, is set by $\alpha_J = 45°$, $\delta_J = 60°$. In this example, I, Ω, and ω undergo the LT shifts of Equations (5.32)–(5.34) due to the spin dipole moment $\boldsymbol{J}$ of the primary; their sizes are suitably rescaled for better visualizing of their effect. The positions on the perturbed trajectory after one, two, and three Keplerian periods T_K are marked as well. At each orbit, the passages at the generic fixed line characterized by ϕ_0 occur always *earlier* than in the Keplerian case by the amount given by the sum of Equation (5.56) and Equation (5.60). It is so because, for the given values of the spin and orbital parameters, $\Delta T_{\rm dra}^{\rm LT} + \Delta T_{\rm sid\,II}^{\rm LT} < 0$, as per Equation (5.56) and Equation (5.60).

Figure 5.5 confirms the analytical results of Equation (5.56) and Equation (5.60). Indeed, over three orbital revolutions, the projection of a generic LT perturbed orbit in the fundamental plane $\{x, y\}$ crosses a fixed direction in the latter set by a certain value ϕ_0 always after a time interval equal to $T_{\rm sid}^{\rm LT} = T_{\rm dra}^{\rm LT} + \Delta T_{\rm sid\,II}^{\rm LT}$ for each orbit. With the particular choice of the primary's spin and the orbital parameters used in the picture, $T_{\rm sid}^{\rm LT}$ turns out to be *shorter* than T_K, in agreement with Equation (5.56) and Equation (5.60).

Furthermore, Figure 5.6 plots the final part of the time series of the cosine of the angle ϕ, normalized to its initial value $\cos\phi_0$, versus time t, in units of T_K, for a numerically integrated fictitious test particle with and without Equation (5.18) starting from the same generic initial position. It can be seen that it comes back

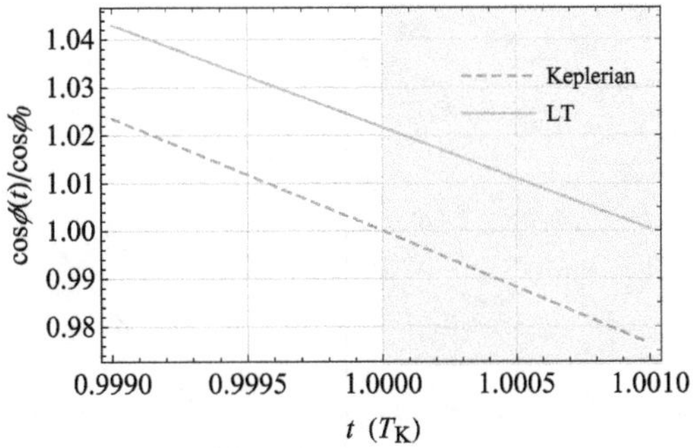

Figure 5.6 Plot of the numerically produced time series of the cosine $\cos\phi\,(t)$ of the azimuthal angle $\phi\,(t)$ normalized to its initial value $\cos\phi_0$ versus time t, in units of $T_{\rm K}$, obtained by integrating the equations of motion of a fictitious test particle with and without the LT acceleration of Equation (5.18) for an elliptical ($e = 0.665$) orbit arbitrarily oriented in space ($I = 40^\circ$, $\Omega = 45^\circ$, $\omega = 310^\circ$) starting from, say, $f_0 = 50^\circ$; the semimajor axis is $a = 6R_{\rm e}$. The physical parameters of the Earth are adopted, apart from the spin axis position set by $\alpha_J = 45^\circ$, $\delta_J = 60^\circ$. The LT acceleration is suitably rescaled in such a way that $\left|\Delta T^{\rm LT}_{\rm sid}\right|/T_{\rm K} = 0.001$. The time needed to $\cos\phi\,(t)$ to assume again its initial value $\cos\phi_0$ is *longer* than in the Keplerian case by the amount $\Delta T^{\rm LT}_{\rm sid} = +0.001T_{\rm K}$, shown by the shaded area, in agreement with the sum of Equation (5.56) and Equation (5.60).

to the same position on the fixed direction chosen in the reference plane, that is, it is $\cos\phi/\cos\phi_0 = +1$ again, just after $T^{\rm LT}_{\rm sid} = T^{\rm LT}_{\rm dra} + \Delta T^{\rm LT}_{\rm sid\,II}$, differing from $T_{\rm K}$ by a (*positive*) amount in agreement with Equation (5.56) and Equation (5.60) for the particular choice of the generic values of the spin and the orbital parameters adopted in the numerical integrations.

5.5 The Gravitomagnetic Clock Effect

According to Equation (5.59), the time difference characterizing the usual scenario for the gravitomagnetic clock effect consisting of two counter-orbiting test particles which move along identical circular orbits in their primary's equatorial plane, illustrated in Figures 5.7–5.8, amounts to (Iorio, 2024b)

$$\delta T_{\rm gvm} = 16\pi\frac{J}{Mc^2}, \tag{5.65}$$

which is four times larger than what is usually[14] found in the literature[15] (Vladimirov et al., 1987; Cohen and Mashhoon, 1993; Mashhoon, 1997; You, 1998; Tartaglia, 2000b; Mashhoon et al., 1999; Mashhoon and Santos, 2000; Mashhoon et al., 2001).

[14] See section 4 of Lichtenegger et al. (2006) for an alternative clock effect in agreement with the present treatment, although less general, and the analysis in Jiale Li et al. (2024).

[15] Mitskevich and Pulido Garcia (1970), incorrectly, obtained the opposite sign.

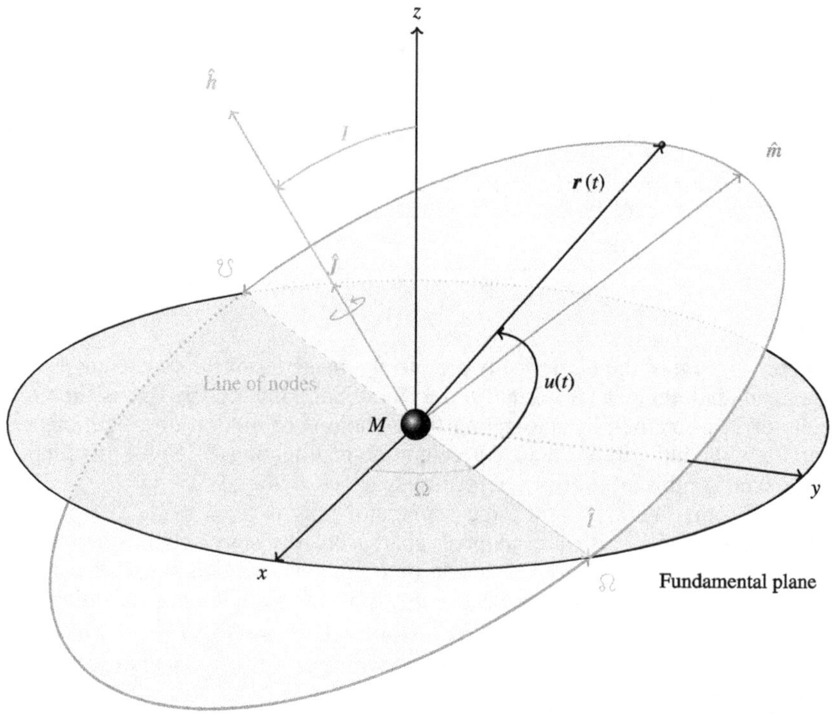

Figure 5.7 Prograde circular equatorial orbit arbitrarily oriented in space with, say, $I = 30°$, $\Omega = 45°$. The orbital plane is aligned with the equator of the central body, and the test particle moves along the *same* sense of rotation of the latter, so that Jh $= +1$.

The standard approach in deriving the aforementioned form of the gravitomagnetic clock effect is to calculate the time interval required for a test particle to come back to some fixed reference direction in the orbital plane from which it began its motion, assumed circular throughout the overall variation of the azimuthal angle $\varphi(t)$ reckoned from such a line and spanning an interval of 2π, when Equation (5.18) is added to the Newtonian inverse-square one. $\hat{\boldsymbol{J}}$ is assumed to be known, so that one can align the reference z axis with it, and the reference $\{x, y\}$ plane coincides with the equatorial one of the source.[16] The usual calculation starts by equating the centripetal acceleration $\dot{\varphi}^2 r_0$, where r_0 is the common constant radius of both the circular orbits followed by the counter-revolving test particles, to the sum of the Newtonian monopole plus Equation (5.20). Indeed, from

$$\left(\frac{d\varphi}{dt}\right)^2 r_0 = \frac{\mu}{r_0^2} \mp \frac{2\, n_{\mathrm{K}}\, G J}{c^2\, r_0^2} \tag{5.66}$$

[16] It should be noted that, even in such a case, Equation (5.64) leads to the same result as Equation (5.65).

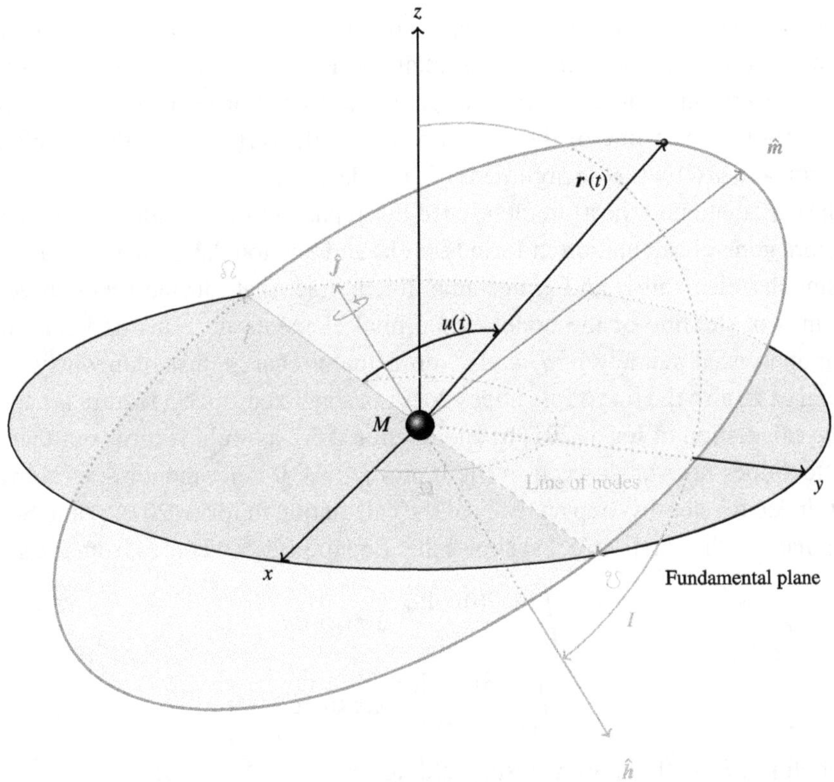

Figure 5.8 Retrograde circular equatorial orbit arbitrarily oriented in space with, say, $I = 150°$, $\Omega = 225°$. The orbital plane is aligned with the equator of the central body, and the test particle moves along the *opposite* sense of rotation of the latter, so that $\mathtt{Jh} = -1$.

one gets

$$\frac{d\varphi}{dt} = \pm n_{\rm K}\sqrt{1 \mp \frac{2\,n_{\rm K}\,J}{M\,c^2}} \tag{5.67}$$

which yields

$$\frac{dt}{d\varphi} = \pm\frac{1}{n_{\rm K}\sqrt{1 \mp \frac{2\,n_{\rm K}\,J}{M\,c^2}}} \simeq \pm\frac{1}{n_{\rm K}}\left(1 \pm \frac{n_{\rm K}\,J}{M\,c^2}\right). \tag{5.68}$$

By integrating Equation (5.68) with respect to φ from 0 to $+2\pi$ for the prograde motion and from 0 to -2π for the retrograde one, the LT orbital period

$$T_{\pm}^{\rm LT} = \frac{2\pi}{n_{\rm K}} \pm 2\pi\,\frac{J}{M\,c^2} \tag{5.69}$$

is obtained. Note that φ is a polar angle counted from some fixed reference polar axis in the orbital plane aimed to instantaneously locate the test particle along its circular orbit; thus, for a generally oriented equatorial orbit, it is straightforward to identify the fixed line of the nodes, as per Equation (5.42), with the reference direction and $\varphi\,(t)$ with the argument of latitude $u\,(t)$.

The explanation in the aforementioned discrepancy likely resides in the fact that the more general calculation in Iorio (2016b) and Section 2.3.2, made by using the nonsingular elements q and k, accounts for the fact that, during two consecutive crossings of the line of the nodes, the orbital elements in terms of which $\mathrm{d}t/\mathrm{d}u$ is parameterized, namely p, q, and k, do actually change instantaneously. In the general case, also the line of the nodes does not stay fixed; such a feature is captured by the calculation in Iorio (2016b) and Section 2.3.2 as well. It turns out that such an effect does not vanish even in the limit $q,\ k \rightarrow 0$ corresponding to a circular orbit. Indeed, a step-by-step analysis of the calculation in Iorio (2016b) and Section 2.3.2 made with Equation (5.18) shows that Equation (5.59) comes from the sum of

$$\int_0^{2\pi} \frac{\partial (\mathrm{d}t/\mathrm{d}u)}{\partial q}\,\Delta q\,(u)\,\mathrm{d}u, \tag{5.70}$$

$$\int_0^{2\pi} \frac{\partial (\mathrm{d}t/\mathrm{d}u)}{\partial k}\,\Delta k\,(u)\,\mathrm{d}u \tag{5.71}$$

which, for $q,\ k \rightarrow 0$, do not vanish, yielding

$$4\,\frac{J}{M\,c^2}\int_0^{2\pi} \cos u\,\left(\cos u - \cos u_0\right)\,\mathrm{d}u = +4\pi\,\frac{J}{M\,c^2}, \tag{5.72}$$

$$4\,\frac{J}{M\,c^2}\int_0^{2\pi} \sin u\,\left(\sin u - \sin u_0\right)\,\mathrm{d}u = +4\pi\,\frac{J}{M\,c^2}. \tag{5.73}$$

Instead, it turns out that $\left[\partial\,(\mathrm{d}t/\mathrm{d}u)\,/\partial p\right]\Delta p\,(u) = 0$ since, in the limit $q,\ k \rightarrow 0$, the instantaneous variation $\Delta p\,(u)$ of the semilatus rectum p vanishes. Also the term due to the change of the line of nodes containing $\mathrm{d}\Omega/\mathrm{d}t$ is zero for an equatorial orbit because of Equation (5.22) and Equations (D.17)–(D.19) of Appendix D. The opposite sign in Equations (5.72)–(5.73) is obtained for the retrograde motion.

Instead, the integration based on Equation (5.66) is performed by considering only φ as variable during an orbital revolution, all the rest being kept fixed.

5.6 The Radial Velocity of a Spectroscopic Binary

From Equation (2.88) and Equations (5.30)–(5.32), it turns out that the semiamplitude K of the radial velocity curve does experience a nonvanishing LT net shift per orbit induced by the LT net shift of the inclination. From Equations (2.89)–(2.91), one finally has

$$\overline{\Delta K}^{\mathrm{LT}} = \gamma_{\mathrm{A,B}} \frac{4\pi G J \cos I_{J1}}{c^2 a^2 \left(1 - e^2\right)^2}, \tag{5.74}$$

where it is meant that $\boldsymbol{J}$ should be replaced by Equation (5.38) for an ordinary binary like, for example, an exoplanet and its parent star. Indeed, for, say, a Sunlike main sequence star and a Jupiter-type planet, by using the figures in Table J.1 and Table J.3 one gets

$$\left(1 + \frac{3}{4}\frac{M_{♃}}{M_\odot}\right) J_\odot = 1.9 \times 10^{41}\,\mathrm{kg\,m^2/s}, \tag{5.75}$$

$$\left(1 + \frac{3}{4}\frac{M_\odot}{M_{♃}}\right) J_{♃} = 5.4 \times 10^{41}\,\mathrm{kg\,m^2/s}. \tag{5.76}$$

However, caution is needed since several main sequence stars hosting exoplanets are much faster rotators than the Sun, so that their spin angular momentum may be up to $\simeq 2-3$ orders of magnitude larger[17] than the solar one. Moreover, tidal effects often synchronize the rotational frequency of close exoplanets with the orbital one. Instead, for a star with the same physical properties as the Sun and the SMBH in Sgr A*, whose angular momentum $J_\bullet$ is calculated with[18] Equation (5.16) and $M_\bullet = 4.1 \times 10^6\,M_\odot$, $\chi_g = 0.5$ (Peißker et al., 2022), one has

$$\left(1 + \frac{3}{4}\frac{M_\bullet}{M_\odot}\right) J_\odot = 5.7 \times 10^{47}\,\mathrm{kg\,m^2/s}, \tag{5.77}$$

$$\left(1 + \frac{3}{4}\frac{M_\odot}{M_\bullet}\right) J_\bullet \simeq 7 \times 10^{54}\,\mathrm{kg\,m^2/s}. \tag{5.78}$$

Furthermore, $\gamma_{\mathrm{A,B}}$ is the ratio of the mass of what, out of A and B, is the unseen companion to the system's total mass M_{b}. Equation (5.74) holds for a pair of bodies of comparable masses M_{A} and M_{B} and angular momenta J_{A} and J_{B}.

In typical tight exoplanetary systems characterized by fast orbits even covered in a fraction of a day,[19] tidal forces tend to either circularize them and align both the system's orbital angular momentum and the individual spins of the star and the planet (Goldreich, 1966; Kasting et al., 1993; Murray and Dermott, 1999; Heller et al., 2011). Thus, Equation (5.74), calculated with Equation (5.38), vanishes,[20] as per Equation (D.1).

Nonetheless, the radial velocity curve exhibits a generally nonvanishing LT instantaneous shift which can be analytically worked out according to Equations (2.77)–(2.78) by using Equations (2.83)–(2.87) along with Equations (5.24)–(5.29)

[17] The case is exactly the same with, for example, WASP–33 and Kelt–9 (Stephan et al., 2022).

[18] Recent estimates for its spin parameter point towards the larger value $\chi_g = 0.9$ (Daly et al., 2024).

[19] As an example, K2–137b (Smith et al., 2018) orbits its parent star in just 0.18 d = 4.3 hr.

[20] It should be recalled that the unit vectors $\hat{\boldsymbol{l}}, \hat{\boldsymbol{m}}, \hat{\boldsymbol{h}}$ are mutually perpendicular, as per Equations (2.9)–(2.10) and Equation (2.25).

calculated with Equation (5.38). Replacing f with $f_0+2\pi$ in the resulting expression allows one to obtain the net shift per orbit, which turns out to be

$$\overline{\Delta V}^{\mathrm{LT}} = \gamma_{\mathrm{A,B}} \frac{4\pi G J}{c^2 a^2 \left(1-e^2\right)^2} \left[2\mathtt{Jh} \sin I \left(e \sin\omega + \sin u_0\right) \right.$$
$$\left. + \cos I \left[\mathtt{Jl} \left(e\cos\omega + \cos u_0\right) + \mathtt{Jm} \left(e \sin\omega + \sin u_0\right)\right]\right], \tag{5.79}$$

where it is intended that $\boldsymbol{J}$ is replaced with Equation (5.38). In the aforementioned tidally induced star–planet scenario, Equation (5.79) reduces to

$$\overline{\Delta V}^{\mathrm{LT}} = \frac{m_{\mathrm{c}}}{M_{\mathrm{b}}} \frac{8\pi G \sin I \sin u_0}{c^2 a^2} \left[\left(1 + \frac{3}{4}\frac{m_{\mathrm{c}}}{M_\star}\right) J_\star + \left(1 + \frac{3}{4}\frac{M_\star}{m_{\mathrm{c}}}\right) J_{\mathrm{c}}\right]. \tag{5.80}$$

Figure 5.9 shows the plot of the time-dependent analytical signature for a fictitious tight exoplanetary system made of a Sunlike star and a gaseous giant planet whose mass and radius are assumed to be those of Jupiter, and whose angular momentum is calculated according to Equations (5.1)–(5.5) in the hypothesis that tides have synchronized its orbital and diurnal rotations. Furthermore, all three angular momenta of the system are aligned, although spin-orbit misalignments are

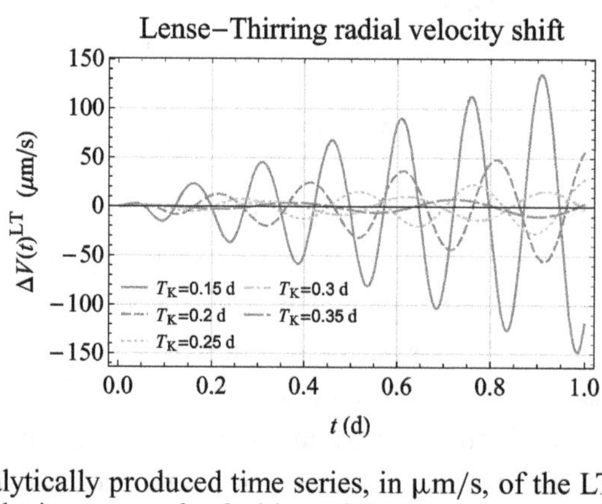

Figure 5.9 Analytically produced time series, in μm/s, of the LT shift $\Delta V(t)^{\mathrm{LT}}$ of the radial velocity curve of a fictitious tight exoplanetary system. It is made of a main sequence star with the same mass, radius and angular momentum as the Sun, and a gaseous giant planet p with the same mass and radius as Jupiter; its angular momentum is calculated with Equations (5.1)–(5.5) by assuming that its rotational frequency is synchronized with the orbital one due to tides. Indeed, it is assumed that tidal effects have either circularized the orbit, so that $e = 0$, and reciprocally aligned all the angular momenta of the system. Different values of the orbital period T_{K} are assumed, so that the star–planet relative distance, in units of $R_\star + R_{\mathrm{p}}$, ranges from 1.08 to 1.9. The other relevant orbital parameters, chosen arbitrarily, are $I = 45°$, $\Omega = 30°$, $\omega = 50°$, $\eta = 20°$. Since p cannot be considered a test particle, Equation (5.38) is used in Equations (5.24)–(5.29) along with Equations (2.83)–(2.87) in Equation (2.77) and Equation (2.78).

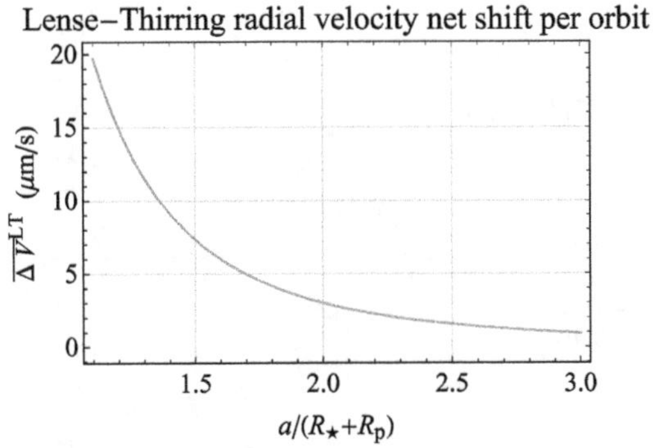

Figure 5.10 LT net shift per orbit $\overline{\Delta V}^{\rm LT}$, in μm/s, of the radial velocity of a fictitious tight exoplanetary system as a function of the star–planet distance according to Equation (5.80). The binary is made of a main sequence star with the same mass, radius, and angular momentum as the Sun, and a gaseous giant planet p with the same mass and radius as Jupiter; its angular momentum is calculated with Equations (5.1)–(5.5) by assuming that its rotational frequency is synchronized with the orbital one due to tides. The other relevant orbital parameters, chosen arbitrarily, are $I = 45°, \omega = 50°, \eta = 20°$. It is assumed that tidal effects have either circularized the orbit, so that $e = 0$, and reciprocally aligned all the angular momenta of the system.

not infrequent in exoplanets (Schlaufman, 2010). By varying the orbital period $T_{\rm K}$ from 0.15 to 0.35 d in such a way that the star–planet relative distance, in units of $R_\star + R_{\rm p}$, ranges from 1.08 to 1.9, it turns out that the peak-to-peak amplitude of the LT shift can reach the level of about $\simeq 10 - 100$ micrometres per second (μm/s) over 1 day.

Figure 5.10 plots the net shift per orbit of Equation (5.80) as a function of the star–planet distance. It turns out that the largest value, occurring for $a = 1.1\left(R_\star + R_{\rm p}\right)$, is 20 μm/s, dropping to about $\simeq$ 1 μm/s for $a/\left(R_\star + R_{\rm p}\right) = 3$.

Figure 5.11 depicts the net shift per orbit $\overline{\Delta K}^{\rm LT}$ of the radial velocity semiamplitude for the recently discovered S star S4716 (Peißker et al., 2022) orbiting Sgr A* in about 4 yr. It is plotted according to Equation (5.74) as a function of the angles $i_\bullet$ and $\zeta_\bullet$ determining the spatial orientation of the BH's spin axis, as per Equation (5.23). It will be noted that $\left|\overline{\Delta K}^{\rm LT}\right|$, which can also vanish for some values of $\zeta_\bullet, i_\bullet$, is $\lesssim 0.1$ km/s; it is $\simeq$ 2–3 orders of magnitude smaller than the uncertainties quoted in Peißker et al. (2022).

5.7 The Characteristic Timescales of Transiting Exoplanets

Here, the 1pN LT net shifts per orbit of the characteristic timescales of transiting exoplanets are calculated in their full generality. In the following sections, it is intended that $\boldsymbol{J}$ has to be replaced with Equation (5.38).

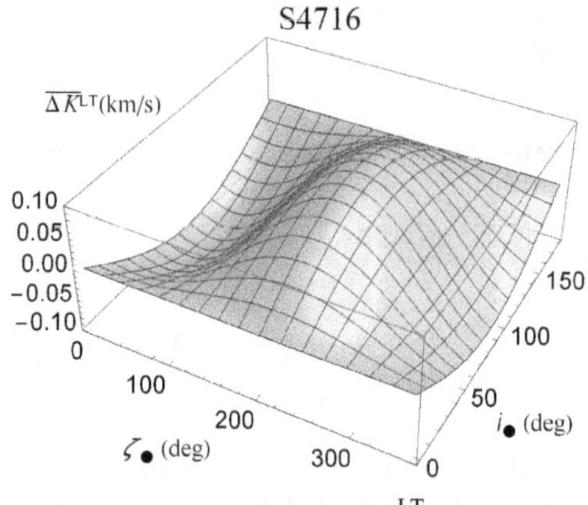

Figure 5.11 Plot of the LT net shift per orbit $\overline{\Delta K}^{\rm LT}$, in km/s, of the radial velocity semiamplitude of the recently discovered S star S4716 (Peißker et al., 2022) as a function of the angles $i_\bullet$ and $\zeta_\bullet$ characterizing the orientation of the BH's spin axis $\hat{\boldsymbol{J}}_\bullet$ in space, assumed to be completely unconstrained. The relevant orbital parameters of S4716 are $T_{\rm K} = 4.02\,{\rm yr}$, $e = 0.756$, $I = 161.24°$, $\Omega = 151.54°$ (Peißker et al., 2022). For the BH, the values $M_\bullet = 4.1 \times 10^6\, M_\odot$, $\chi_g = 0.5$ are used for its mass and spin parameter (Peißker et al., 2022).

5.7.1 The Total Transit Duration δt_D

From Equations (2.102)–(2.105) and the shifts obtainable from Equations (5.30)–(5.35), it turns out that

$$\begin{aligned}\overline{\Delta\delta t_D}^{\rm LT} = & \frac{4\pi J}{c^2 M_{\rm b} R_\star \left(1-e^2\right)\sqrt{(1+\vartheta)^2 - b^2}\,(1+e\sin\omega)^3\left(1+e^2+2e\sin\omega\right)^{3/2}} \\ & \Big\{a^2\left(1-e^2\right)^2 \mathtt{Jl}\sin 2I\,(1+e\sin\omega)\left(1+e^2\right. \\ & \left. +\,2e\sin\omega\right) + 2e\cos\omega\left(2\mathtt{Jh} + \mathtt{Jm}\cot I\right)\Big[R_\star^2\,(1+\vartheta)^2\,(1+e\sin\omega)^3 \\ & - a^2\left(1-e^2\right)^2\cos^2 I\left(2+e^2+3e\sin\omega\right)\Big]\Big\}. \end{aligned} \tag{5.81}$$

It should be noted that Equation (5.81) does not vanish for circular orbits, since its first nonvanishing term is just of zeroth order in e.

5.7.2 The Ingress/Egress Transit Duration $\delta t_{\rm i/e}$

From Equations (2.108)–(2.111) and the shifts obtainable from Equations (5.30)–(5.35), it turns out that

$$\begin{aligned}\overline{\Delta\delta t_{\rm i/e}}^{\rm LT} = & \frac{2\pi J}{c^2 M_{\rm b}}\left(\frac{\mathtt{Jl}\,a\left(1-e^2\right)\sin 2I}{R_\star\,(1+e\sin\omega)^2\sqrt{1+e^2+2e\sin\omega}}\left[\frac{1}{\sqrt{(1+\vartheta)^2-b^2}}\right.\right. \\ & \left. - \frac{1}{\sqrt{(1-\vartheta)^2-b^2}}\right]\end{aligned}$$

$$
\begin{aligned}
&- \frac{2R_\star e\cos\omega\left(2\mathtt{Jh}+\cot I\,\mathtt{Jm}\right)}{a\left(1-e^2\right)\left(1+e^2+2e\sin\omega\right)^{3/2}}\left\{\sqrt{(1-\vartheta)^2-b^2}\right.\\
&-\sqrt{(1+\vartheta)^2-b^2}\\
&+\frac{a^2\left(1-e^2\right)^2\cos^2 I\left(1+e^2+2e\sin\omega\right)}{R_\star^2\left(1+e\sin\omega\right)^3}\left[\frac{1}{\sqrt{(1+\vartheta)^2-b^2}}\right.\\
&\left.\left.\left.-\frac{1}{\sqrt{(1-\vartheta)^2-b^2}}\right]\right\}\right).
\end{aligned}
\tag{5.82}
$$

It should be noted that Equation (5.82) does not vanish for circular orbits, being its first nonvanishing term just of zeroth order in e.

5.7.3 The Full Width at Half Maximum Primary Transit Duration δt_H

From Equations (2.115)–(2.118) and the shifts obtainable from Equations (5.30)–(5.35), it turns out that

$$
\begin{aligned}
\overline{\Delta\delta t_H}^{\mathrm{LT}} &= \frac{2\pi J}{c^2 M_\mathrm{b}}\left(\frac{\mathtt{Jl}\,a\left(1-e^2\right)\sin 2I}{R_\star\left(1+e\sin\omega\right)^2\sqrt{1+e^2+2e\sin\omega}}\left[\frac{1}{\sqrt{(1+\vartheta)^2-b^2}}\right.\right.\\
&\left.+\frac{1}{\sqrt{(1-\vartheta)^2-b^2}}\right]\\
&- \frac{2R_\star e\cos\omega\left(2\mathtt{Jh}+\cot I\,\mathtt{Jm}\right)}{a\left(1-e^2\right)\left(1+e^2+2e\sin\omega\right)^{3/2}}\left\{-\sqrt{(1-\vartheta)^2-b^2}\right.\\
&-\sqrt{(1+\vartheta)^2-b^2}\\
&+\frac{a^2\left(1-e^2\right)^2\cos^2 I\left(1+e^2+2e\sin\omega\right)}{R_\star^2\left(1+e\sin\omega\right)^3}\left[\frac{1}{\sqrt{(1+\vartheta)^2-b^2}}\right.\\
&\left.\left.\left.+\frac{1}{\sqrt{(1-\vartheta)^2-b^2}}\right]\right\}\right).
\end{aligned}
\tag{5.83}
$$

It should be noted that Equation (5.83) does not vanish for circular orbits, since its first nonvanishing term is just of zeroth order in e.

5.7.4 The Time of Inferior Conjunction t_{cj}

From Equations (2.121)–(2.124) and the shifts obtainable from Equations (5.30)–(5.35), it turns out that

$$\overline{\Delta t_{\mathrm{cj}}}^{\mathrm{LT}} = \frac{4\pi J\left(2\mathtt{Jh} + \cot I\,\mathtt{Jm}\right)}{c^2 M_{\mathrm{b}}\left(1 + e\sin\omega\right)^2}. \tag{5.84}$$

It can be noted that Equation (5.84) is independent of a.

5.7.5 A Numerical Evaluation

Figure 5.12 displays the plots of Equations (5.81)–(5.84) for a Jovian-type exoplanet transiting in front of its Sun-like host star, seen edge-on, as functions of the minimum distance r_{min}, ranging from 1.1 to 20 times the sum of the radii of the star and the planet, for various values of e and by assuming that both the stellar and planetary angular momenta are aligned with the orbital angular momentum. It will be noted that the largest effect occurs for the time of inferior conjunction whose

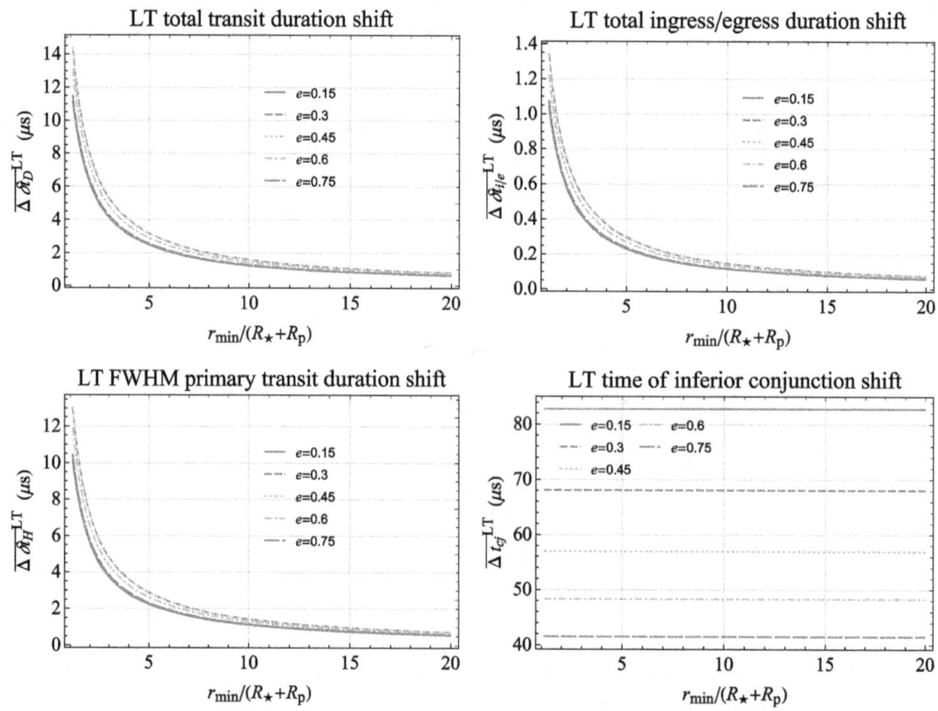

Figure 5.12 Plots of the net shifts per orbit $\overline{\Delta\delta t_D}^{\mathrm{LT}}$, $\overline{\Delta\delta t_{\mathrm{i/e}}}^{\mathrm{LT}}$, $\overline{\Delta\delta t_H}^{\mathrm{LT}}$, $\overline{\Delta t_{\mathrm{cj}}}^{\mathrm{LT}}$ of a fictitious Sun-Jupiter exoplanetary system, seen edge-on, as functions of the minimum star-planet distance r_{min}, in units of $R_\star + R_{\mathrm{p}}$, for different values of the eccentricity e according to Equations (5.81)–(5.84). The values $I = 90^\circ, \omega = 50^\circ$ are used for the relevant orbital parameters; all the three angular momenta are assumed to be aligned. The units are μs.

shift per orbit, which is independent of $r_{\rm min}$, is at the $\simeq$ 40–80 μs level. Instead, the variations of the other timescales are of the order of $\simeq$ 1–10 μs.

5.8 The Sky-Projected Spin-Orbit Angle of Transiting Exoplanets

According to Equation (2.126), the rate of change of the sky-projected spin-orbit angle λ of a transiting exoplanet is equal to the node precession, provided that the angular momentum of the host star can be considered as constant.

By expressing the stellar spin axis $\hat{\boldsymbol{J}}_\star$ in terms of its inclination $i_\star$ to the line of sight and its azimuth $\zeta_\star$ in the plane of the sky as

$$\hat{\boldsymbol{J}}_\star = \begin{cases} \sin i_\star \cos \zeta_\star \\ \sin i_\star \sin \zeta_\star \ , \\ \cos i_\star \end{cases} \tag{5.85}$$

it turns out, from Equation (5.33), Equation (5.85), and Equation (D.2), that the LT precession of λ is

$$\left\langle \frac{d\lambda}{dt} \right\rangle^{\rm LT} = \left\langle \frac{d\Omega}{dt} \right\rangle^{\rm LT} = \frac{2GJ_\star \cos i_\star}{c^2 a^3 \left(1 - e^2\right)^{3/2}}. \tag{5.86}$$

In Equation (5.86), only the angular momentum of the host star appears since it turns out that its contribution to Equation (5.38) is some orders of magnitude larger than that of a typical Jovian-type exoplanet.

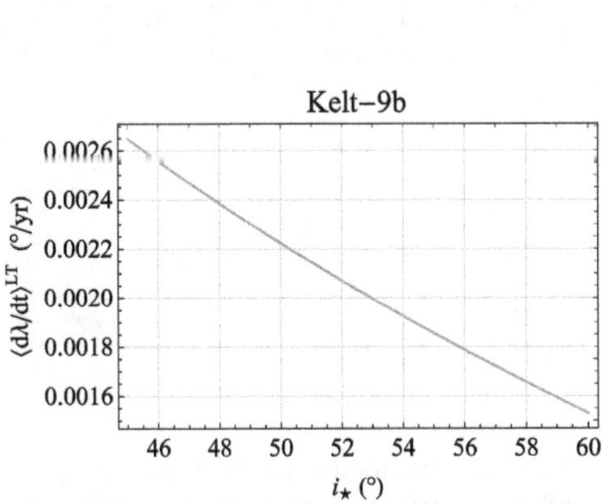

Figure 5.13 Plot of the LT averaged precession, in °/yr, of the sky-projected spin-orbit angle of the transiting exoplanet Kelt–9b as a function of the tilt $i_\star$ to the line of sight of the spin axis of its host star according to Equation (5.86). The physical parameters of the star and the planet are retrieved from Tables A.1 and A.2 of Albrecht et al. (2022) as well as the range of values for $i_\star$. The stellar angular momentum is calculated with Equations (5.1)–(5.7).

In Figure 5.13, Equation (5.86) is plotted as a function of $i_\star$ for the exoplanet Kelt–9b. It can be noted that the LT secular rate of λ ranges from $\simeq 0.001$ to $\simeq 0.002°/\mathrm{yr}$ within the allowed observational interval for $i_\star$.

5.9 The Variation of the Times of Arrival of Binary Pulsars

For a binary pulsar, the instantaneous shift of $\delta\widetilde{t}_{\mathrm{psr}}$ can be obtained from Equation (2.77) calculated with Equations (2.130)–(2.134), Equations (5.24)–(5.28), and Equation (5.51). By replacing f with $f_0 + 2\pi$ in the resulting expression, the net variation per orbit is inferred; it turns out to be (Iorio, 2017)

$$\overline{\Delta\delta\widetilde{t}_{\mathrm{psr}}}^{\mathrm{LT}} = \frac{m_{\mathrm{c}}}{M_{\mathrm{b}}}\frac{4\pi J\left[-\cos u_0\left(\mathtt{Jm}\cos I + 2\mathtt{Jh}\sin I\right) + \mathtt{Jl}\cos I\sin u_0\right]}{c^3 a^2 n_{\mathrm{K}}\sqrt{1-e^2}\left(1+e\cos f_0\right)}, \tag{5.87}$$

where it is intended that $\boldsymbol{J}$ has to be replaced by Equation (5.38) written for the pulsar and its companion. It may happen that the angular momentum of the latter is much larger than that of the pulsar itself, as for the PSR J1141–6545 system (Antoniadis et al., 2011; Iorio, 2020a; Venkatraman Krishnan et al., 2020).

Figure 5.14 displays the instantaneous shifts $\Delta\delta\widetilde{t}_{\mathrm{psr}}(t)$ of a fictitious binary pulsar whose companion is a white dwarf obtained by varying T_{K} and e in such a way

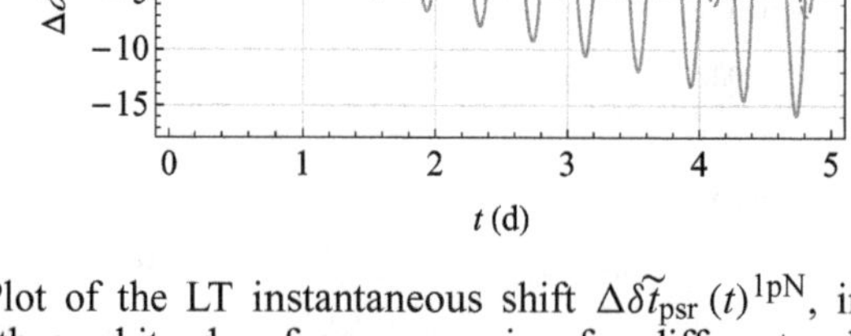

Figure 5.14 Plot of the LT instantaneous shift $\Delta\delta\widetilde{t}_{\mathrm{psr}}(t)^{1\mathrm{pN}}$, in μs, of a binary pulsar with a white dwarf as companion for different values of T_{K} and e in such a way that $r_{\min} = 1.8 \times 10^6$ km. The relevant orbital parameters are $I = 75°$, $\Omega = 50°$, $\omega = 42.457°$, $\eta = 20°$. For the stellar corpses, the values $M_{\mathrm{psr}} = 1.27\,M_\odot$, $M_{\mathrm{wd}} = 1.02\,M_\odot$, $J_{\mathrm{psr}} \simeq 4 \times 10^{40}\,\mathrm{kg\,m^2/s}$, $J_{\mathrm{wd}} \simeq 2.2 \times 10^{43}\,\mathrm{kg\,m^2/s}$, taken from those of PSR J1141–6545 (Antoniadis et al., 2011; Iorio, 2020a; Venkatraman Krishnan et al., 2020), are used. It is assumed that both the individual angular momenta and the orbital angular momentum are aligned.

that the periastron distance is kept fixed to, say, $r_{\min} = 1.8 \times 10^6$ km. For the masses and the angular momenta, assumed aligned with the orbital angular momentum, the values of PSR J1141–6545 (Antoniadis et al., 2011; Iorio, 2020a; Venkatraman Krishnan et al., 2020) are taken. It turns out that the peak–to–peak amplitudes of the signals may be as large as $\simeq$ 20–30 μs over 5 d.

It may be noted that, when the three angular momenta of the system are mutually aligned, Equation (5.87) reduces to

$$\overline{\Delta\delta\tilde{t}_{\text{psr}}}^{\text{LT}} = \frac{m_{\text{c}}}{M_{\text{b}}} \frac{8\pi J \sin I}{c^3 a^2 n_{\text{K}} \sqrt{1 - e^2}\,(1 + e\cos f_0)}, \tag{5.88}$$

where the replacement

$$J \rightarrow \left(1 + \frac{3}{4}\frac{m_{\text{c}}}{m_{\text{psr}}}\right) J_{\text{psr}} + \left(1 + \frac{3}{4}\frac{m_{\text{psr}}}{m_{\text{c}}}\right) J_{\text{c}} \tag{5.89}$$

has to be taken.

5.10 The Astrometric Angles RA and dec.

For a test particle and a massive primary, as in the case of the S stars and Sgr A*, the instantaneous shifts of the RA and dec. can be obtained from Equation (2.77) calculated with Equations (2.137)–(2.143), Equations (5.24)–(5.29), and Equation (5.51). By replacing f with $f_0 + 2\pi$ in the resulting expressions, the net variations per orbit are inferred; they turn out to be

$$\overline{\Delta\alpha}^{\text{LT}} = -\frac{16\pi G J\left[2\mathtt{Jh}\cos I + \cos u_0 \sin I\left(-\mathtt{Jm}\cos u_0 + \mathtt{Jl}\sin u_0\right)\right]}{c^2 n_{\text{K}} a^3 \left(1 - e^2\right)^{3/2}\left(3 + \cos 2I + 2\sin^2 I \cos 2u_0\right)}, \tag{5.90}$$

$$\overline{\Delta\delta}^{\text{LT}} = -\frac{4\pi G J\left[\cos u_0\left(\mathtt{Jm}\cos I + 2\mathtt{Jh}\sin I\right) - \mathtt{Jl}\cos I \sin u_0\right]}{c^2 n_{\text{K}} a^3 \left(1 - e^2\right)^{3/2}\sqrt{1 - \sin^2 I \sin^2 u_0}}. \tag{5.91}$$

Figures 5.15 and 5.16 display the instantaneous shifts $\Delta\alpha\,(t)\,, \Delta\delta\,(t)$ of the RA and the dec. of a fictitious S star obtained by varying T_{K} and e in such a way that the stellar perinigricon distance is kept fixed to $r_{\min} = 1250\,\mathcal{R}_\bullet$. The SMBH's angular momentum $J_\bullet$ is calculated with Equation (5.16), and the orientation of its spin axis is set equal to, say, $i_\bullet = 30^\circ, \zeta_\bullet = 150^\circ$. It turns out that the spin-induced astrometric signatures can be as large as $\lesssim 5$–$50''$, depending on the star's orbital period and eccentricity.

5.11 The Two-Body Range and Range Rate

Here, Equations (5.24)–(5.29) are used in Equation (2.147) and Equation (2.152) to analytically calculate the time series of the range and range rate shifts for A = Mercury and B = Earth, both moving in the gravitomagnetic field of the Sun.

Figures 5.17–5.18 plot the resulting signals, obtained by introducing the dependence on time t through the mean anomaly by means of Equation (2.4) and Equations (2.158)–(2.159), over 2 yr, which is the expected duration of the extended

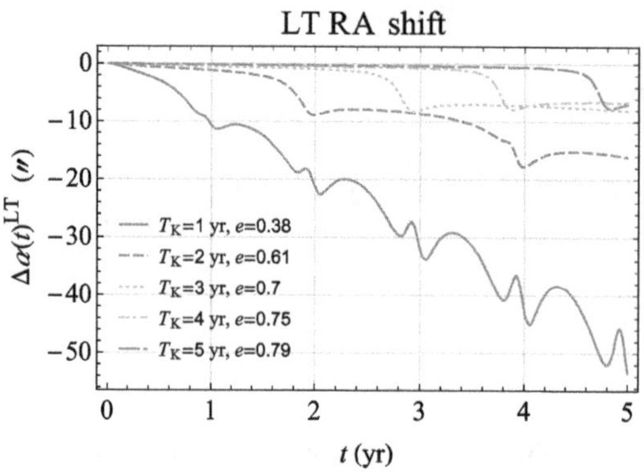

Figure 5.15 Plot of the LT instantaneous shift $\Delta\alpha\,(t)^{\rm LT}$, in ″, of the RA of a fictitious S star for different values of $T_{\rm K}$ and e in such a way that $r_{\rm min} = 1250\,\mathcal{R}_\bullet$. The relevant stellar orbital parameters are $I = 161.24°$, $\Omega = 151.54°$, $\omega = 0.073°$, $\eta = 20°$. For the BH, the values $M_\bullet = 4.1 \times 10^6\,M_\odot$, $\chi_g = 0.5$ are used for its mass and spin parameter (Peißker et al., 2022). Furthermore, the orientation of its spin axis is set by, say, $i_\bullet = 30°$, $\zeta_\bullet = 150°$.

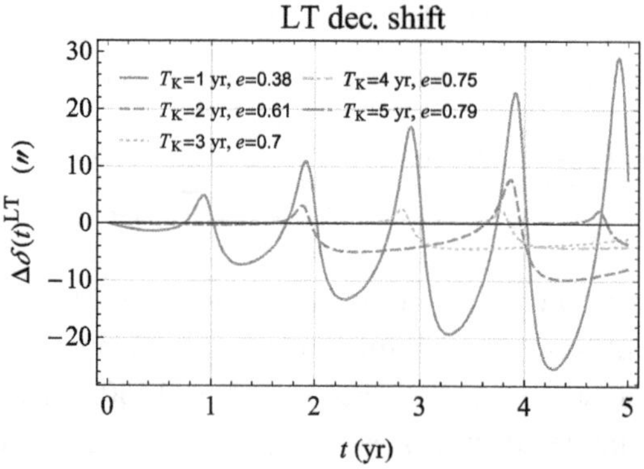

Figure 5.16 Plot of the LT instantaneous shift $\Delta\delta\,(t)^{\rm LT}$, in ″, of the dec. of a fictitious S star for different values of $T_{\rm K}$ and e in such a way that $r_{\rm min} = 1250\,\mathcal{R}_\bullet$. The relevant stellar orbital parameters are $I = 161.24°$, $\Omega = 151.54°$, $\omega = 0.073°$, $\eta = 20°$. For the BH, the values $M_\bullet = 4.1 \times 10^6\,M_\odot$, $\chi_g = 0.5$ are used for its mass and spin parameter (Peißker et al., 2022). Furthermore, the orientation of its spin axis is set by, say, $i_\bullet = 30°$, $\zeta_\bullet = 150°$.

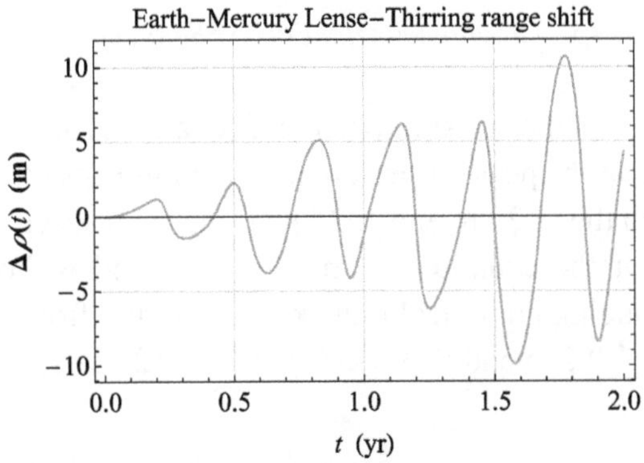

Figure 5.17 Analytically produced time series, in m, of the two-body LT range shift $\Delta\rho\,(t)$ due to the Sun's angular momentum $\boldsymbol{J}_\odot$ for A=Mercury and B=Earth plotted over 2 yr. It was worked out by calculating Equations (2.149)–(2.151), entering Equations (2.147)–(2.148) with Equations (5.24)–(5.29), Equation (5.51) and Equation (2.4). The initial values of the Keplerian orbital elements of both planets, referred to the International Celestial Reference Frame (ICRF), were retrieved from the WEB interface Horizons System at https://ssd.jpl.nasa.gov/horizons/, maintained by the Jet Propulsion Laboratory (JPL) of the National Aeronautics and Space Administration (NASA), and accessed 12th February 2024. The values $J_\odot = 1.90 \times 10^{41}\ \mathrm{kg\,m^2/s}$ (Pijpers, 1998) and $\alpha_{J_\odot} = 286^\circ.13, \delta_{J_\odot} = 63^\circ.87$ (Seidelmann et al., 2007) were adopted for the magnitude and the orientation of the Sun's angular momentum; see Table J.1.

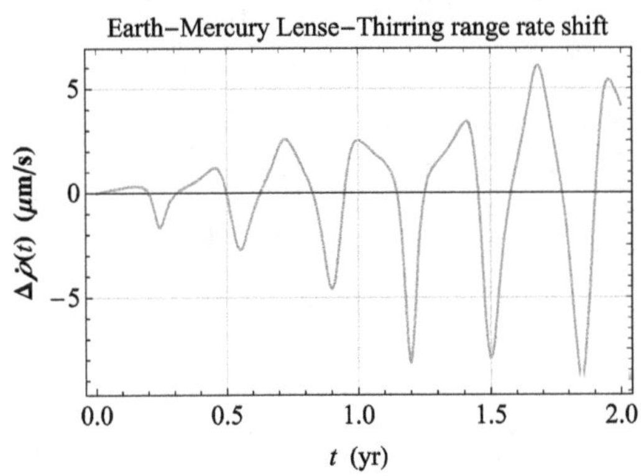

Figure 5.18 Analytically produced time series, in μm/s, of the two-body LT range rate shift $\Delta\dot{\rho}\,(t)$ due to the Sun's angular momentum $\boldsymbol{J}_\odot$ for A=Mercury and B=Earth plotted over 2 yr. It was worked out by calculating Equations (2.155)–(2.157), entering Equations (2.152)–(2.154) with Equations (5.24)–(5.29), Equation (5.51), and Equation (2.4). The initial values of the Keplerian orbital elements of both planets, referred to the International Celestial Reference Frame (ICRF), were retrieved from the WEB interface Horizons System at https://ssd.jpl.nasa.gov/horizons/, maintained by the Jet Propulsion Laboratory (JPL) of the National Aeronautics and Space Administration (NASA), and accessed 12th February 2024. The values $J_\odot = 1.90 \times 10^{41}\ \mathrm{kg\,m^2/s}$ (Pijpers, 1998) and $\alpha_{J_\odot} = 286^\circ.13, \delta_{J_\odot} = 63^\circ.87$ (Seidelmann et al., 2007) were adopted for the magnitude and the orientation of the Sun's angular momentum; see Table J.1.

phase of the BepiColombo mission (Benkhoff et al., 2010, 2021) from[21] 2026 to 2028. It turns out that the peak-to-peak nominal amplitudes of the range and range rate shifts can reach the $\simeq 20$ m and $\simeq 15\ \mu\mathrm{m/s}$ level, respectively.

An integration of the equations of motion with and without Equation (5.18) generated corresponding numerical time series differing from the aforementioned analytical ones by $\lesssim 0.2$ m and $\lesssim 3\ \mu\mathrm{m/s}$ level over 2 years, respectively.

[21] See www.esa.int/Science_Exploration/Space_Science/BepiColombo/BepiColombo_factsheet. Accessed 19th January 2024.

6

1pN Gravitomagnetic Effects: Spin Octupole

To the 1pN order, the gravitomagnetic Panhans–Soffel (PS) spin octupole acceleration $\boldsymbol{A}^{\mathrm{PS}}$ felt by a test particle moving about an oblate spheroid of constant density that is rigidly and uniformly rotating is (Panhans and Soffel, 2014)

$$\boldsymbol{A}^{\mathrm{PS}} = \frac{\boldsymbol{v}}{c^2} \times \boldsymbol{B}^{\mathrm{oct}}, \tag{6.1}$$

where the gravitomagnetic octupole field $\boldsymbol{B}^{\mathrm{oct}}$ can be calculated as (Panhans and Soffel, 2014)

$$\boldsymbol{B}^{\mathrm{oct}} = -\nabla W^{\mathrm{oct}}, \tag{6.2}$$

with the gravitomagnetic octupolar potential W^{oct} given by (Panhans and Soffel, 2014)

$$W^{\mathrm{oct}} = \frac{6GJR_{\mathrm{e}}^2\varepsilon^2}{7r^4}\mathcal{P}_3\left(r_J\right). \tag{6.3}$$

In Equation (6.3),

$$\varepsilon := \sqrt{1-\left(\frac{R_{\mathrm{po}}}{R_{\mathrm{e}}}\right)^2} \tag{6.4}$$

is the body's ellipticity written in terms of its polar and equatorial radius R_{po} and R_{e}, respectively, and $\mathcal{P}_3\left(r_J\right)$ is the Legendre polynomial of degree 3 in r_J which is defined as in Equation (5.19).

From Equations (6.1)–(6.3), the PS 1pN gravitomagnetic spin octupole acceleration can be expressed as

$$\boldsymbol{A}^{\mathrm{PS}} = \frac{3GJR_{\mathrm{e}}^2\varepsilon^2}{7c^2r^5}\boldsymbol{v} \times \left[5r_J\left(7r_J^2-3\right)\hat{\boldsymbol{r}} + 3\left(1-5r_J^2\right)\hat{\boldsymbol{J}}\right]. \tag{6.5}$$

It should be noted that the spin dipole moment in Panhans and Soffel (2014), namely $\boldsymbol{J}$, yields the usual LT acceleration of Equation (5.18). Other studies on relativistic multipoles can be found, for example, in Meichsner and Soffel (2015); Schanner and Soffel (2018); Frutos-Alfaro and Soffel (2018).

By projecting Equation (6.5) onto the unit vectors $\hat{\boldsymbol{r}}, \hat{\boldsymbol{\tau}}, \hat{\boldsymbol{h}}$ defined in Equations (2.23)–(2.25) and by using Equations (2.7)–(2.8) along with Equation (2.11), its radial, transverse, and normal components are obtained; they turn out to be

$$A_r^{\mathrm{PS}} = \frac{9Gn_{\mathrm{K}}R_{\mathrm{e}}^2 J\varepsilon^2 \,\mathtt{Jh}\, (1+e\cos f)^6 \left[1 - 5\left(\mathtt{Jl}\cos u + \mathtt{Jm}\sin u\right)^2\right]}{7c^2a^4\left(1-e^2\right)^{11/2}}, \tag{6.6}$$

$$A_\tau^{\mathrm{PS}} = -\frac{9eGn_{\mathrm{K}}R_{\mathrm{e}}^2 J\varepsilon^2 \,\mathtt{Jh}\, (1+e\cos f)^5 \sin f \left[1 - 5\left(\mathtt{Jl}\cos u + \mathtt{Jm}\sin u\right)^2\right]}{7c^2a^4\left(1-e^2\right)^{11/2}}, \tag{6.7}$$

$$\begin{aligned} A_h^{\mathrm{PS}} = {} & \frac{3Gn_{\mathrm{K}}R_{\mathrm{e}}^2 J\varepsilon^2 (1+e\cos f)^5}{7c^2a^4\left(1-e^2\right)^{11/2}} \left\{15\left(1+e\cos f\right)\left(\mathtt{Jl}\cos u + \mathtt{Jm}\sin u\right)\right. \\ & - 35\left(1+e\cos f\right)\left(\mathtt{Jl}\cos u + \mathtt{Jm}\sin u\right)^3 \\ & - 3\left[\mathtt{Jl}\left(e\cos\omega + \cos u\right) + \mathtt{Jm}\left(e\sin\omega + \sin u\right)\right] \\ & + 15\left(\mathtt{Jl}\cos u + \mathtt{Jm}\sin u\right)^2\left[\mathtt{Jl}\left(e\cos\omega + \cos u\right)\right. \\ & \left.\left. + \mathtt{Jm}\left(e\sin\omega + \sin u\right)\right]\right\}. \end{aligned} \tag{6.8}$$

The quantities $\mathtt{Jl}$, $\mathtt{Jm}$, and $\mathtt{Jh}$ entering Equations (6.6)–(6.8) are defined in Equations (D.1)–(D.3) of Appendix D.

6.1 The Averaged Orbital Precessions

Inserting Equations (6.6)–(6.8) in the machinery of Equations (2.13)–(2.18) yields the averaged precessions of the orbital elements. They turn out to be

$$\left\langle\frac{da}{dt}\right\rangle^{\mathrm{PS}} = 0, \tag{6.9}$$

$$\left\langle\frac{de}{dt}\right\rangle^{\mathrm{PS}} = \frac{45eGJR_{\mathrm{e}}^2\varepsilon^2\,\mathtt{Jh}\left(\widehat{T}_3\sin 2\omega - 2\widehat{T}_6\cos 2\omega\right)}{28c^2a^5\left(1-e^2\right)^{5/2}}, \tag{6.10}$$

$$\begin{aligned} \left\langle\frac{dI}{dt}\right\rangle^{\mathrm{PS}} = {} & -\frac{9GJR_{\mathrm{e}}^2\varepsilon^2}{56c^2a^5\left(1-e^2\right)^{7/2}}\left\{2\left(2+3e^2\right)\mathtt{Jl}\left(-4\widehat{T}_1 + 5\widehat{T}_2\right)\right. \\ & + 5e^2\left[\mathtt{Jl}\left(-2\widehat{T}_1 + 3\mathtt{Jl}^2 + \mathtt{Jm}^2\right)\cos 2\omega\right. \\ & \left.\left. + 2\mathtt{Jm}\left(-\widehat{T}_1 + 2\mathtt{Jl}^2 + \mathtt{Jm}^2\right)\sin 2\omega\right]\right\}, \end{aligned} \tag{6.11}$$

$$\left\langle \frac{d\Omega}{dt} \right\rangle^{\rm PS} = -\frac{9GJR_{\rm e}^2\varepsilon^2 \csc I}{56c^2a^5\left(1-e^2\right)^{7/2}} \left\{2\left(2+3e^2\right) \mathtt{Jm}\left(-4\widehat{T}_1+5\widehat{T}_2\right)\right.$$
$$+ 5e^2\left[-\mathtt{Jm}\left(-2\widehat{T}_1+\mathtt{Jl}^2+3\mathtt{Jm}^2\right)\cos 2\omega\right.$$
$$\left.\left.+ 2\mathtt{Jl}\left(-\widehat{T}_1+\mathtt{Jl}^2+2\mathtt{Jm}^2\right)\sin 2\omega\right]\right\}, \tag{6.12}$$

$$\left\langle \frac{d\omega}{dt} \right\rangle^{\rm PS} = \frac{9GJR_{\rm e}^2\varepsilon^2}{56c^2a^5\left(1-e^2\right)^{7/2}} \left\{4\left(3+2e^2\right) \mathtt{Jh}\left(-2\widehat{T}_1+5\widehat{T}_2\right)\right.$$
$$+2\left(2+3e^2\right)\mathtt{Jm}\left(-4\widehat{T}_1+5\widehat{T}_2\cot I\right)$$
$$+5\left[2\left(1+2e^2\right)\mathtt{Jh}\widehat{T}_3\right.$$
$$\left.- e^2\mathtt{Jm}\left(-2+\mathtt{Jl}^2+3\mathtt{Jm}^2\right)\cot I\right]\cos 2\omega$$
$$+10\mathtt{Jl}\left[2\left(1+2e^2\right)\widehat{T}_5\right.$$
$$\left.\left.+e^2\left(-1+\mathtt{Jl}^2+2\mathtt{Jm}^2\right)\cot I\right]\sin 2\omega\right\}, \tag{6.13}$$

$$\left\langle \frac{d\eta}{dt} \right\rangle^{\rm PS} = -\frac{9GJR_{\rm e}^2\varepsilon^2\mathtt{Jh}}{28c^2a^5\left(1-e^2\right)^2} \left\{5\widehat{T}_3\cos 2\omega\right.$$
$$\left.+2\left[-2\widehat{T}_1+5\left(\widehat{T}_2+\widehat{T}_6\sin 2\omega\right)\right]\right\}. \tag{6.14}$$

The coefficients $\widehat{T}_j, j = 1, 2, \ldots 6$ in Equations (6.9)–(6.14), characterizing the orientation of the orbital plane with respect to the primary's symmetry axis $\hat{\boldsymbol{J}}$, are explicitly shown in Equations (D.4)–(D.9) of Appendix D.

6.1.1 Some Special Orbital Configurations

Here, some peculiar orbital configurations, characterized in Appendix D, are examined.

Let, first, the satellite's orbital plane be assumed coincident with the body's equatorial one, irrespectively of the orientation of the latter in the adopted reference frame, namely, for generic values of α_J, δ_J. Then, according to Equations (D.17)–(D.25) of Appendix D, Equations (6.9)–(6.14) reduce to

$$\left\langle \frac{da}{dt} \right\rangle^{\rm PS} = 0, \tag{6.15}$$

$$\left\langle \frac{de}{dt} \right\rangle^{\rm PS} = 0, \tag{6.16}$$

$$\left\langle\frac{dI}{dt}\right\rangle^{\mathrm{PS}} = 0, \tag{6.17}$$

$$\left\langle\frac{d\Omega}{dt}\right\rangle^{\mathrm{PS}} = 0, \tag{6.18}$$

$$\left\langle\frac{d\omega}{dt}\right\rangle^{\mathrm{PS}} = \mp\frac{9GJR_{\mathrm{e}}^2\varepsilon^2\left(3+2e^2\right)}{7c^2a^5\left(1-e^2\right)^{7/2}}, \tag{6.19}$$

$$\left\langle\frac{d\eta}{dt}\right\rangle^{\mathrm{PS}} = \pm\frac{9GJR_{\mathrm{e}}^2\varepsilon^2}{7c^2a^5\left(1-e^2\right)^2}. \tag{6.20}$$

Equations (6.19)–(6.20) are genuine secular trends; the upper signs in them refer to the prograde motion ($\mathtt{Jh} = +1$), while the lower signs are for the retrograde motion ($\mathtt{Jh} = -1$).

Let, now, the body's spin axis, irrespective of its orientation in the adopted coordinate system, namely, for generic values of α_J, δ_J, lie somewhere in the satellite's orbital plane between $\hat{\boldsymbol{l}}$ and $\hat{\boldsymbol{m}}$. Then, according to Equations (D.26)–(D.34) of Appendix D, Equations (6.9)–(6.14) can be written as

$$\left\langle\frac{da}{dt}\right\rangle^{\mathrm{PS}} = 0, \tag{6.21}$$

$$\left\langle\frac{de}{dt}\right\rangle^{\mathrm{PS}} = 0, \tag{6.22}$$

$$\left\langle\frac{dI}{dt}\right\rangle^{\mathrm{PS}} = -\frac{9GJR_{\mathrm{e}}^2\varepsilon^2\cos\delta_J\left\{4+6e^2+5e^2\cos\left[2\left(\delta_J-\omega\right)\right]\right\}}{56c^2a^5\left(1-e^2\right)^{7/2}}, \tag{6.23}$$

$$\left\langle\frac{d\Omega}{dt}\right\rangle^{\mathrm{PS}} = -\frac{9GJR_{\mathrm{e}}^2\varepsilon^2\sin\delta_J\left\{4+6e^2+5e^2\cos\left[2\left(\delta_J-\omega\right)\right]\right\}}{56c^2a^5\left(1-e^2\right)^{7/2}}, \tag{6.24}$$

$$\left\langle\frac{d\omega}{dt}\right\rangle^{\mathrm{PS}} = 0, \tag{6.25}$$

$$\left\langle\frac{d\eta}{dt}\right\rangle^{\mathrm{PS}} = 0. \tag{6.26}$$

Equations (6.23)–(6.24), in addition to secular trends, include also long-period signatures due to the evolution of pericentre which is mainly driven by the zonal harmonics of the Newtonian component of the multipolar field of the central body (See Section 7.1).

7

Newtonian Effects: Mass Quadrupole(s)

The pK acceleration due to the first even zonal harmonic coefficient J_2 of the multipolar expansion of the exterior Newtonian gravitational potential of a massive body endowed with axial symmetry is

$$\boldsymbol{A}^{J_2} = \frac{3\mu J_2 R_e^2}{2r^4}\left[\left(5r_J^2 - 1\right)\hat{\boldsymbol{r}} - 2r_J\hat{\boldsymbol{J}}\right], \tag{7.1}$$

where r_J is defined as in Equation (5.19). In most cases, its orbital effects represent a major source of systematic bias for the pN features of motion one may be interested in since the former often have the same temporal patterns as the latter ones along with usually much larger nominal magnitudes. The knowledge of J_2 and/or of the orientation of the primary's symmetry axis $\hat{\boldsymbol{J}}$ is often imperfect to such a level that it leaves relevant mismodelled signatures. Thus, the perturbations induced by Equation (7.1) have to be carefully investigated in order to devise strategies to circumvent them or, at least, reduce their impact as much as possible.

Equation (7.1) is obtained as[1]

$$\boldsymbol{A}^{J_2} = -\boldsymbol{\nabla}\Delta U_2(\boldsymbol{r}), \tag{7.2}$$

where ΔU_2 is the term of degree[2] $\ell = 2$ of the expansion in multipoles of the Newtonian gravitational potential of an axisymmetric body (Montenbruck and Gill, 2000; Murray and Dermott, 1999; Bertotti et al., 2003; Capderou, 2005; Roy, 2005; Kopeikin et al., 2011; Poisson and Will, 2014)

$$\mathcal{U}(\boldsymbol{r}) = U_N(r) + \sum_{\ell=2}^{\infty}\Delta U_\ell(\boldsymbol{r}) = -\frac{\mu}{r}\left[1 - \sum_{\ell=2}^{\infty}\left(\frac{R_e}{r}\right)^{\ell}\mathcal{P}_\ell(r_J)\right], \tag{7.3}$$

where $\mathcal{P}_\ell(r_J)$ is the Legendre polynomial of degree ℓ in r_J (Olver et al., 2010).

1 In some textbooks, the Newtonian potential U_N is conventionally defined as positive; in this case, the acceleration is the gradient of ΔU_2.

2 If the axisymmetric primary exhibits 'north–south symmetry', only the even zonal harmonics J_ℓ, $\ell = 2, 4, 6, \ldots$ enter Equation (7.3).

For a material body in hydrostatic equilibrium like, for example, a main sequence star or a planet, it is, in general (Ragozzine and Wolf, 2009; Correia and Rodríguez, 2013),

$$J_2 = \frac{k_2}{3}\left(q_c - \frac{q_t}{2}\right), \tag{7.4}$$

where q_c is defined as in Equation (5.4), and

$$q_t := -3\left(\frac{R_e}{a'}\right)^3 \frac{M'}{M}. \tag{7.5}$$

In Equation (7.5), M' is the finite mass of a nearby tide-raising companion about which the former one revolves in an orbit whose relative semimajor axis is a'. It should be remarked that Equation (7.4) is used also for white dwarfs (Boshkayev et al., 2017; Mathew and Nandy, 2017), with, say (Iorio, 2020a),

$$k_2^{\rm wd} \simeq 0.228 \tag{7.6}$$

for the pulsar's companion (Antoniadis et al., 2011) in the binary system PSR J1141–6545 (Kaspi et al., 2000).

In dealing with astrophysical compact objects like white dwarfs, neutron stars, and BHs, it is convenient to express Equation (7.1) in terms of the negative definite dimensional quadrupole mass moment Q_2 with the replacement (Laarakkers and Poisson, 1999; Will, 2014)

$$J_2 \rightarrow -\frac{Q_2}{MR_e^2}. \tag{7.7}$$

According to Laarakkers and Poisson (1999), for a neutron star it is

$$Q_2^{\rm ns} = \xi_{\rm ns}\frac{M_{\rm ns}^3 G^2}{c^4}, \tag{7.8}$$

where $|\xi_{\rm ns}|$ ranges from 0.074 to 3.507 for a variety of equations of state (EOSs) and $M_{\rm ns} = 1.4M_\odot$; compare table 4 of Laarakkers and Poisson (1999). Then, for a typical neutron star, Equation (7.8) yields

$$9 \times 10^{35} \lesssim \left|Q_2^{\rm ns}\right| \lesssim 4 \times 10^{37}\ {\rm kg\,m^2}. \tag{7.9}$$

As a comparison, Equation (7.4), calculated for an isolated white dwarf with $M_{\rm wd} = 1.02\,M_\odot$, $R_e^{\rm wd} = 5400$ km (Antoniadis et al., 2011) and Equation (7.6), yields[3]

$$\left|Q_2^{\rm wd}\right| \simeq 4.5 \times 10^{42}\ {\rm kg\,m^2}. \tag{7.10}$$

Thus, the quadrupole mass moment of a white dwarf may be several orders of magnitude larger than that of a neutron star.

[3] It is computed by setting $q_t = 0$ $q_c = 1$ in Equation (7.4) (Iorio, 2020a).

For a Kerr BH, from Equation (5.17) calculated with $\ell = 2$, it turns out (Carter, 1971; Robinson, 1975)

$$\mathbb{M}_\bullet^2 := Q_2^\bullet = -\frac{J_\bullet^2}{c^2 M_\bullet}, \tag{7.11}$$

where $J_\bullet$ is given by Equation (5.16). Then, for a rotating Kerr BH, the quadrupole mass moment is of the order of $\mathcal{O}\left(1/c^4\right)$.

For a binary system made of bodies with comparable masses $M_{\rm A}$ and $M_{\rm B}$ and quadrupole mass moments $Q_2^{\rm A}$ and $Q_2^{\rm B}$, Equation (7.1) can be written, by means of Equation (7.7), as (Barker and O'Connell, 1975)

$$\begin{aligned} \boldsymbol{A}^{Q_2} &= \frac{3G}{2r^4}\left(1+\frac{M_{\rm B}}{M_{\rm A}}\right) Q_2^{\rm A}\left[\left(1-5r_{J_{\rm A}}^2\right)\hat{\boldsymbol{r}} + 2r_{J_{\rm A}}\hat{\boldsymbol{J}}_{\rm A}\right] \\ &+ \frac{3G}{2r^4}\left(1+\frac{M_{\rm A}}{M_{\rm B}}\right) Q_2^{\rm B}\left[\left(1-5r_{J_{\rm B}}^2\right)\hat{\boldsymbol{r}} + 2r_{J_{\rm B}}\hat{\boldsymbol{J}}_{\rm B}\right], \end{aligned} \tag{7.12}$$

where r and $\hat{\boldsymbol{r}}$ refer to the relative orbital motion.

By projecting Equation (7.1) onto the unit vectors $\hat{\boldsymbol{r}}, \hat{\boldsymbol{\tau}}, \hat{\boldsymbol{h}}$ defined in Equations (2.23)–(2.25) and by using Equations (2.7)–(2.8) along with Equation (2.11), its radial, transverse, and normal components are obtained; they turn out to be

$$A_r^{J_2} = \frac{3\mu J_2 R_{\rm e}^2 \left(1+e\cos f\right)^4}{2a^4\left(1-e^2\right)^4}\left[-\widehat{T}_1 + 3\left(\frac{\widehat{T}_2}{2} + \frac{\widehat{T}_3\cos 2u}{2} + \widehat{T}_6\sin 2u\right)\right], \tag{7.13}$$

$$A_\tau^{J_2} = \frac{3\mu J_2 R_{\rm e}^2 \left(1+e\cos f\right)^4}{a^4\left(1-e^2\right)^4}\left(\frac{\widehat{T}_3\sin 2u}{2} - \widehat{T}_6\cos 2u\right), \tag{7.14}$$

$$A_h^{J_2} = -\frac{3\mu J_2 R_{\rm e}^2 \left(1+e\cos f\right)^4}{a^4\left(1-e^2\right)^4}\left(\widehat{T}_4\cos u + \widehat{T}_5\sin u\right), \tag{7.15}$$

where the coefficients $\widehat{T}_j, j = 1, 2, \ldots 6$ are defined in Equations (D.4)–(D.9) of Appendix D. For an arbitrary orbital configuration and a generic direction of $\hat{\boldsymbol{J}}$, it is not possible to spot any particular feature of Equations (7.13)–(7.15), apart from a radial part of $\boldsymbol{A}^{J_2}$ directed inward which is always present. For an equatorial orbit, the latter, which is proportional to Equation (D.4) of Appendix D, is the only nonvanishing component of the J_2-driven acceleration, as per Equations (D.20)–(D.25) of Appendix D; it enhances Equation (2.1). In the case of a polar orbit, $\boldsymbol{A}^{J_2}$ lies entirely in the orbital plane since, according to Equations (D.29)–(D.34) of Appendix D, only $\widehat{T}_3$ and $\widehat{T}_6$ are generally non-zero, apart from $\widehat{T}_1$. In the particular case in which $\hat{\boldsymbol{J}}$ is known and oriented along, say, the reference z-axis, all the three r, τ, h components of $\boldsymbol{A}^{J_2}$ are, in general, non-zero.

7.1 The Orbital Shifts of the Keplerian Orbital Elements

The J_2-driven instantaneous shifts $\Delta\kappa\,(f)$ of the Keplerian orbital elements $\kappa = a, e, I, \Omega, \omega, \eta$ due to Equation (7.1) can be calculated as per Equation (2.12) by using Equations (7.13)–(7.15) in Equations (2.13)–(2.18). They turn out to be

$$\Delta a\,(f)^{J_2} = -\frac{J_2 R_\mathrm{e}^2}{16a\left(1-e^2\right)^3}\,\mathcal{A}^{J_2}, \tag{7.16}$$

$$\Delta e\,(f)^{J_2} = \frac{J_2 R_\mathrm{e}^2}{32a^2\left(1-e^2\right)^2}\,\mathcal{E}^{J_2}, \tag{7.17}$$

$$\Delta I\,(f)^{J_2} = -\frac{J_2 R_\mathrm{e}^2}{4a^2\left(1-e^2\right)^2}\,\mathcal{I}^{J_2}, \tag{7.18}$$

$$\Delta \Omega\,(f)^{J_2} = -\frac{J_2 R_\mathrm{e}^2 \csc I}{4a^2\left(1-e^2\right)^2}\,\mathcal{N}^{J_2}, \tag{7.19}$$

$$\Delta \omega\,(f)^{J_2} = \frac{J_2 R_\mathrm{e}^2}{32a^2 e\left(1-e^2\right)^2}\,\mathcal{G}^{J_2}, \tag{7.20}$$

$$\Delta \eta\,(f)^{J_2} = \frac{J_2 R_\mathrm{e}^2}{32a^2 e\left(1-e^2\right)^{3/2}}\,\mathcal{H}^{J_2}, \tag{7.21}$$

where

$$\mathcal{A}^{J_2} := \sum_{j=1}^{6} \mathcal{A}_j^{J_2}\,\widehat{T}_j, \tag{7.22}$$

$$\mathcal{E}^{J_2} := \sum_{j=1}^{6} \mathcal{E}_j^{J_2}\,\widehat{T}_j, \tag{7.23}$$

$$\mathcal{I}^{J_2} := \sum_{j=1}^{6} \mathcal{I}_j^{J_2}\,\widehat{T}_j, \tag{7.24}$$

$$\mathcal{N}^{J_2} := \sum_{j=1}^{6} \mathcal{N}_j^{J_2}\,\widehat{T}_j, \tag{7.25}$$

$$\mathcal{G}^{J_2} := \sum_{j=1}^{6} \mathcal{G}_j^{J_2}\,\widehat{T}_j, \tag{7.26}$$

$$\mathcal{H}^{J_2} := \sum_{j=1}^{6} \mathcal{H}_j^{J_2}\,\widehat{T}_j. \tag{7.27}$$

The coefficients $\mathcal{A}_1^{J_2}, \ldots \mathcal{H}_6^{J_2}$ entering Equations (7.22)–(7.27) are explicitly shown in Appendix F.

By calculating Equations (7.16)–(7.21) with the replacement $f \rightarrow f_0 + 2\pi$ and dividing the result by T_K, one obtains the averaged precessions

$$\left\langle \frac{da}{dt} \right\rangle^{J_2} = 0, \tag{7.28}$$

$$\left\langle \frac{de}{dt} \right\rangle^{J_2} = 0, \tag{7.29}$$

$$\left\langle \frac{dI}{dt} \right\rangle^{J_2} = -\frac{3n_K J_2 R_e^2 \widehat{T}_4}{2a^2 \left(1 - e^2\right)^2}, \tag{7.30}$$

$$\left\langle \frac{d\Omega}{dt} \right\rangle^{J_2} = -\frac{3n_K J_2 R_e^2 \widehat{T}_5}{2a^2 \left(1 - e^2\right)^2 \sin I}, \tag{7.31}$$

$$\left\langle \frac{d\omega}{dt} \right\rangle^{J_2} = \frac{3n_K J_2 R_e^2 \left(2\widehat{T}_1 - 3\widehat{T}_2 + 2\widehat{T}_5 \cot I\right)}{4a^2 \left(1 - e^2\right)^2}, \tag{7.32}$$

$$\left\langle \frac{d\eta}{dt} \right\rangle^{J_2} = \frac{3n_K J_2 R_e^2 \left(2\widehat{T}_1 - 3\widehat{T}_2\right)}{4a^2 \left(1 - e^2\right)^{3/2}}. \tag{7.33}$$

Equations (7.28)–(7.33) coincide with the definite integrals of $(1/T_K)\, d\kappa/df$ from f_0 to $f_0 + 2\pi$.

Using Equations (2.27)–(2.32) with[4]

$$\langle \mathfrak{R} \rangle^{J_2} = \frac{\mu J_2 R_e^2 \left(2\widehat{T}_1 - 3\widehat{T}_2\right)}{4a^3 \left(1 - e^2\right)^{3/2}}, \tag{7.34}$$

obtained by averaging Equation (C.21) of Appendix C over one orbit, yields just Equations (7.28)–(7.33).

7.1.1 Some Special Orbital Configurations

Here, some peculiar orbital configurations, characterized in Appendix D, are examined.

Let, first, the satellite's orbital plane be assumed coincident with the body's equatorial one, irrespective of the orientation of the latter in the adopted reference

[4] The scheme outlined in Appendix C is followed.

frame, that is, for generic values of $\alpha_J,\ \delta_J$. Then, according to Equations (D.17)–(D.25) of Appendix D, Equations (7.28)–(7.33) reduce to

$$\left\langle \frac{da}{dt} \right\rangle^{J_2} = 0, \tag{7.35}$$

$$\left\langle \frac{de}{dt} \right\rangle^{J_2} = 0, \tag{7.36}$$

$$\left\langle \frac{dI}{dt} \right\rangle^{J_2} = 0, \tag{7.37}$$

$$\left\langle \frac{d\Omega}{dt} \right\rangle^{J_2} = 0, \tag{7.38}$$

$$\left\langle \frac{d\omega}{dt} \right\rangle^{J_2} = \frac{3n_{\rm K}J_2R_{\rm e}^2}{2a^2\left(1-e^2\right)^2}, \tag{7.39}$$

$$\left\langle \frac{d\eta}{dt} \right\rangle^{J_2} = \frac{3n_{\rm K}J_2R_{\rm e}^2}{2a^2\left(1-e^2\right)^{3/2}}. \tag{7.40}$$

Equations (7.39)–(7.40) are genuine secular trends.

Let, now, the body's spin axis, irrespective of its orientation in the adopted coordinate system, that is, for generic values of $\alpha_J,\ \delta_J$, lie somewhere in the satellite's orbital plane between $\hat{\boldsymbol{l}}$ and $\hat{\boldsymbol{m}}$. Then, according to Equations (D.26)–(D.34) of Appendix D, Equations (7.28)–(7.33) can be written as

$$\left\langle \frac{da}{dt} \right\rangle^{J_2} = 0, \tag{7.41}$$

$$\left\langle \frac{de}{dt} \right\rangle^{J_2} = 0, \tag{7.42}$$

$$\left\langle \frac{dI}{dt} \right\rangle^{J_2} = 0, \tag{7.43}$$

$$\left\langle \frac{d\Omega}{dt} \right\rangle^{J_2} = 0, \tag{7.44}$$

$$\left\langle \frac{d\omega}{dt} \right\rangle^{J_2} = -\frac{3n_{\rm K}J_2R_{\rm e}^2}{4a^2\left(1-e^2\right)^2}, \tag{7.45}$$

$$\left\langle \frac{d\eta}{dt} \right\rangle^{J_2} = -\frac{3n_{\rm K}J_2R_{\rm e}^2}{4a^2\left(1-e^2\right)^{3/2}}. \tag{7.46}$$

Equations (7.45)–(7.46) are genuine secular trends.

7.1.2 The Contribution of Φ to the Mean Anomaly

When Equation (7.1) enters the equations of motion, the analytical expression of the term Φ entering the shift of the mean anomaly, calculated with Equation (2.80), turns out to be

$$\Phi\left(f\right)^{J_2} = \frac{3J_2R_{\mathrm{e}}^2}{2a^2\left(1-e^2\right)^3}\mathcal{F}^{J_2}, \tag{7.47}$$

with

$$\mathcal{F}^{J_2} = \sum_{j=1}^{6}\mathcal{F}_j^{J_2}\widehat{T}_j. \tag{7.48}$$

The coefficients $\mathcal{F}_j^{J_2}, j = 1, 2, \ldots 6$ entering Equation (7.48) are explicitly shown in Appendix F.

The net change per orbit of Φ can be obtained with the replacement $f \rightarrow f_0 + 2\pi$ in Equation (7.47), getting

$$\begin{aligned}\overline{\Phi}^{J_2} = &-\frac{3\pi J_2R_{\mathrm{e}}^2}{2a^2\left(1-e^2\right)^3}\left\{\left(2\widehat{T}_1 - 3\widehat{T}_2\right)\left[\left(1-e^2\right)^{3/2} - \left(1+e\cos f_0\right)^3\right]\right.\\ &\left.+\left(1+e\cos f_0\right)^3\left(3\widehat{T}_3\cos 2u_0 + 6\widehat{T}_6\sin 2u_0\right)\right\}.\end{aligned} \tag{7.49}$$

7.2 The Anomalistic Period

When the perturbation due to J_2 is taken into account, the anomalistic period can be calculated by means of Equations (7.13)–(7.15) as explained in Section 2.3.1. It turns out to be

$$\Delta T_{\mathrm{ano}}^{J_2} = \frac{3\pi J_2R_{\mathrm{e}}^2\left(1+e\cos f_0\right)^3}{2\left(1-e^2\right)^3\sqrt{\mu a}}\left(-2\widehat{T}_1 + 3\widehat{T}_2 + 3\widehat{T}_3\cos 2u_0 + 6\widehat{T}_6\sin 2u_0\right). \tag{7.50}$$

Figure 7.1, obtained for generic values of the Keplerian orbital parameters, confirms the analytical result of Equation (7.50); over three orbital revolutions, the test particle always reaches the precessing line of apsides after a time interval equal to T_{ano}^{J2} for each orbit. For the particular choice of the values of the primary's spin and orbital parameters, it turns out to be *longer* than T_{K}, in agreement with Equation (7.50).

Furthermore, Figure 7.2 plots the final part of the time series of the cosine $\hat{\boldsymbol{r}}\boldsymbol{\cdot}\hat{\boldsymbol{C}}$ of the angle between the position vector $\boldsymbol{r}$ and the Laplace–Runge–Lenz unit vector $\hat{\boldsymbol{C}}$ versus time t, in units of T_{K}, for a numerically integrated fictitious test particle with

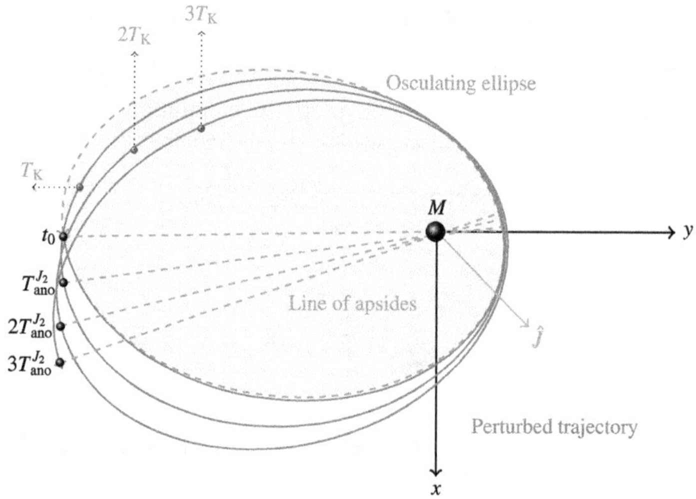

Figure 7.1 Perturbed J_2 trajectory (continuous curve) and its osculating Keplerian ellipse (dashed curve) at the initial instant of time t_0 characterized by $e = 0.7$, $I = 30°$, $\Omega = 45°$, $\omega = 50°$, $f_0 = 180°$ as seen from the z-axis. The orientation of the spin axis $\hat{\boldsymbol{J}}$ of the central body is set by $\alpha_J = 45°$, $\delta_J = 60°$. In this example, I, Ω, ω, and η undergo the Newtonian shifts of Equations (7.30)–(7.33) due to the quadrupole mass moment J_2 of the primary; their magnitudes are suitably rescaled by enhancing them for a better visualization. The positions on the perturbed trajectory after one, two, and three Keplerian periods T_K are marked as well. At each orbit, the passages at the drifting line of apsides always occur *later* than in the Keplerian case by the amount given by Equation (7.50), which is *positive* for the given values of the spin and orbital parameters.

and without Equation (7.1) starting in both cases from, say, the moving pericentre, namely, for $\hat{\boldsymbol{r}}_0 \cdot \hat{\boldsymbol{C}}_0 = +1$. It can be seen that it comes back to the same position on the precessing line of apsides, that is, it is $\hat{\boldsymbol{r}} \cdot \hat{\boldsymbol{C}} = +1$ again, just after $T^{J_2}_{\mathrm{ano}} = T_K + \Delta T^{J_2}_{\mathrm{ano}}$ differing from T_K by a (*positive*) amount in agreement with Equation (7.50) for the particular choice of the generic values of the spin and the orbital parameters adopted in the numerical integrations.

7.3 The Draconitic Period

The draconitic period, calculated when the perturbation due to J_2 is taken into account as explained in Section 2.3.2, turns out to be

$$\Delta T^{J_2}_{\mathrm{dra}} = \frac{3\pi J_2 R_e^2}{2\sqrt{\mu a\left(1-e^2\right)}}\left[\frac{1}{(1+e\cos\omega)^2}\left(-2\widehat{T}_1+3\widehat{T}_2-2\widehat{T}_5\cot I\right)\right.$$
$$\left.+\frac{(1+e\cos f_0)^3}{\left(1-e^2\right)^{5/2}}\left(-2\widehat{T}_1+3\widehat{T}_2+3\widehat{T}_3\cos 2u_0+6\widehat{T}_6\sin 2u_0\right)\right]. \quad (7.51)$$

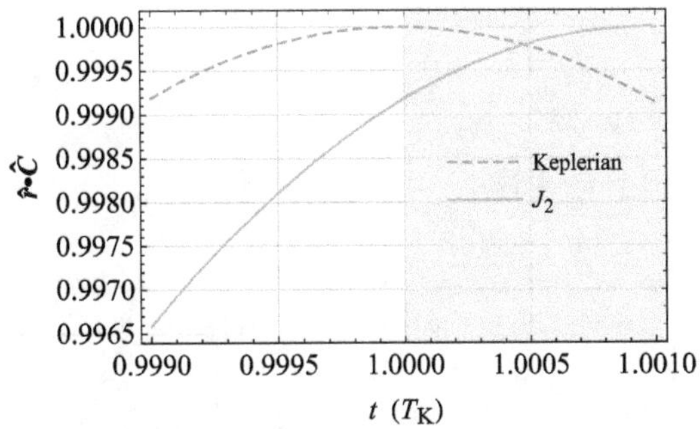

Figure 7.2 Plot of the numerically produced time series of the cosine $\hat{\boldsymbol{r}} \cdot \hat{\boldsymbol{C}}$ of the angle between the position vector $\boldsymbol{r}$ and the Laplace–Runge–Lenz vector $\boldsymbol{C}$ versus time t, in units of $T_{\rm K}$, obtained by integrating the equations of motion of a fictitious test particle with and without the J_2 acceleration of Equation (7.1) for an elliptical ($e = 0.665$) orbit arbitrarily oriented in space ($I = 40°$, $\Omega = 45°$, $\omega = 50°$) starting from the periapsis ($f_0 = 0$), that is, $\hat{\boldsymbol{r}}_0 \cdot \hat{\boldsymbol{C}}_0 = +1$; the semimajor axis is $a = 6R_{\rm e}$. The physical parameters of the Earth are adopted, apart from the spin axis position set by $\alpha_J = 45°$, $\delta_J = 60°$. The J_2 acceleration is suitably rescaled in such a way that $\left|\Delta T_{\rm ano}^{J_2}\right| / T_{\rm K} = 0.001$. The time needed to come back to the initial position on the (moving) line of apsides, so that $\hat{\boldsymbol{r}} \cdot \hat{\boldsymbol{C}} = +1$ again, is *longer* than in the Keplerian case by the amount $\Delta T_{\rm ano}^{J_2} = +0.001 T_{\rm K}$, shown by the shaded area, in agreement with Equation (7.50).

It can be noted that Equation (7.51) is not defined for $I \to 0$ because of the term

$$\widehat{T}_5 \cot I = \cot I \left[\sin I \sin \delta_J + \cos I \cos \delta_J \sin\left(\alpha_J - \Omega\right)\right] \\ \times \left[\cos I \sin \delta_J - \cos \delta_J \sin I \sin\left(\alpha_J - \Omega\right)\right], \qquad (7.52)$$

as is expected since, in this case, the line of nodes is no longer defined.

Figure 7.3, obtained for generic values of the Keplerian orbital parameters, confirms the analytical result of Equation (7.51); over three orbital revolutions, the test particle reaches always the precessing line of nodes after a time interval equal to $T_{\rm dra}^{J_2}$ after each orbit. For the particular choice of the values of the primary's spin and orbital parameters, it is *shorter* than $T_{\rm K}$, in agreement with Equation (7.51).

Furthermore, Figure 7.4 plots the final part of the time series of the cosine $\hat{\boldsymbol{r}} \cdot \hat{\boldsymbol{l}}$ of the angle between the position vector $\boldsymbol{r}$ and the node unit vector $\hat{\boldsymbol{l}}$ versus time t, in units of $T_{\rm K}$, for a numerically integrated fictitious test particle with and without Equation (7.1), starting in both cases from, say, the moving ascending node, namely, for $\hat{\boldsymbol{r}}_0 \cdot \hat{\boldsymbol{l}}_0 = +1$. It can be seen that it comes back to the same position on the precessing line of nodes, that is, it is $\hat{\boldsymbol{r}} \cdot \hat{\boldsymbol{l}} = +1$ again, just after $T_{\rm dra}^{J_2} = T_{\rm K} + \Delta T_{\rm dra}^{J_2}$ differing from $T_{\rm K}$ by a (*positive*) amount in agreement with Equation (7.51) for

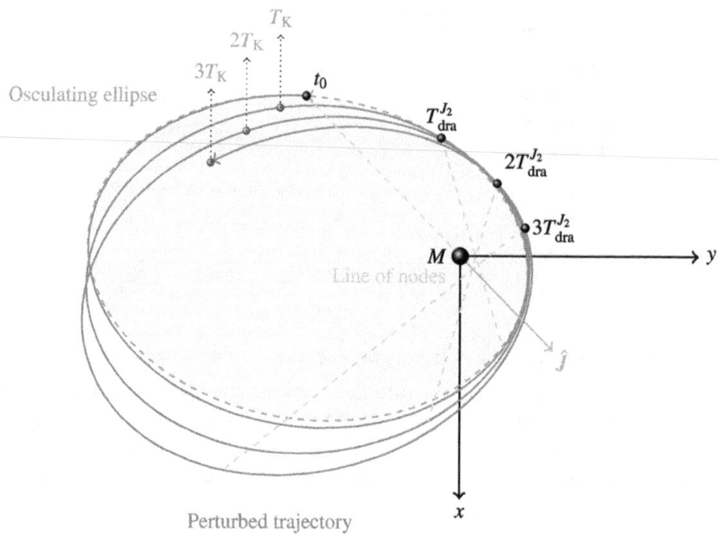

Figure 7.3 Perturbed J_2 trajectory (continuous curve) and its osculating Keplerian ellipse (dashed curve) at the initial instant of time t_0 characterized by $e = 0.7$, $I = 30°$, $\Omega = 45°$, $\omega = 50°$, $f_0 = 180° - \omega$ as seen from the z-axis. The orientation of the spin axis $\hat{\boldsymbol{J}}$ of the central body is set by $\alpha_J = 45°$, $\delta_J = 60°$. In this example, I, Ω, ω, and η undergo the Newtonian shifts of Equations (7.30)–(7.33) due to the quadrupole mass moment J_2 of the primary; their magnitudes are suitably rescaled for better visualizing their effect. The positions on the perturbed trajectory after one, two, and three Keplerian periods T_K are marked as well. At each orbit, the passages at the precessing line of nodes always occur *earlier* than in the Keplerian case by the amount given by Equation (7.51), which is *negative* for the given values of the spin and orbital parameters.

the particular choice of the generic values of the spin and the orbital parameters adopted in the numerical integrations.

7.4 The Sidereal Period

As shown in Section 2.3.3, the sidereal period for a generic perturbed orbit is the sum of the draconitic period, calculated as explained in Section 2.3.2, and the term given by Equation (2.66). For Equation (7.1), Equation (2.66) turns out to be

$$\Delta T^{J_2}_{\text{sid II}} = \frac{3\pi J_2 R_e^2 \cot I}{e^2\sqrt{\mu a\left(1-e^2\right)}}\left\{\widehat{T}_5\left[e^2 + 2\left(-2+e^2+2\sqrt{1-e^2}\right)\cos 2\omega\right]\right.$$
$$\left. - 2\widehat{T}_4\left(-2+e^2+2\sqrt{1-e^2}\right)\sin 2\omega\right\} \simeq \frac{3\pi J_2 R_e^2 \widehat{T}_5 \cot I}{\sqrt{\mu a}} + \mathcal{O}\left(e^2\right). \tag{7.53}$$

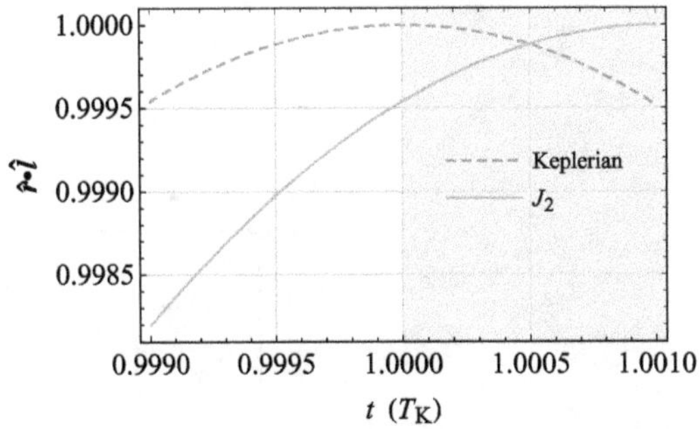

Figure 7.4 Plot of the numerically produced time series of the cosine $\hat{\boldsymbol{r}} \cdot \hat{\boldsymbol{l}}$ of the angle between the position vector $\boldsymbol{r}$ and the node unit vector $\hat{\boldsymbol{l}}$ versus time t, in units of T_K, obtained by integrating the equations of motion of a fictitious test particle with and without the J_2 acceleration of Equation (7.1) for an elliptical ($e = 0.665$) orbit arbitrarily oriented in space ($I = 40°$, $\Omega = 45°$, $\omega = 50°$) starting from the ascending node ☊ ($f_0 = -\omega + 360°$), i.e., $\hat{\boldsymbol{r}}_0 \cdot \hat{\boldsymbol{l}}_0 = +1$; the semimajor axis is $a = 6R_e$. The physical parameters of the Earth are adopted, apart from the spin axis position set by $\alpha_J = 45°$, $\delta_J = 60°$. The J_2 acceleration is suitably rescaled in such a way that $\left|\Delta T_{\rm dra}^{J_2}\right| / T_K = 0.001$. The time needed to come back to the initial position on the (moving) line of nodes, so that $\hat{\boldsymbol{r}} \cdot \hat{\boldsymbol{l}} = +1$ again, is *longer* than in the Keplerian case by the amount $\Delta T_{\rm dra}^{J_2} = +0.001 T_K$, shown by the shaded area, in agreement with Equation (7.51).

For equatorial orbits, Equation (7.53) vanishes because of Equations (D.17)–(D.19) of Appendix D, and the sidereal period reduces to the draconitic one. The oblateness-induced correction of the sidereal period $\Delta T_{\rm sid}^{J_2}$ can be obtained by summing Equation (7.51) and Equation (7.53); for an elliptic orbit, it turns out to be singular in $I = 0$. Instead, in the limit $e \to 0$, it reduces to

$$\begin{aligned}
\Delta T_{\rm sid}^{J_2} = \frac{3\pi J_2 R_e^2}{2\sqrt{\mu a}} \{ & -4 + 6\cos^2\delta_J \cos^2(\alpha_J - \Omega) \\
& + 6\cos\delta_J \cos(\alpha_J - \Omega) \sin 2u_0 \left[\sin I \sin\delta_J \right. \\
& \left. + \cos I \cos\delta_J \sin(\alpha_J - \Omega)\right] \\
& + 6\left[\sin I \sin\delta_J + \cos I \cos\delta_J \sin(\alpha_J - \Omega)\right]^2 \\
& + 3\cos 2u_0 \left[\cos\delta_J \cos(\alpha_J - \Omega) \right. \\
& \left. - \sin I \sin\delta_J - \cos I \cos\delta_J \sin(\alpha_J - \Omega)\right] \\
& \times \left[\cos\delta_J \cos(\alpha_J - \Omega) + \sin I \sin\delta_J \right. \\
& \left. + \cos I \cos\delta_J \sin(\alpha_J - \Omega)\right]\},
\end{aligned} \tag{7.54}$$

which is defined also for that value of the inclination. In such a case, using the true longitude l yields

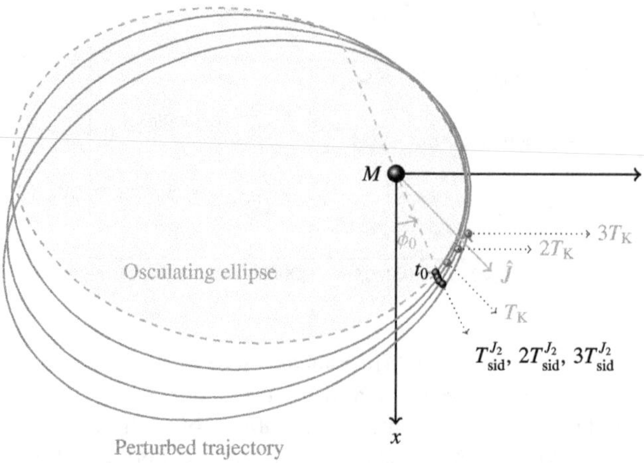

Figure 7.5 Projections of the perturbed J_2 trajectory (continuous curve) and of its osculating Keplerian ellipse (dashed curve) in the reference plane $\{x, y\}$ at the initial instant of time t_0 characterized by the generic initial conditions $e = 0.7, I = 30°, \Omega = 45°, \omega = 50°, f_0 = 285°$. The orientation of the spin axis $\hat{\boldsymbol{J}}$ of the central body, whose projection in the fundamental plane is depicted as well, is set by $\alpha_J = 45°, \delta_J = 60°$. In this example, I, Ω, ω and η undergo the Newtonian shifts of Equations (7.30)–(7.33) due to the quadrupole mass moment J_2 of the primary; their magnitudes are suitably rescaled for better visualizing their effect. The positions on the perturbed trajectory after one, two and three Keplerian periods T_K are marked as well. At each orbit, the passages at the generic fixed line characterized by ϕ_0 occur always *earlier* than in the Keplerian case by the amount given by the sum of Equation (7.51) and Equation (7.53). It is so because, for the given values of the spin and orbital parameters, $\Delta T_{dra}^{J_2} + \Delta T_{sid\,II}^{J_2} < 0$, as per Equation (7.51) and Equation (7.53).

$$
\begin{aligned}
\Delta T_{\text{sid}}^{J_2} = & -\frac{3\pi J_2 R_e^2}{4\left(1-e^2\right)^2\sqrt{\mu a}}\left[\frac{\left(-2+3\cos^2\delta_J\right)}{\left(1+e\cos\varpi\right)^2}\left[2+e^2-2\left(1-e^2\right)^{3/2}\right.\right. \\
& \left. + 4e\cos\varpi + e^2\cos 2\varpi\right] + \frac{1}{2\left(1-e^2\right)}\left(\left(4+e^2\right)\left(1-3\cos 2\delta_J\right)\right. \\
& -e\left[-1+6\cos^2\left(l_0-\alpha_J\right)\cos 2\delta_J\right] \\
& \times\left[3e\cos\left(2l_0-2\varpi\right)+6\cos\left(l_0-\varpi\right)\right. \\
& \left.+2e^2\cos^3\left(l_0-\varpi\right)\right] - 3\cos\left(2l_0-2\alpha_J\right)\left\{\left(2+3e^2\right)\cos 2\delta_J\right. \\
& \left.\left.\left.+2\left[1+e\cos\left(l_0-\varpi\right)\right]^3\right\}\right)\right].
\end{aligned}
\tag{7.55}
$$

In the limit $e \to 0$, Equation (7.55) agrees with Equation (7.54) calculated for $I = 0$.

Figure 7.5 confirms the analytical results of Equation (7.51) and Equation (7.53). Indeed, over three orbital revolutions, the projection of a generic J_2-perturbed orbit

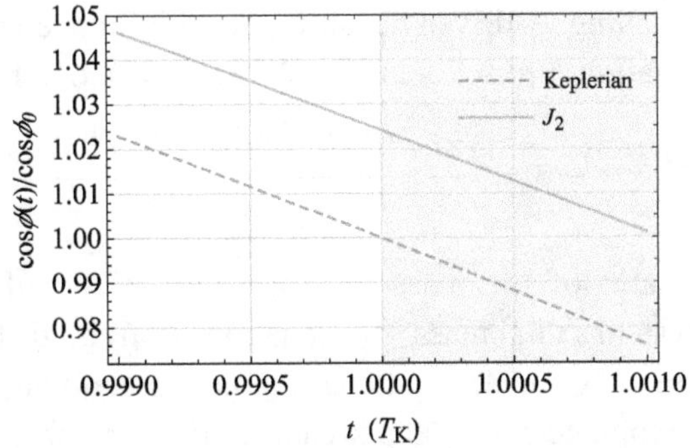

Figure 7.6 Plot of the numerically produced time series of the cosine $\cos\phi\,(t)$ of the azimuthal angle $\phi\,(t)$ normalized to its initial value $\cos\phi_0$ versus time t, in units of T_K, obtained by integrating the equations of motion of a fictitious test particle with and without the J_2 acceleration of Equation (7.1) for an elliptical ($e = 0.665$) orbit arbitrarily oriented in space ($I = 40°$, $\Omega = 45°$, $\omega = 50°$) starting from, say, the ascending node ☊ ($f_0 = -\omega + 360°$); the semimajor axis is $a = 6R_e$. The physical parameters of the Earth are adopted, apart from the spin axis position set by $\alpha_J = 45°$, $\delta_J = 60°$. The J_2 acceleration is suitably rescaled in such a way that $\left|\Delta T_{\mathrm{sid}}^{J_2}\right|/T_K = 0.001$. The time needed to $\cos\phi\,(t)$ to assume again its initial value $\cos\phi_0$ is *longer* than in the Keplerian case by the amount $\Delta T_{\mathrm{sid}}^{J_2} = +0.001\,T_K$, shown by the shaded area, in agreement with the sum of Equation (7.51) and Equation (7.53).

in the fundamental plane $\{x, y\}$ crosses a fixed direction in the latter set by a certain value ϕ_0 always after a time interval equal to $T_{\mathrm{sid}}^{J_2} = T_{\mathrm{dra}}^{J_2} + \Delta T_{\mathrm{sid\,II}}^{J_2}$ after each orbit. For the particular choice of the primary's spin and the orbital parameters used in the picture, $T_{\mathrm{sid}}^{J_2}$ turns out to be *shorter* than T_K, in agreement with Equation (7.51) and Equation (7.53).

Furthermore, Figure 7.6 plots the final part of the time series of the cosine of the angle ϕ, normalized to its initial value $\cos\phi_0$, versus time t, in units of T_K, for a numerically integrated fictitious test particle with and without Equation (7.1) starting from the same generic initial position. It can be seen that it comes back to the same position on the fixed direction chosen in the reference plane, that is, it is $\cos\phi/\cos\phi_0 = +1$ again, just after $T_{\mathrm{sid}}^{J_2} = T_{\mathrm{dra}}^{J_2} + \Delta T_{\mathrm{sid\,II}}^{J_2}$, differing from T_K by a (*positive*) amount in agreement with Equation (7.51) and Equation (7.53) for the particular choice of the generic values of the spin and the orbital parameters adopted in the numerical integrations.

7.5 The Radial Velocity of a Spectroscopic Binary

From Equation (2.88) and the net orbital shifts obtainable from Equations (7.28)–(7.30), it turns out that the semiamplitude K of the radial velocity curve does experience a nonvanishing J_2-driven net shift per orbit induced by the quadrupolar net shift of the inclination. From Equations (2.89)–(2.91), one finally has

$$\overline{\Delta K}^{J_2} = -\gamma_{\mathrm{A,B}} \frac{3\pi n_{\mathrm{K}} \left(J_2^{\mathrm{A}} R_{\mathrm{A}}^2 \widehat{T}_4^{\mathrm{A}} + J_2^{\mathrm{B}} R_{\mathrm{B}}^2 \widehat{T}_4^{\mathrm{B}}\right) \cos I}{a\left(1-e^2\right)^{5/2}}, \tag{7.56}$$

where $\gamma_{\mathrm{A,B}}$ is the ratio of the mass of what, out of A and B, is the unseen companion to the system's total mass M_{b}. Equation (7.56) holds for a pair of bodies of comparable masses M_{A} and M_{B} and quadrupole mass moments Q_2^{A} and Q_2^{B}.

In typical tight exoplanetary systems characterized by fast orbits, sometimes covered in a fraction of a day,[5] tidal forces tend often to either circularize them and align both the system's orbital angular momentum and the spins of the star and the planet (Goldreich, 1966; Kasting et al., 1993; Murray and Dermott, 1999; Heller et al., 2011). Thus, Equation (7.56) vanishes,[6] as per Equation (D.7) of Appendix D.

Nonetheless, the radial velocity curve exhibits a generally nonvanishing oblateness-driven instantaneous shift which can be analytically worked out according to Equations (2.77)–(2.78) by using Equations (2.83)–(2.87) along with Equations (7.16)–(7.21) and Equation (7.47) calculated for both bodies. By replacing f with $f_0 + 2\pi$ in the resulting expression allows one to obtain the net shift per orbit.

In general, the products $J_2^{\star} R_{\star}^2$ and $J_2^{\mathrm{p}} R_{\mathrm{p}}^2$ might be of comparable magnitude for an exoplanet and its Sunlike parent star; for the Sun and Jupiter, they amount to $9.6 \times 10^{10}\ \mathrm{m}^2$ and $7.5 \times 10^{13}\ \mathrm{m}^2$, respectively, as per Table J.1 and Table J.3. However, the same caveat about faster rotators than the Sun and the tidally synchronized diurnal rotation of the hosted exoplanet raised in Section 5.6 holds also in this case. Instead, for the SMBH in Sgr A*, whose quadrupole mass moment $Q_2^{\bullet}$ is calculated according to Equation (5.16) and Equation (7.11) along with $M_{\bullet} = 4.1 \times 10^6\, M_{\odot}$, $\chi_g = 0.5$ (Peißker et al., 2022), and a Sunlike star, the BH's term turns out to be up to eight orders of magnitude larger than the stellar one.

Figure 7.7 shows the plot of the resulting signature for a fictitious tight exoplanetary system made of a Sunlike star and a gaseous giant planet p whose mass and radius are assumed to be those of Jupiter, and whose first even zonal harmonic J_2^{p} is calculated according to Equations (7.4)–(7.5) and Equation (5.4) in the hypothesis that tides have synchronized its orbital and diurnal rotations. Furthermore, all the three angular momenta of the system are aligned. By varying the orbital period T_{K} from 0.15 to 0.35 d in such a way that the star–planet relative distance, in units of $R_{\star} + R_{\mathrm{p}}$, ranges from 1.08 to 1.9, it turns out that the peak-to-peak amplitude of the J_2 nominal shift can reach the level of about $\simeq 10 - 200\ \mathrm{m/s}$ over 1 day.

[5] As an example, K2–137b (Smith et al., 2018) orbits its parent star in just $0.18\ \mathrm{d} = 4.3$ hr.

[6] It should be recalled that the unit vectors $\hat{\boldsymbol{l}}, \hat{\boldsymbol{m}}, \hat{\boldsymbol{h}}$ are mutually perpendicular, as per Equations (2.9)–(2.10) and Equation (2.25).

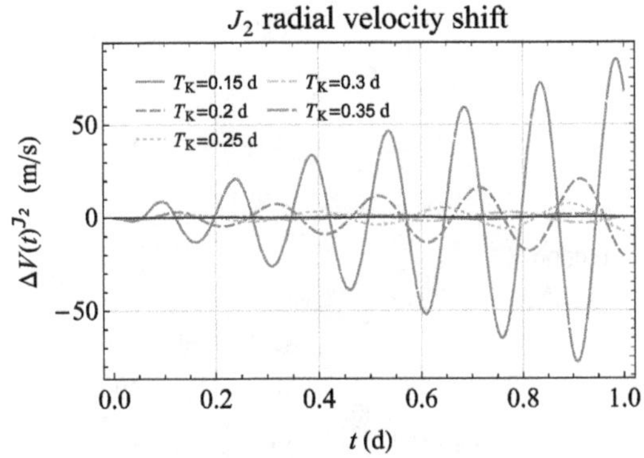

Figure 7.7 Analytically produced time series, in m/s, of the oblateness–driven shift $\Delta V(t)^{J_2}$ of the radial velocity curve of a fictitious tight exoplanetary system. It is made of a main sequence star with the same mass, radius and quadrupole mass moment of the Sun, and a gaseous giant planet p with the same mass and radius of Jupiter; J_2^{p} is calculated with Equations (7.4)–(7.5) and Equation (5.4) by assuming that its rotational frequency is synchronized with the orbital one due to tides. Different values of the orbital period T_{K} are assumed, so that the star–planet relative distance, in units of $R_\star + R_{\mathrm{p}}$, ranges from 1.08 to 1.9. The other relevant orbital parameters, chosen arbitrarily, are $I = 45°, \Omega = 30°, \omega = 50°, \eta = 20°$. It is assumed that tidal effects have either circularized the orbit, so that $e = 0$, and reciprocally aligned all the angular momenta of the system. Since p cannot be considered a test particle, $J_2R_{\mathrm{e}}^2$ in Equations (7.16)–(7.21) and Equation (7.47) is replaced with $J_2^\star R_\star^2 + J_2^{\mathrm{p}} R_{\mathrm{p}}^2$, in agreement with Equation (7.12), to be used along with Equations (2.83)–(2.87) in Equation (2.77) and Equation (2.78).

Figure 7.8 depicts the net shift per orbit $\overline{\Delta K}^{Q_2}$ of the radial velocity semi-amplitude for the S star S4716 (Peißker et al., 2022). It is plotted according to Equation (7.56), calculated with Equation (7.7) and Equation (7.11), as a function of the angles $i_\bullet$ and $\zeta_\bullet$ of the BH's spin axis assumed as independent variables. It turns out that $\left|\overline{\Delta K}^{Q_2}\right|$, which can also vanish for some values of $\zeta_\bullet$, $i_\bullet$, is $\lesssim 0.0002$ km/s; it is several orders of magnitude smaller than the uncertainties quoted in Peißker et al. (2022).

7.6 The Characteristic Timescales of Transiting Exoplanets

Here, the oblateness-driven net shifts per orbit of the characteristic timescales of transiting exoplanets are calculated in their full generality. In the following expressions, it is intended that $J_2R_{\mathrm{e}}^2\widehat{T}_j, j = 1, 2, \ldots 5$ is a shorthand for

$$J_2^\star R_\star^2 \widehat{T}_j^\star + J_2^{\mathrm{p}} R_{\mathrm{p}}^2 \widehat{T}_j^{\mathrm{p}}, j = 1, 2, \ldots 5. \tag{7.57}$$

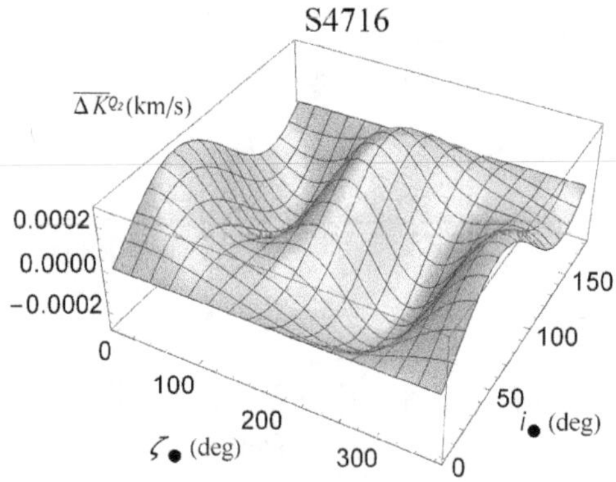

Figure 7.8 Plot of the oblateness-driven net shift per orbit $\overline{\Delta K}^{Q_2}$, in km/s, of the radial velocity semiamplitude of the recently discovered S star S4716 (Peißker et al., 2022) as a function of the angles $i_\bullet$ and $\zeta_\bullet$ characterizing the orientation of the BH's spin axis $\hat{\boldsymbol{J}}_\bullet$ in space, assumed completely unconstrained. The relevant orbital parameters of S4716 are $T_{\rm K} = 4.02\,{\rm yr}$, $e = 0.756$, $I = 161.24°$, $\Omega = 151.54°$ (Peißker et al., 2022). For the BH, the values $M_\bullet = 4.1 \times 10^6\, M_\odot$, $\chi_g = 0.5$ are used for its mass and spin parameter (Peißker et al., 2022).

7.6.1 The Total Transit Duration δt_D

From Equations (2.102)–(2.105) and the shifts obtainable from Equations (7.28)–(7.33), it turns out

$$\begin{aligned}\overline{\Delta\delta t_D}^{J_2} = &\frac{3\pi J_2 R_{\rm e}^2}{n_{\rm K} a^2 R_\star \left(1-e^2\right)^{3/2}\sqrt{(1+\vartheta)^2-b^2}\left(1+e^2+2e\sin\omega\right)^{3/2}(1+e\sin\omega)^3}\\ &\Big\{-a^2\left(1-e^2\right)^2 \widehat{T}_4 \sin 2I\,(1+e\sin\omega)\left(1+e^2+2e\sin\omega\right)\\ &+e\cos\omega\left(2\widehat{T}_1-3\widehat{T}_2+2\widehat{T}_5\cot I\right)\Big[-R_\star^2(1+\vartheta)^2(1+e\sin\omega)^3\\ &+a^2\left(1-e^2\right)^2\cos^2 I\left(2+e^2+3e\sin\omega\right)\Big]\Big\}. \qquad (7.58)\end{aligned}$$

It should be noted that Equation (7.58) does not vanish for circular orbits, being its first nonvanishing term just of zeroth order in e.

7.6.2 The Ingress/Egress Transit Duration $\delta t_{\rm i/e}$

From Equations (2.108)–(2.111) and the shifts obtainable from Equations (7.28)–(7.33), it turns out

$$\overline{\Delta\delta t_{\mathrm{i/e}}}^{J_2} = \frac{3\pi J_2 R_{\mathrm{e}}^2}{2n_{\mathrm{K}} R_\star a^3 \left(1-e^2\right)^{3/2}\left(1+e^2+2e\sin\omega\right)^{3/2}} \left(-\frac{a^2\left(1-e^2\right)^2 \widehat{T}_4 \sin 2I\left(1+e^2+2e\sin\omega\right)}{(1+e\sin\omega)^2}\left[\frac{1}{\sqrt{(1+\vartheta)^2-b^2}} - \frac{1}{\sqrt{(1-\vartheta)^2-b^2}}\right] + R_\star^2 e\cos\omega\left(2\widehat{T}_1-3\widehat{T}_2+2\widehat{T}_5\cot I\right)\left\{\sqrt{(1-\vartheta)^2-b^2} - \sqrt{(1+\vartheta)^2-b^2} + \frac{a^2\left(1-e^2\right)^2\cos^2 I\left(1+e^2+2e\sin\omega\right)}{R_\star^2(1+e\sin\omega)^3} \left[\frac{1}{\sqrt{(1+\vartheta)^2-b^2}} - \frac{1}{\sqrt{(1-\vartheta)^2-b^2}}\right]\right\}\right). \tag{7.59}$$

It should be noted that Equation (7.59) does not vanish for circular orbits, being its first nonvanishing term just of zeroth order in e.

7.6.3 The Full Width at Half Maximum Primary Transit Duration δt_H

From Equations (2.115)–(2.118) and the shifts obtainable from Equations (7.28)–(7.33), it turns out that

$$\overline{\Delta\delta t_H}^{J_2} = \frac{3\pi J_2 R_{\mathrm{e}}^2}{2n_{\mathrm{K}} R_\star a^3 \left(1-e^2\right)^{3/2}\left(1+e^2+2e\sin\omega\right)^{3/2}} \left(-\frac{a^2\left(1-e^2\right)^2 \widehat{T}_4 \sin 2I\left(1+e^2+2e\sin\omega\right)}{(1+e\sin\omega)^2}\left[\frac{1}{\sqrt{(1+\vartheta)^2-b^2}} + \frac{1}{\sqrt{(1-\vartheta)^2-b^2}}\right] + R_\star^2 e\cos\omega\left(2\widehat{T}_1-3\widehat{T}_2+2\widehat{T}_5\cot I\right)\left\{-\sqrt{(1-\vartheta)^2-b^2} - \sqrt{(1+\vartheta)^2-b^2} + \frac{a^2\left(1-e^2\right)^2\cos^2 I\left(1+e^2+2e\sin\omega\right)}{R_\star^2(1+e\sin\omega)^3} \left[\frac{1}{\sqrt{(1+\vartheta)^2-b^2}} + \frac{1}{\sqrt{(1-\vartheta)^2-b^2}}\right]\right\}\right). \tag{7.60}$$

It should be noted that Equation (7.60) does not vanish for circular orbits, since its first nonvanishing term is only of zeroth order in e.

7.6.4 The Time of Inferior Conjunction t_{cj}

From Equations (2.121)–(2.124) and the shifts obtainable from Equations (7.28)–(7.33), it turns out

$$\overline{\Delta t_{\mathrm{cj}}}^{J_2} = -\frac{3\pi J_2 R_{\mathrm{e}}^2}{2 n_{\mathrm{K}} a^2 \left(1-e^2\right)^{3/2} \left(1+e\sin\omega\right)^2}\left\{-2\left(1-e^2\right)\widehat{T}_5\cot I + \left(2\widehat{T}_1 - 3\widehat{T}_2\right)\left[2-e^2+e\sin\omega\left(2+e\sin\omega\right)\right]\right\}. \tag{7.61}$$

7.6.5 A Numerical Evaluation

Figure 7.9 displays the plots of Equations (7.58)–(7.61) for a Jovian-type exoplanet transiting in front of its Sunlike host star, seen edge-on, as functions of

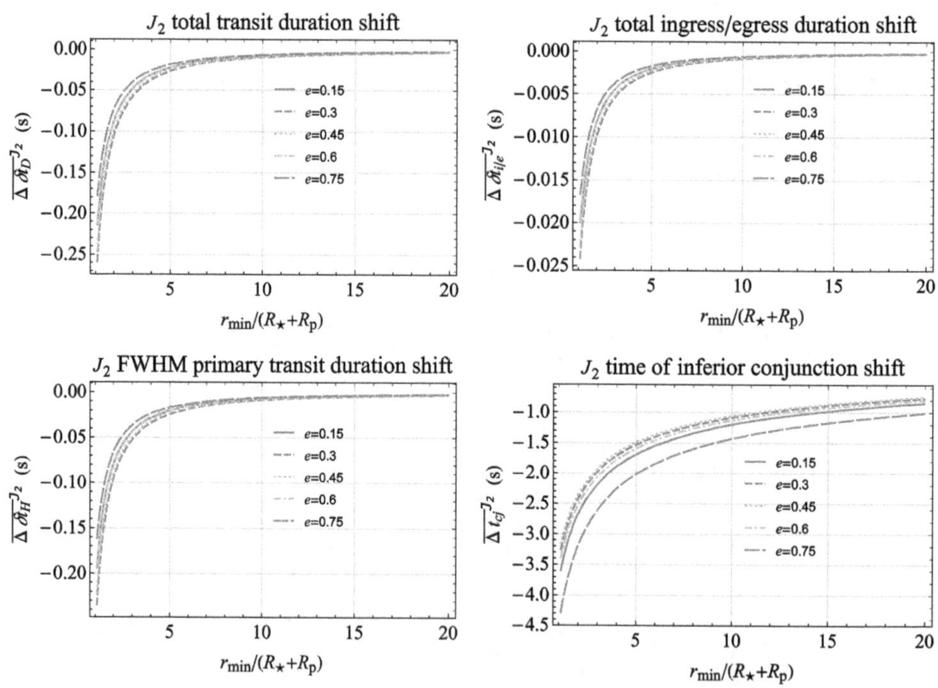

Figure 7.9 Plots of the net shifts per orbit $\overline{\Delta\delta t_D}^{J_2}$, $\overline{\Delta\delta t_{\mathrm{i/e}}}^{J_2}$, $\overline{\Delta\delta t_H}^{J_2}$, $\overline{\Delta t_{\mathrm{cj}}}^{J_2}$ of a fictitious Sun–Jupiter exoplanetary system, seen edge-on, as functions of the minimum star–planet distance $r_{\min}$, in units of $R_\star + R_{\mathrm{p}}$, for different values of the e according to Equations (7.58)–(7.61). The values $I = 90^\circ, \omega = 50^\circ$ are used for the relevant orbital parameters; all the three angular momenta are assumed to be aligned. The units are s.

the minimum distance $r_{\min}$, ranging from 1.1 to $20\,R_\star + R_{\rm p}$, for various values of e and by assuming that both the stellar and planetary angular momenta are aligned with the orbital angular momentum. It can be noted that the largest effect occurs for the time of inferior conjunction whose shift per orbit is at the $\simeq 1 - 4$ s level. Instead, the variations of the other timescales are of the order of $\simeq 0.01 - 0.1$ s.

7.7 The Sky-Projected Spin-Orbit Angle of Transiting Exoplanets

According to Equation (2.126), the rate of change of the sky-projected spin-orbit angle λ of a transiting exoplanet is equal to the node precession, provided that the angular momentum of the host star can be considered as constant.

By expressing the stellar spin axis as in Equation (5.85), it turns out, from Equation (7.31), Equation (5.85), Equations (D.2)–(D.3), and Equation (D.8) of Appendix D, that the J_2-driven precession of λ is

$$\left\langle \frac{d\lambda}{dt} \right\rangle^{J_2} = \left\langle \frac{d\Omega}{dt} \right\rangle^{J_2} = \frac{3 n_{\rm K} J_2^\star R_\star^2 \cos\lambda \sin 2i_\star}{4a^2\left(1-e^2\right)^2}. \tag{7.62}$$

In Equation (7.62), only the quadrupole mass moment of the host star, assumed to be as fast rotating as Kelt–9, appears since it turns out that its product by the stellar radius squared is some orders of magnitude larger than that of a typical close–in, tidally synchronized Jovian-type exoplanet.

In Figure 7.10, Equation (7.62) is plotted as a function of $i_\star$ for the exoplanet Kelt–9b. It can be noted that the J_2-induced secular rate of λ ranges from $\simeq 0.3$ to $\simeq 0.5^\circ$/yr within the allowed observational interval for $i_\star$.

7.8 The Variation of the Times of Arrival of Binary Pulsars

For a binary pulsar, the instantaneous shift of $\delta\widetilde{t}_{\rm psr}$ can be obtained from Equation (2.77) calculated with Equations (2.130)–(2.134), Equations (7.16)–(7.21), and Equation (7.47). By replacing f with $f_0 + 2\pi$ in the resulting expression, the net variation per orbit is inferred.

Figure 7.11 displays the instantaneous shifts $\Delta\delta\widetilde{t}_{\rm psr}\left(t\right)^{Q_2}$ of a fictitious binary pulsar whose companion is a white dwarf obtained by varying $T_{\rm K}$ and e in such a way that the periastron distance is kept fixed to, say, $r_{\min} = 1.8 \times 10^6$ km. For the masses and the quadrupole mass moments, the values of PSR J1141–6545 (Antoniadis et al., 2011; Iorio, 2020a; Venkatraman Krishnan et al., 2020) are taken. It is assumed that both the rotational axes of the pulsar and the white dwarf are aligned with the orbital angular momentum. It turns out that the peak-to-peak amplitudes of the signals may be as large as $\simeq 0.4 - 0.8$ ms over 5 d.

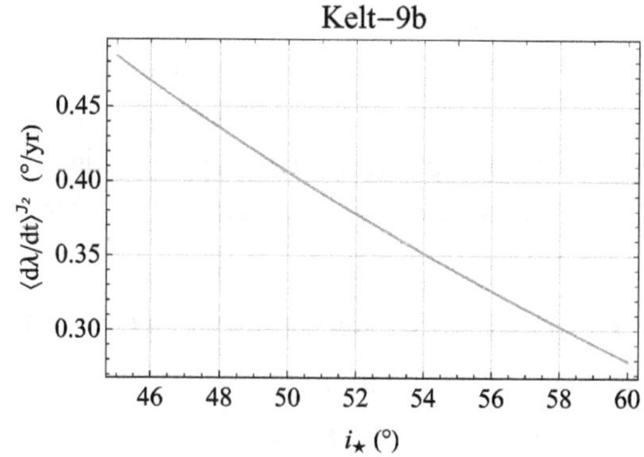

Figure 7.10 Plot of the J_2–induced averaged precession, in °/yr, of the sky-projected spin-orbit angle λ of the transiting exoplanet Kelt–9b as a function of the tilt $i_\star$ to the line of sight of the spin axis of its host star according to Equation (7.62). The physical parameters of the star and the planet are retrieved from Tables A.1 and A.2 of Albrecht et al. (2022) as well as the range of the admitted values for $i_\star$. The stellar quadrupole mass moment $J_2^\star$ is calculated with Equation (7.4).

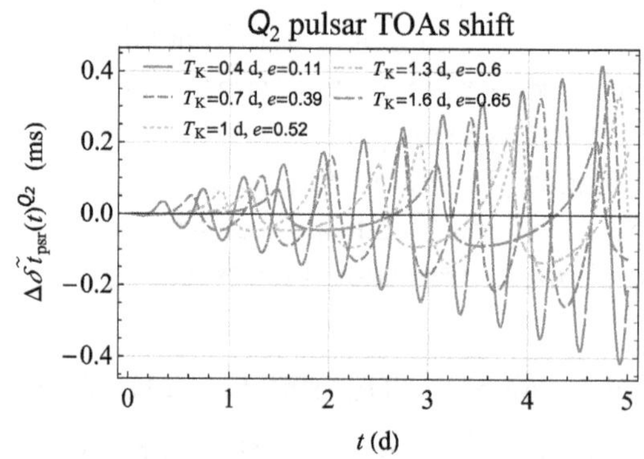

Figure 7.11 Plot of the quadrupole-induced instantaneous shift $\Delta\delta\widetilde{t}_{\mathrm{psr}}\,(t)^{Q_2}$, in ms, of a binary pulsar with a white dwarf as companion for different values of T_{K} and e in such a way that $r_{\min} = 1.8 \times 10^6$ km. The relevant orbital parameters are $I = 75°$, $\omega = 42.457°$, $\Omega = 50°$, $\eta = 20°$. For the stellar corpses, the values $M_{\mathrm{psr}} = 1.27\,M_\odot, M_{\mathrm{wd}} = 1.02\,M_\odot$, $\left|Q_2^{\mathrm{psr}}\right| \simeq 3 \times 10^{37}\,\mathrm{kg\,m^2}$, $\left|Q_2^{\mathrm{wd}}\right| \simeq 4.5 \times 10^{42}\,\mathrm{kg\,m^2}$, taken from those of PSR J1141–6545 (Antoniadis et al., 2011; Iorio, 2020a; Venkatraman Krishnan et al., 2020), are used. It is assumed that both the individual angular momenta and the orbital angular momentum are aligned.

7.9 The Astrometric Angles RA and dec.

For a test particle and a massive primary, as in the case of the S stars and Sgr A*, the instantaneous shifts of the RA and dec. can be obtained from Equation (2.77) calculated with Equations (2.137)–(2.143), Equations (7.16)–(7.21), and

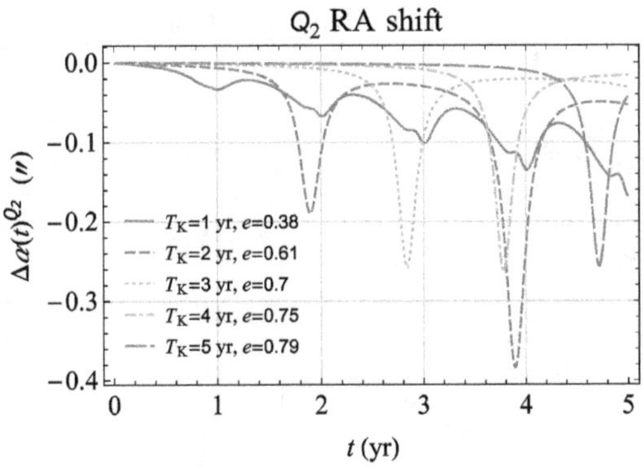

Figure 7.12 Plot of the oblateness-driven instantaneous shift $\Delta\alpha\,(t)^{Q_2}$, in $''$, of the RA of a fictitious S star for different values of $T_{\rm K}$ and e in such a way that $r_{\rm min} = 1250\,\mathcal{R}_\bullet$. The relevant stellar orbital parameters are $I = 161.24^\circ$, $\Omega = 151.54^\circ$, $\omega = 0.073^\circ$, $\eta = 20^\circ$. For the BH, the values $M_\bullet = 4.1\times10^6\,M_\odot$, $\chi_g = 0.5$ are used for its mass and spin parameter (Peißker et al., 2022). Furthermore, the orientation of its spin axis is set by, say, $i_\bullet = 30^\circ, \zeta_\bullet = 150^\circ$.

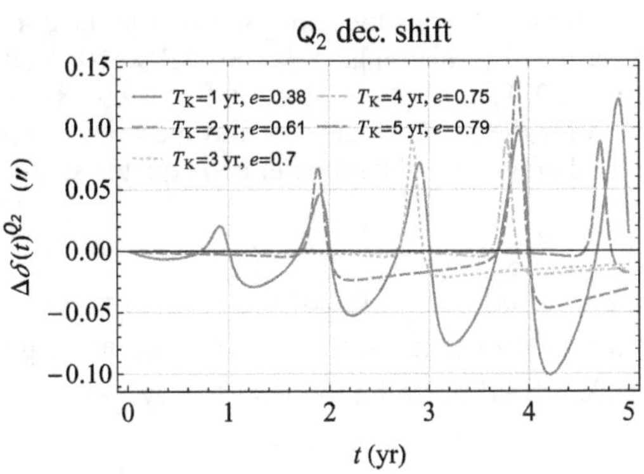

Figure 7.13 Plot of the oblateness-driven instantaneous shift $\Delta\delta\,(t)^{Q_2}$, in $''$, of the dec. of a fictitious S star for different values of $T_{\rm K}$ and e in such a way that $r_{\rm min} = 1250\,\mathcal{R}_\bullet$. The relevant stellar orbital parameters are $I = 161.24^\circ$, $\Omega = 151.54^\circ$, $\omega = 0.073^\circ$, $\eta = 20^\circ$. For the BH, the values $M_\bullet = 4.1\times10^6\,M_\odot$, $\chi_g = 0.5$ are used for its mass and spin parameter (Peißker et al., 2022). Furthermore, the orientation of its spin axis is set by, say, $i_\bullet = 30^\circ, \zeta_\bullet = 150^\circ$.

Equation (7.47). By replacing f with $f_0 + 2\pi$ in the resulting expressions, the net variations per orbit are inferred.

Figures 7.12 and 7.13 display the instantaneous shifts $\Delta\alpha\,(t)\,, \Delta\delta\,(t)$ of the RA and the dec. of a fictitious S star obtained by varying $T_{\rm K}$ and e in such a way that the stellar perinigricon distance is kept fixed to $r_{\rm min} = 1250\,\mathcal{R}_\bullet$. The SMBH's quadrupole mass moment $Q_2^\bullet$ is calculated with Equation (5.16) and Equation (7.11), and

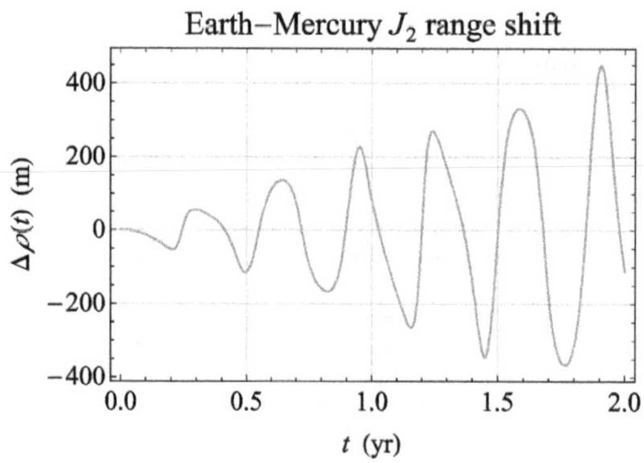

Figure 7.14 Analytically produced time series, in m, of the two-body range shift $\Delta\rho\,(t)$ due to the Sun's quadrupole mass moment $J_2^{\odot}$ for A=Mercury and B=Earth plotted over 2 yr. It was worked out by calculating Equations (2.149)–(2.151) entering Equations (2.147)–(2.148) with Equations (7.16)–(7.21), Equation (7.47), and Equation (2.4). The initial values of the Keplerian orbital elements of both planets, referred to as the International Celestial Reference Frame (ICRF), were retrieved from the WEB interface Horizons System at https://ssd.jpl.nasa.gov/horizons/, maintained by the Jet Propulsion Laboratory (JPL) of the National Aeronautics and Space Administration (NASA), and accessed 12 February 2024. The values $J_2^{\odot} = 2.2 \times 10^{-7}$ (Park et al., 2017; Mecheri and Meftah, 2021) and $\alpha_{J_{\odot}} = 286^{\circ}.13, \delta_{J_{\odot}} = 63^{\circ}.87$ (Seidelmann et al., 2007) were adopted for the Sun's oblateness and the orientation of its spin axis. For the solar standard gravitational parameter and equatorial radius, see Table J.1.

the orientation of its spin axis is set equal to, say, $i_{\bullet} = 30^{\circ}, \zeta_{\bullet} = 150^{\circ}$. It turns out that the Q_2-induced astrometric signatures can be as large as $\lesssim 0.01 - 0.1''$, depending on the star's orbital period and eccentricity.

7.10 The Two-Body Range and Range Rate

Here, Equations (7.16)–(7.20), along with Equation (7.21) and Equation (7.47) for the shift of the mean anomaly, are used in Equation (2.147) and Equation (2.152) to analytically calculate the time series of the range and range rate shifts for A = Mercury and B = Earth due to the Sun's oblateness $J_2^{\odot}$.

Figures 7.14–7.15 plot the resulting signals, obtained by introducing the dependence on time t through the mean anomaly by means of Equation (2.4) and Equations (2.158)–(2.159), over 2 yr, which is the expected duration of the extended phase of the BepiColombo mission (Benkhoff et al., 2010, 2021) from 2026 to 2028.[7] It turns out that the peak-to-peak nominal amplitudes of the range and range rate shifts can reach the $\simeq 800$ m and $\simeq 700\ \mu$m/s level, respectively.

[7] See www.esa.int/Science_Exploration/Space_Science/BepiColombo/BepiColombo_factsheet. Accessed 19 January 2024.

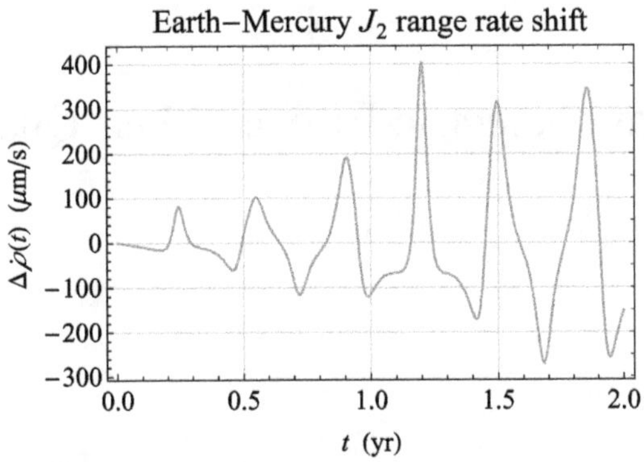

Figure 7.15 Analytically produced time series, in μm/s, of the two-body range rate shift $\Delta\dot{\rho}\,(t)$ due to the Sun's quadrupole mass moment $J_2^{\odot}$ for A=Mercury and B=Earth plotted over 2 yr. It was worked out by calculating Equations (2.155)–(2.157) entering Equations (2.152)–(2.154) with Equations (7.16)–(7.21), Equation (7.47), and Equation (2.4). The initial values of the Keplerian orbital elements of both planets, referred to as the International Celestial Reference Frame (ICRF), were retrieved from the WEB interface Horizons System at https://ssd.jpl.nasa.gov/horizons/, maintained by the Jet Propulsion Laboratory (JPL) of the National Aeronautics and Space Administration (NASA), and accessed 12 February 2024. The values $J_2^{\odot} = 2.2 \times 10^{-7}$ (Park et al., 2017; Mecheri and Meftah, 2021) and $\alpha_{J_\odot} = 286^{\circ}.13, \delta_{J_\odot} = 63^{\circ}.87$ (Seidelmann et al., 2007) were adopted for the Sun's oblateness and the orientation of its spin axis. For the solar standard gravitational parameter and equatorial radius, see Table J.1.

An integration of the equations of motion with and without Equation (7.1) generated corresponding numerical time series differing from the aforementioned analytical ones by $\lesssim 3$ m and $\lesssim 6\,\mu\text{m/s}$ level over 2 years, respectively.

8

1pN Gravitoelectric Effects: Mass Quadrupole

The oblateness of the primary impacts the orbital motion of a test particle not only at the Newtonian level, as shown in Chapter 7, but also to the 1pN order (Soffel et al., 1987; Soffel, 1989; Heimberger et al., 1989; Brumberg, 1991; Huang and Liu, 1992; Will, 2014; Iorio, 2015b, 2023e, 2024a). It occurs both *directly* because of a specific pK acceleration $\boldsymbol{A}^{J_2/c^2}$ of the order of $\mathcal{O}\left(J_2/c^2\right)$ entering the equations of motion, and *indirectly* due to the interplay between the 1pN monopole acceleration of Equation (3.1) and the Newtonian quadrupolar one of Equation (7.1).

In view of their generally small sizes, such effects have not yet been studied in detail nor have they been put to the test in any astronomical scenarios.

Proposals for spacecraft-based missions in the field of Earth and Jupiter recently appeared in the literature (Iorio, 2013a, 2019a,g, 2024a); in Iorio et al. (2019), it was suggested to look at binary pulsars whose companion is a fast rotating main sequence star as well.

To this aim, it should be remarked that the mixed signatures, which nonetheless contribute to the overall orbital evolution to the order of $\mathcal{O}\left(J_2/c^2\right)$, may not be measurable independently of other dynamical features of motion. Indeed, they do not come from any new pK acceleration, still unmodelled in the softwares used worldwide to process data. If it were so, it could be possible, at least in principle, to include it in the dynamical models and estimate some dedicated solve-for parameters in the usual least-square approach. On the other hand, the standard pK accelerations of Equation (3.1) and Equation (7.1) are accurately modelled; thus, just very tiny signatures, due to the current level of mismodelling in the latter ones, would impact the post-fit residuals produced in data analyses.

8.1 The Direct Effects

By inserting the Newtonian potential of Equation (7.3) up to degree $\ell = 2$ in the 1pN Lagrangian of Equation (C.10) in Appendix C, a new acceleration of the order of $\mathcal{O}\left(J_2/c^2\right)$ arises from the Lagrange equations of motion

$$\frac{d\nabla_v L^{J_2/c^2}}{dt} = \nabla L^{J_2/c^2}. \tag{8.1}$$

It can be cast into the form (Will, 2014)

$$\begin{aligned} \boldsymbol{A}^{J_2/c^2} = \frac{\mu J_2 R_e^2}{c^2 r^4} & \left\{ \frac{3}{2}\left[\left(5r_J^2 - 1\right)\hat{\boldsymbol{r}} - 2r_J\hat{\boldsymbol{J}}\right]\left(v^2 - \frac{4\mu}{r}\right) \right. \\ & \left. - 6\left[\left(5r_J^2 - 1\right)v_r - 2r_J v_J\right]\boldsymbol{v} - \frac{2\mu}{r}\left(3r_J^2 - 1\right)\hat{\boldsymbol{r}} \right\}, \end{aligned} \tag{8.2}$$

where r_J is defined as in Equation (5.19), and

$$v_J := \boldsymbol{v} \cdot \hat{\boldsymbol{J}} \tag{8.3}$$

is the projection of the velocity $\boldsymbol{v}$ on the direction of $\boldsymbol{J}$; for previous derivations in a reference frame whose z-axis is aligned with $\boldsymbol{J}$, see Soffel et al. (1987), Soffel (1989), Brumberg (1991), and Huang and Liu (1992). Here, it is proposed to dub Equation (8.2) as Brumberg–Soffel (BS) acceleration.

By projecting Equation (8.2) onto the unit vectors $\hat{\boldsymbol{r}}, \hat{\boldsymbol{\tau}}, \hat{\boldsymbol{h}}$ defined in Equations (2.23)–(2.25) and by using Equations (2.7)–(2.8) along with Equation (2.11), its radial, transverse, and normal components are obtained; they turn out to be

$$\begin{aligned} A_r^{J_2/c^2} = & -\frac{\mu^2 J_2 R_e^2 (1 + e\cos f)^4}{4c^2 a^5 \left(1 - e^2\right)^5} \left(\left(-2\widehat{T}_1 + 3T_2\right)\left(-13 - 3e^2\right.\right. \\ & - \left. 10e\cos f + 6e^2\cos 2f\right) + 3\widehat{T}_3\left\{e^2\cos 2\omega - \left(13 + 3e^2\right)\cos 2u\right. \\ & + \left. e\left[5e\cos(4f + 2\omega) - 9\cos(f + 2\omega) - \cos(3f + 2\omega)\right]\right\} \\ & + 6\widehat{T}_6\left\{e^2\sin 2\omega - \left(13 + 3e^2\right)\sin 2u + e\left[5e\sin(4f + 2\omega)\right.\right. \\ & - \left.\left.\left. 9\sin(f + 2\omega) - \sin(3f + 2\omega)\right]\right\}\right), \end{aligned} \tag{8.4}$$

$$\begin{aligned} A_\tau^{J_2/c^2} = & -\frac{3\mu^2 J_2 R_e^2 (1 + e\cos f)^4}{2c^2 a^5 \left(1 - e^2\right)^5} \left[\left(-3 + e^2 - 2e\cos f\right)\left(2\widehat{T}_6\cos 2u\right.\right. \\ & - \left.\widehat{T}_3\sin 2u\right) + (1 + e\cos f)\left(-8\widehat{T}_6\cos 2u + 4\widehat{T}_3\sin 2u\right. \\ & + e\left\{2\widehat{T}_6\left[\cos(f + 2\omega) - 5\widehat{T}_1\cos(3f + 2\omega)\right]\right. \\ & + \left(-4\widehat{T}_1 + 6\widehat{T}_2\right)\sin f - \widehat{T}_3\sin(f + 2\omega) \\ & + \left.\left.\left. 5\widehat{T}_3\sin(3f + 2\omega)\right\}\right)\right], \end{aligned} \tag{8.5}$$

$$A_h^{J_2/c^2} = -\frac{3\mu^2 J_2 R_e^2 (1+e\cos f)^4 \left(-3+e^2-2e\cos f\right)}{c^2 a^5 \left(1-e^2\right)^5}\left(\widehat{T}_4\cos u+\widehat{T}_5\sin u\right). \tag{8.6}$$

The coefficients $\widehat{T}_j, j = 1,2,\ldots 6$ entering Equations (8.4)–(8.6) are listed in Equations (D.4)–(D.9) of Appendix D.

The orbital effects of the order of $\mathcal{O}\left(J_2/c^2\right)$ induced by Equation (8.2) are dubbed as *direct*.

8.1.1 The Net Orbital Precessions

The averaged orbital precessions induced by Equation (8.2) were calculated in their full generality in Iorio (2015b) and, in a more compact form, in Iorio (2024a). For other derivations obtained by orienting $\hat{\boldsymbol{J}}$ along the z-axis of the reference frame chosen, see Soffel et al. (1987), Brumberg (1991), and Huang and Liu (1992).

By inserting Equations (8.4)–(8.6) in the machinery of Equations (2.13)–(2.18) yields the averaged precessions of the orbital elements for an arbitrary orientation of $\hat{\boldsymbol{J}}$. They turn out to be

$$\left\langle\frac{da}{dt}\right\rangle^{J_2/c^2} = -\frac{9e^2\left(6+e^2\right) n_{\mathrm{K}}\mu J_2 R_e^2\left(\widehat{T}_3\sin 2\omega-2\widehat{T}_6\cos 2\omega\right)}{8c^2a^2\left(1-e^2\right)^4}, \tag{8.7}$$

$$\left\langle\frac{de}{dt}\right\rangle^{J_2/c^2} = -\frac{21e\left(2+e^2\right) n_{\mathrm{K}}\mu J_2 R_e^2\left(\widehat{T}_3\sin 2\omega-2\widehat{T}_6\cos 2\omega\right)}{16c^2a^3\left(1-e^2\right)^3}, \tag{8.8}$$

$$\left\langle\frac{dI}{dt}\right\rangle^{J_2/c^2} = \frac{3n_{\mathrm{K}}\mu J_2 R_e^2\left[\widehat{T}_4\left(6+e^2\cos 2\omega\right)+e^2\widehat{T}_5\sin 2\omega\right]}{4c^2a^3\left(1-e^2\right)^3}, \tag{8.9}$$

$$\left\langle\frac{d\Omega}{dt}\right\rangle^{J_2/c^2} = -\frac{3n_{\mathrm{K}}\mu J_2 R_e^2\csc I\left[-e^2\widehat{T}_4\sin 2\omega+\widehat{T}_5\left(-6+e^2\cos 2\omega\right)\right]}{4c^2a^3\left(1-e^2\right)^3}, \tag{8.10}$$

$$\begin{aligned}\left\langle\frac{d\omega}{dt}\right\rangle^{J_2/c^2} = &-\frac{3n_{\mathrm{K}}\mu J_2 R_e^2}{16c^2a^3\left(1-e^2\right)^3}\left\{\left(-8+3e^2\right)\left(-2\widehat{T}_1+3\widehat{T}_2\right)\right.\\ &+14\widehat{T}_3\cos 2\omega+4\left[e^2\widehat{T}_4\cot I\sin 2\omega\right.\\ &\left.\left.+\widehat{T}_5\cot I\left(6-e^2\cos 2\omega\right)+7\widehat{T}_6\sin 2\omega\right]\right\},\end{aligned} \tag{8.11}$$

$$\begin{aligned}\left\langle\frac{d\eta}{dt}\right\rangle^{J_2/c^2} = &\frac{n_{\mathrm{K}}\mu J_2 R_e^2}{16c^2a^3\left(1-e^2\right)^{5/2}}\left[\left(80+73e^2\right)\left(-2\widehat{T}_1+3\widehat{T}_2\right)\right.\\ &\left.+42\left(1+2e^2\right)\left(\widehat{T}_3\cos 2\omega+2\widehat{T}_6\sin 2\omega\right)\right].\end{aligned} \tag{8.12}$$

Some Special Orbital Configurations

Here, some peculiar orbital configurations, characterized in Appendix D, are examined (Iorio, 2024a).

Let, first, the satellite's orbital plane be assumed coincident with the body's equatorial one, irrespective of the orientation of the latter in the adopted reference frame, namely for generic values of α_J, δ_J. Then, according to Equations (D.17)–(D.25) of Appendix D, Equations (8.7)–(8.12) reduce to

$$\left\langle\frac{da}{dt}\right\rangle^{J_2/c^2} = 0, \tag{8.13}$$

$$\left\langle\frac{de}{dt}\right\rangle^{J_2/c^2} = 0, \tag{8.14}$$

$$\left\langle\frac{dI}{dt}\right\rangle^{J_2/c^2} = 0, \tag{8.15}$$

$$\left\langle\frac{d\Omega}{dt}\right\rangle^{J_2/c^2} = 0, \tag{8.16}$$

$$\left\langle\frac{d\omega}{dt}\right\rangle^{J_2/c^2} = -\frac{3n_{\mathrm{K}}\mu J_2 R_{\mathrm{e}}^2\left(8-3e^2\right)}{8c^2a^3\left(1-e^2\right)^3}, \tag{8.17}$$

$$\left\langle\frac{d\eta}{dt}\right\rangle^{J_2/c^2} = -\frac{n_{\mathrm{K}}\mu J_2 R_{\mathrm{e}}^2\left(80+73e^2\right)}{8c^2a^3\left(1-e^2\right)^{5/2}}. \tag{8.18}$$

Equations (8.17)–(8.18) are genuine secular trends.

Let, now, the body's spin axis, irrespective of its orientation in the adopted coordinate system, namely for generic values of α_J, δ_J, lie somewhere in the satellite's orbital plane between $\hat{l}$ and $\hat{m}$. Then, according to Equations (D.26)–(D.34) of Appendix D, Equations (8.7)–(8.12) can be written as

$$\left\langle\frac{da}{dt}\right\rangle^{J_2/c^2} = \frac{9e^2\left(6+e^2\right)n_{\mathrm{K}}\mu J_2 R_{\mathrm{e}}^2\sin\left[2\left(\delta_J-\omega\right)\right]}{8c^2a^2\left(1-e^2\right)^4}, \tag{8.19}$$

$$\left\langle\frac{de}{dt}\right\rangle^{J_2/c^2} = \frac{21e\left(2+e^2\right)n_{\mathrm{K}}\mu J_2 R_{\mathrm{e}}^2\sin\left[2\left(\delta_J-\omega\right)\right]}{16c^2a^3\left(1-e^2\right)^3}, \tag{8.20}$$

$$\left\langle\frac{dI}{dt}\right\rangle^{J_2/c^2} = 0, \tag{8.21}$$

$$\left\langle\frac{d\Omega}{dt}\right\rangle^{J_2/c^2} = 0, \tag{8.22}$$

$$\left\langle \frac{d\omega}{dt} \right\rangle^{J_2/c^2} = -\frac{3 n_{\mathrm{K}} \mu J_2 R_{\mathrm{e}}^2 \left\{-8 + 3e^2 + 14 \cos\left[2\left(\delta_J - \omega\right)\right]\right\}}{16 c^2 a^3 \left(1 - e^2\right)^3}, \tag{8.23}$$

$$\left\langle \frac{d\eta}{dt} \right\rangle^{J_2/c^2} = \frac{n_{\mathrm{K}} \mu J_2 R_{\mathrm{e}}^2 \left\{80 + 73e^2 + 42\left(1 + 2e^2\right) \cos\left[2\left(\delta_J - \omega\right)\right]\right\}}{16 c^2 a^3 \left(1 - e^2\right)^{5/2}}. \tag{8.24}$$

Equations (8.19)–(8.20) and Equations (8.23)–(8.24), in addition to secular trends, also include long-period signatures due to the evolution of pericentre which is dominated by the Newtonian perturbation due to J_2 (see Section 7.1).

8.2 The Mixed Effects

If both Equation (3.1) and Equation (7.1) are simultaneously taken into account, indirect, or mixed, effects of the order of $\mathcal{O}\left(J_2/c^2\right)$ arise as well. Their net shifts per orbit and their averaged precessions can be calculated as explained in Section 2.2.3.

8.2.1 The Net Orbital Shifts

Equation (2.39), applied to Equation (3.1) and Equation (7.1) and calculated with Equations (3.6)–(3.10) and Equations (7.16)–(7.20), allows one to obtain the total mixed shifts per orbit of order $\mathcal{O}\left(J_2/c^2\right)$ with the replacement $f \rightarrow f_0 + 2\pi$ in the resulting expressions. For previous derivations with another orbital parametrization and with the z-axis aligned with $\hat{\boldsymbol{J}}$, see Heimberger et al. (1989). The Keplerian orbital elements and the same reference frame were used by Will (2014). The mixed orbital effects were computed in their full generality in Iorio (2015b), and, in a more compact form, in Iorio (2023e). They are (Iorio, 2023e)

$$\overline{\Delta a}_{\mathrm{mix}}^{J_2/c^2} = \frac{9\pi J_2 R_{\mathrm{e}}^2 \mu}{4 c^2 a^2 \left(1 - e^2\right)^4} \widetilde{\mathcal{A}}^{J_2/c^2}, \tag{8.25}$$

$$\overline{\Delta e}_{\mathrm{mix}}^{J_2/c^2} = -\frac{3\pi J_2 R_{\mathrm{e}}^2 \mu}{8 c^2 a^3 \left(1 - e^2\right)^3} \widetilde{\mathcal{E}}^{J_2/c^2}, \tag{8.26}$$

$$\overline{\Delta I}_{\mathrm{mix}}^{J_2/c^2} = -\frac{3\pi J_2 R_{\mathrm{e}}^2 \mu}{c^2 a^3 \left(1 - e^2\right)^3} \widetilde{\mathcal{I}}^{J_2/c^2}, \tag{8.27}$$

$$\overline{\Delta \Omega}_{\mathrm{mix}}^{J_2/c^2} = -\frac{3\pi J_2 R_{\mathrm{e}}^2 \mu \csc I}{c^2 a^3 \left(1 - e^2\right)^3} \widetilde{\mathcal{N}}^{J_2/c^2}, \tag{8.28}$$

$$\overline{\Delta \omega}_{\mathrm{mix}}^{J_2/c^2} = -\frac{3\pi J_2 R_{\mathrm{e}}^2 \mu \csc I}{16 c^2 a^3 e \left(1 - e^2\right)^3} \widetilde{\mathcal{P}}^{J_2/c^2}, \tag{8.29}$$

$$\overline{\Delta\eta}^{J_2/c^2}_{\text{mix}} = -\frac{3\pi J_2 R^2 \mu}{16 c^2 a^3 e \left(1 - e^2\right)^{7/2}} \widetilde{\mathcal{H}}^{J_2/c^2}, \tag{8.30}$$

where

$$\widetilde{\mathcal{A}}^{J_2/c^2} := \sum_{j=1}^{6} \widetilde{\mathcal{A}}_j^{J_2/c^2} \widehat{T}_j, \tag{8.31}$$

$$\widetilde{\mathcal{E}}^{J_2/c^2} := \sum_{j=1}^{6} \widetilde{\mathcal{E}}_j^{J_2/c^2} \widehat{T}_j, \tag{8.32}$$

$$\widetilde{\mathcal{I}}^{J_2/c^2} := \sum_{j=1}^{6} \widetilde{\mathcal{I}}_j^{J_2/c^2} \widehat{T}_j, \tag{8.33}$$

$$\widetilde{\mathcal{N}}^{J_2/c^2} := \sum_{j=1}^{6} \widetilde{\mathcal{N}}_j^{J_2/c^2} \widehat{T}_j, \tag{8.34}$$

$$\widetilde{\mathcal{P}}^{J_2/c^2} := \sum_{j=1}^{6} \widetilde{\mathcal{P}}_j^{J_2/c^2} \widehat{T}_j, \tag{8.35}$$

$$\widetilde{\mathcal{H}}^{J_2/c^2} := \sum_{j=1}^{6} \widetilde{\mathcal{H}}_j^{J_2/c^2} \widehat{T}_j. \tag{8.36}$$

The explicit forms of the coefficients $\widetilde{\mathcal{A}}_1^{J_2/c^2}, \ldots \widetilde{\mathcal{H}}_6^{J_2/c^2}$ entering Equations (8.31)–(8.36) are displayed in Appendix G.

As explained in Section 2.2.3, the ratios of Equations (8.25)–(8.30) to T_{K} do not yield the total mixed averaged orbital precessions of the order of $\mathcal{O}\left(J_2/c^2\right)$; see Section 8.2.2 for the calculation of the required additional contributions.

Some Special Orbital Configurations

Here, some peculiar orbital configurations, characterized in Appendix D, are examined (Iorio, 2023e).

Let, first, the satellite's orbital plane be assumed coincident with the body's equatorial one, irrespective of the orientation of the latter in the adopted reference frame, namely for generic values of α_J, δ_J. Then, according to Equations (D.17)–(D.25) of Appendix D, Equations (8.25)–(8.30) reduce to

$$\overline{\Delta a}^{J_2/c^2}_{\text{mix}} = 0, \tag{8.37}$$

$$\overline{\Delta e}^{J_2/c^2}_{\text{mix}} = 0, \tag{8.38}$$

$$\overline{\Delta I}^{J_2/c^2}_{\text{mix}} = 0, \tag{8.39}$$

$$\overline{\Delta\Omega}^{J_2/c^2}_{\text{mix}} = 0, \tag{8.40}$$

$$\overline{\Delta\omega}^{J_2/c^2}_{\text{mix}} = \frac{3\pi J_2 R_\text{e}^2 \mu \left(44 + 17e^2 - 64e\cos f_0\right)}{4c^2a^3\left(1-e^2\right)^3}, \tag{8.41}$$

$$\begin{aligned}\overline{\Delta\eta}^{J_2/c^2}_{\text{mix}} = &\frac{3\pi J_2 R_\text{e}^2 \mu}{4c^2a^3\left(1-e^2\right)^{7/2}}\left(-88 + 16\sqrt{1-e^2}\right.\\ &+ e^2\left[63 - 5e^2 + 24\sqrt{1-e^2}\right]\\ &+ e\left\{3e^2\left[7 + 4\sqrt{1-e^2}\right] + 8\left[-17 + 6\sqrt{1-e^2}\right]\right\}\cos f_0\\ &+ 8e^2\left[-5 + 3\sqrt{1-e^2}\right]\cos 2f_0\\ &\left.+ e^3\left[-5 + 4\sqrt{1-e^2}\right]\cos 3f_0\right).\end{aligned} \tag{8.42}$$

Let, now, the body's spin axis, irrespective of its orientation in the adopted coordinate system, namely for generic values of α_J, δ_J, lie somewhere in the satellite's orbital plane between $\hat{l}$ and $\hat{m}$. Then, according to Equations (D.26)–(D.34) of Appendix D, Equations (8.25)–(8.30) can be written as

$$\begin{aligned}\overline{\Delta a}^{J_2/c^2}_{\text{mix}} = &-\frac{9\pi J_2 R_\text{e}^2 \mu}{4c^2a^2\left(1-e^2\right)^4}\left\{e^3\sin\left(f_0 + 2\delta_J - 2\omega\right)\right.\\ &+ e^2\left(12 + e^2\right)\sin\left(2\delta_J - 2\omega\right)\\ &- 2\left[4 + 6e^2 + 3e\left(4 + e^2\right)\cos f_0\right]\sin\left(2f_0 - 2\delta_J + 2\omega\right)\\ &\left.- 6e^2\sin\left(4f_0 - 2\delta_J + 2\omega\right) - e^3\sin\left(5f_0 - 2\delta_J + 2\omega\right)\right\},\end{aligned} \tag{8.43}$$

$$\begin{aligned}\overline{\Delta e}^{J_2/c^2}_{\text{mix}} = &\frac{3\pi J_2 R_\text{e}^2 \mu}{8c^2a^3\left(1-e^2\right)^3}\left\{4\left[3\sin\left(f_0 - 2\delta_J + 2\omega\right)\right.\right.\\ &\left.+ 7\sin\left(3f_0 - 2\delta_J + 2\omega\right)\right]\\ &+ e\left[-3e\sin\left(f_0 + 2\delta_J - 2\omega\right) - \left(20 + 19e^2\right)\sin\left(2\delta_J - 2\omega\right)\right.\\ &+ 60\sin\left(2f_0 - 2\delta_J + 2\omega\right)\\ &+ 18\sin\left(4f_0 - 2\delta_J + 2\omega\right) + 33e\sin\left(f_0 - 2\delta_J + 2\omega\right)\\ &\left.\left.+ 17e\sin\left(3f_0 - 2\delta_J + 2\omega\right) + 3e\sin\left(5f_0 - 2\delta_J + 2\omega\right)\right]\right\},\end{aligned} \tag{8.44}$$

$$\overline{\Delta I}^{J_2/c^2}_{\text{mix}} = 0, \tag{8.45}$$

$$\overline{\Delta\Omega}^{J_2/c^2}_{\text{mix}} = 0, \tag{8.46}$$

$$\overline{\Delta\omega}_{\text{mix}}^{J_2/c^2} = -\frac{3\pi J_2 R_{\text{e}}^2 \mu}{8c^2 a^3 e\left(1-e^2\right)^3}\left((-12+45e^2)\cos\left(f_0-2\delta_J+2\omega\right)\right.$$
$$+\left(28+19e^2\right)\cos\left(3f_0-2\delta_J+2\omega\right)$$
$$+e\left\{2\left(-10+9e^2\right)\cos\left(2\delta_J-2\omega\right)+60\cos\left(2f_0-2\delta_J+2\omega\right)\right.$$
$$+18\cos\left(4f_0-2\delta_J+2\omega\right)+3e\cos\left(5f_0-2\delta_J+2\omega\right)$$
$$\left.\left.+44+17e^2-e\left[64\cos f_0+3\cos\left(f_0+2\delta_J-2\omega\right)\right]\right\}\right), \tag{8.47}$$

$$\overline{\Delta\eta}_{\text{mix}}^{J_2/c^2} = -\frac{3\pi J_2 R_{\text{e}}^2 \mu}{16c^2 a^3 e\left(1-e^2\right)^{7/2}}\left[-2e\left\{88+5e^4-16\sqrt{1-e^2}\right.\right.$$
$$-3e^2\left(21+8\sqrt{1-e^2}\right)$$
$$-e\left[3e^2\left(7+4\sqrt{1-e^2}\right)+8\left(-17+6\sqrt{1-e^2}\right)\right]\cos f_0$$
$$+e^2\left[8\left(5-3\sqrt{1-e^2}\right)\cos 2f_0\right.$$
$$\left.\left.+e\left(5-4\sqrt{1-e^2}\right)\cos 3f_0\right]\right\}$$
$$+\cos 2\delta_J\left\{3e^2\left(2-7e^2\right)\cos\left(f_0-2\omega\right)\right.$$
$$-2e\left(-20+7e^2+13e^4\right)\cos 2\omega+12e\left[-14-11e^2\right.$$
$$\left.+8\sqrt{1-e^2}\left(1+e\cos f_0\right)^3\right]\cos 2u_0$$
$$-18e\left(2+3e^2\right)\cos\left(4f_0+2\omega\right)$$
$$-3\left(-8+74e^2+9e^4\right)\cos\left(f_0+2\omega\right)$$
$$-\left(4+e^2\right)\left(14+31e^2\right)\cos\left(3f_0+2\omega\right)$$
$$\left.-3e^2\left(2+3e^2\right)\cos\left(5f_0+2\omega\right)\right\}$$
$$-\sin 2\delta_J\left(3e^2\left[2+e^2\left(-7+4\sqrt{1-e^2}\right)\right]\sin\left(f_0-2\omega\right)\right.$$
$$+2e\left[-20+13e^4+e^2\left(7-36\sqrt{1-e^2}\right)\right]\sin 2\omega$$
$$+12e\left[14-8\sqrt{1-e^2}+e^2\left(11-12\sqrt{1-e^2}\right)\right]\sin 2u_0$$
$$+18e\left[2+e^2\left(3-4\sqrt{1-e^2}\right)\right]\sin\left(4f_0+2\omega\right)$$
$$+\left\{-24+3e^2\left[74-48\sqrt{1-e^2}\right.\right.$$
$$\left.\left.+e^2\left(9-12\sqrt{1-e^2}\right)\right]\right\}\sin\left(f_0+2\omega\right)$$
$$+\left\{56+e^2\left[138-144\sqrt{1-e^2}\right.\right.$$
$$\left.\left.+e^2\left(31-36\sqrt{1-e^2}\right)\right]\right\}\sin\left(3f_0+2\omega\right)$$
$$\left.\left.+3e^2\left[2+e^2\left(3-4\sqrt{1-e^2}\right)\right]\sin\left(5f_0+2\omega\right)\right)\right]. \tag{8.48}$$

8.2.2 The Net Orbital Precessions

As explained in Section 2.2.3, the calculation of the total mixed orbital precessions is not limited just to the ratios of Equations (8.25)–(8.30) to $T_{\rm K}$; further contributions are needed (Iorio, 2023e), which are calculated as follows.

By using the 1pN gravitoelectric anomalistic period of Equation (3.24) with the J_2 net shifts per orbit obtainable from Equations (7.28)–(7.33) and the J_2 anomalistic period of Equation (7.50) with the 1pN gravitoelectric net shifts per orbit obtainable from Equations (3.13)–(3.18), one finally gets

$$\left\langle\frac{da}{dt}\right\rangle_{\rm mix}^{J_2/c^2} = 0, \tag{8.49}$$

$$\left\langle\frac{de}{dt}\right\rangle_{\rm mix}^{J_2/c^2} = 0, \tag{8.50}$$

$$\begin{aligned}\left\langle\frac{dI}{dt}\right\rangle_{\rm mix}^{J_2/c^2} &= \frac{9n_{\rm K}J_2R_{\rm e}^2\mu\widehat{T}_4}{4c^2a^3\left(1-e^2\right)^4}\left[6+7e^2+2e^4\right.\\ &\left.+2e\left(7+3e^2\right)\cos f_0+5e^2\cos 2f_0\right],\end{aligned} \tag{8.51}$$

$$\begin{aligned}\left\langle\frac{d\Omega}{dt}\right\rangle_{\rm mix}^{J_2/c^2} &= \frac{9n_{\rm K}J_2R_{\rm e}^2\mu\widehat{T}_5\csc I}{4c^2a^3\left(1-e^2\right)^4}\left[6+7e^2+2e^4\right.\\ &\left.+2e\left(7+3e^2\right)\cos f_0+5e^2\cos 2f_0\right],\end{aligned} \tag{8.52}$$

$$\begin{aligned}\left\langle\frac{d\omega}{dt}\right\rangle_{\rm mix}^{J_2/c^2} &= \frac{9n_{\rm K}J_2R_{\rm e}^2\mu}{8c^2a^3\left(1-e^2\right)^4}\left\{\left[6+7e^2+2e^4+2e\left(7+3e^2\right)\cos f_0\right.\right.\\ &\left.+5e^2\cos 2f_0\right]\left(-2\widehat{T}_1+3\widehat{T}_2-2\widehat{T}_5\cot I\right)\\ &+2\left(1+e\cos f_0\right)^3\left[2\widehat{T}_1-3\left(\widehat{T}_2+\widehat{T}_3\cos 2u_0\right)\right.\\ &\left.\left.-6\widehat{T}_6\sin 2u_0\right]\right\},\end{aligned} \tag{8.53}$$

$$\begin{aligned}\left\langle\frac{d\eta}{dt}\right\rangle_{\rm mix}^{J_2/c^2} &= -\frac{9n_{\rm K}J_2R_{\rm e}^2\mu}{8c^2a^3\left(1-e^2\right)^{7/2}}\left\{\left(2\widehat{T}_1-3\widehat{T}_2\right)\left[6+7e^2+2e^4\right.\right.\\ &\left.+2e\left(7+3e^2\right)\cos f_0+5e^2\cos 2f_0\right]\\ &+2\left(5-2\sqrt{1-e^2}\right)\left(1+e\cos f_0\right)^3\left[2\widehat{T}_1\right.\\ &\left.\left.-3\left(\widehat{T}_2+\widehat{T}_3\cos 2u_0\right)-6\widehat{T}_6\sin 2u_0\right]\right\}.\end{aligned} \tag{8.54}$$

Equations (8.49)–(8.54) add to the ratios of Equations (8.25)–(8.30) to the Keplerian orbital period $T_{\rm K}$ in order to give the total mixed orbital precessions of the order of $\mathcal{O}\left(J_2/c^2\right)$, as pointed out in Section 2.2.3.

9

pK Tidal Effects: Distant 3rd Body

If a gravitationally bound two-body (restricted) system $\mathcal{S}$, made of a primary of mass M and a test particle orbiting it, freely moves in the exterior, generally non-uniform gravitational field of a distant, external massive object[1] of mass $M^{'} \gg M$, tidal effects affect the internal dynamics of the former to the Newtonian and the pN levels (Iorio, 2014b). It is assumed that

$$n_{\mathrm{K}} \gg n_{\mathrm{K}}^{'}, \tag{9.1}$$

that is, the orbital period $T_{\mathrm{K}}^{'}$ of the motion of $\mathcal{S}$ about $M^{'}$ is much longer than the time T_{K} needed for the test particle to make a full revolution around M.

Here, the primed quantities refer to the orbit of $\mathcal{S}$ around $M^{'}$, while the unprimed ones describe the motion of the test particle about M within $\mathcal{S}$.

For full treatments of orbital effects within hierarchical triple systems made of bodies of comparable masses, including also those exerted by the inner binary on the distant companion, to the Newtonian and the 1pN orders, see, for example, Will (2014), Lim and Rodriguez (2020), Kuntz et al. (2021), and Kuntz (2022), and references therein.

9.1 Tidal Orbital Effects in a Kinematically Non-rotating Frame

It should be pointed out that, in the following, the quasi-inertial[2] local reference frame $\mathcal{K}$ attached to $\mathcal{S}$ is considered both dynamically and kinematically non-rotating (Brumberg and Kopeikin, 1989a; Damour et al., 1994; Kopeikin et al., 2011) over the characteristic timescales of the system under consideration. It means that the frequencies of the unavoidable de Sitter–Fokker (de Sitter, 1916b; Schouten, 1918; Fokker, 1921) and Pugh–Schiff precessions (Pugh, 1959; Schiff,

[1] In principle, it may have its own Newtonian and pN mass and spin multipole moments; here, apart from $M^{'}$, only its angular momentum $\boldsymbol{J}^{'}$ is taken into account to the pN level.

[2] Here, the use of 'quasi' refers just to the fact that tidal effects occur in such a reference frame which, thus, cannot be considered as *exactly* inertial.

1960) experienced by its axes due to the fact that they are parallel transported (Fermi, 1922; Levi-Civita, 1926; Synge, 1927) along the geodesic worldline of M through the deformed spacetime of $M^{'}$, are much smaller than both n_{K} and $n^{'}_{\mathrm{K}}$, or that the former ones are corrected for.

The pK tidal acceleration due to the field of $M^{'}$ experienced by the test particle in its motion around M is of the form (Mashhoon et al., 1989)

$$A_i^{\mathrm{tid}} = -\sum_{j=1}^{3} \mathsf{T}_{ij} x_j,\ i = 1, 2, 3, \tag{9.2}$$

where $x_1 := x$, $x_2 := y$, $x_3 := z$. The elements of the tidal matrix T_{ij}, $i, j = 1, 2, 3$

$$\mathsf{T}_{ij} = \mathsf{R}_{0i0j},\ i, j = 1, 2, 3 \tag{9.3}$$

are the 'electric' tetrad components of the curvature Riemann tensor evaluated onto the geodesic of the observer comoving with $\mathcal{S}$, and dimensionally are the reciprocal of a time squared. Equation (9.3) is the sum of the Newtonian and pN tidal matrices, which are all traceless and symmetric (Mashhoon et al., 1989). In general, a tidal acceleration experienced by a slowly moving test particle due to an external curved spacetime metric can be written as Equation (9.2) (Misner et al., 2017). In particular, in the linearized weak-field and slow-motion approximation of GTR, it is worth noticing that Equation (9.2) holds also in the case of a plane gravitational wave which, propagating along an arbitrary spatial direction, encounters a gravitationally bound two-body system.[3] For the orbital effects of a low-frequency gravitational wave on a two-body system, see, for example, Bertotti (1973), Rudenko (1975), Mashhoon (1978), Mashhoon (1979), Futamase and Matsuda (1979), Turner (1979), Grishchuk and Polnarev (1980), Mashhoon et al. (1981), Linet (1982a,b), Nelson and Chau (1982), Ivashchenko (1987), Kochkin and Sbytov (1987), Chicone et al. (1996a,b), and Iorio (2014a).

The radial, transverse, and normal components of Equation (9.2) turn out to be (Iorio, 2014b)

$$\begin{aligned}
\frac{A_r^{\mathrm{tid}}}{r} = {} & (\cos I \sin u \sin\Omega - \cos u \cos\Omega) \\
& \{\sin f\,[\mathsf{T}_{21}\cos I\cos\omega\cos\Omega + (\mathsf{T}_{22}+\mathsf{T}_{33})\sin\omega\cos\Omega + \mathsf{T}_{31}\cos\omega\sin I] \\
& + \cos f\,[\mathsf{T}_{11}\cos\omega\cos\Omega + (\mathsf{T}_{21}\cos I\cos\Omega + \mathsf{T}_{31}\sin I)\sin\omega] \\
& + [\mathsf{T}_{21}\cos u + (\mathsf{T}_{22}+\mathsf{T}_{33})\cos I\sin u]\sin\Omega\} \\
& + (\cos I\cos\Omega\sin u + \cos u\sin\Omega)\,\{\cos u\,(\mathsf{T}_{21}\cos\Omega + \mathsf{T}_{22}\sin\Omega) \\
& + \sin u\,[\mathsf{T}_{32}\sin I + \cos I\,(\mathsf{T}_{22}\cos\Omega - \mathsf{T}_{21}\sin\Omega)]\} \\
& + \sin I\sin u\,\{\cos u\,(\mathsf{T}_{31}\cos\Omega + \mathsf{T}_{32}\sin\Omega) \\
& + \sin u\,[\mathsf{T}_{33}\sin I + \cos I\,(\mathsf{T}_{32}\cos\Omega - \mathsf{T}_{31}\sin\Omega)]\},
\end{aligned} \tag{9.4}$$

[3] See the discussion in, for example, section 2 of Iorio (2014a).

$$
\begin{aligned}
-\frac{4A_\tau^{\mathrm{tid}}}{r} = {} & [-\mathrm{T}_{33} + (2\mathrm{T}_{22} + \mathrm{T}_{33})\cos 2\Omega \\
& - 2\mathrm{T}_{21}\sin 2\Omega]\cos^2 I \sin 2u + 2\{2\sin I\,(\mathrm{T}_{32}\cos\Omega - \mathrm{T}_{31}\sin\Omega)\sin 2u \\
& + [2\mathrm{T}_{21}\cos 2\Omega + (2\mathrm{T}_{22} + \mathrm{T}_{33})\sin 2\Omega]\}\cos I\cos 2u \\
& + 4\sin I\,(\mathrm{T}_{31}\cos\Omega + \mathrm{T}_{32}\sin\Omega)\cos 2u + [(2 - \cos 2I)\,\mathrm{T}_{33} \\
& + (2\mathrm{T}_{22} + \mathrm{T}_{33})\cos 2\Omega - 2\mathrm{T}_{21}\sin 2\Omega]\sin 2u, \qquad (9.5)
\end{aligned}
$$

$$
\begin{aligned}
-\frac{A_h^{\mathrm{tid}}}{r} = {} & 4\cos I\,(\mathrm{T}_{31}\cos\Omega + \mathrm{T}_{32}\sin\Omega)\cos u \\
& + 2\sin I\,[-2\mathrm{T}_{21}\cos 2\Omega - (2\mathrm{T}_{22} + \mathrm{T}_{33})\sin 2\Omega]\cos u \\
& + 4\cos 2I\,(\mathrm{T}_{32}\cos\Omega - \mathrm{T}_{31}\sin\Omega)\sin u \\
& + \sin 2I\,[3\mathrm{T}_{33} - (2\mathrm{T}_{22} + \mathrm{T}_{33})\cos 2\Omega + 2\mathrm{T}_{21}\sin 2\Omega]\sin u, \qquad (9.6)
\end{aligned}
$$

where r is given by Equation (2.11), and the coefficients $\mathrm{T}_{ij},\ i,j = 1,2,3$ depend only on the orbital parameters of the motion about M', on and on the orientation of the spin axis $\hat{\boldsymbol{J}}'$ of the latter.

In view of the hypothesis of Equation (9.1), it can be assumed that the coefficients $\mathrm{T}_{lo},\ l,o = 1,2,3$ are constant during T_{K}. Thus, calculating Equations (2.13)–(2.18) with Equations (9.4)–(9.6) and using the eccentric anomaly E as a fast variable of integration for computational purposes straightforwardly yields (Iorio, 2014b)

$$\left\langle\frac{da}{dt}\right\rangle^{\mathrm{tid}} = 0, \qquad (9.7)$$

$$\left\langle\frac{de}{dt}\right\rangle^{\mathrm{tid}} = \frac{5e\sqrt{1-e^2}}{8n_{\mathrm{K}}}\mathcal{E}^{\mathrm{tid}}, \qquad (9.8)$$

$$\left\langle\frac{dI}{dt}\right\rangle^{\mathrm{tid}} = -\frac{1}{8n_{\mathrm{K}}\sqrt{1-e^2}}\mathcal{I}^{\mathrm{tid}}, \qquad (9.9)$$

$$\left\langle\frac{d\Omega}{dt}\right\rangle^{\mathrm{tid}} = -\frac{\csc I}{16n_{\mathrm{K}}\sqrt{1-e^2}}\mathcal{N}^{\mathrm{tid}}, \qquad (9.10)$$

$$\left\langle\frac{d\omega}{dt}\right\rangle^{\mathrm{tid}} = \frac{1}{4n_{\mathrm{K}}\sqrt{1-e^2}}\sum_{i=1}^{2}k_i^{\mathrm{tid}}\mathcal{P}_i^{\mathrm{tid}}, \qquad (9.11)$$

$$\left\langle\frac{d\eta}{dt}\right\rangle^{\mathrm{tid}} = \frac{1}{16n_{\mathrm{K}}}\mathcal{H}^{\mathrm{tid}}. \qquad (9.12)$$

The explicit expressions of the coefficients $\mathcal{E}^{\mathrm{tid}}, \ldots \mathcal{H}^{\mathrm{tid}}$ entering Equations (9.8)–(9.12) can be found in Equations (H.1)–(H.8) of Appendix H. Equations (9.7)–(9.12) hold also when the tidal field is due to a passing plane gravitational wave whose frequency is much lower than the orbital one.

9.1.1 The Newtonian Effects

The Newtonian tidal matrix is (Mashhoon et al., 1989)

$$\mathsf{T}_{ij}^{\mathrm{N}} = \frac{\mu'}{r'^3}\left(\delta_{ij} - 3\hat{r}'_i\hat{r}'_j\right), \tag{9.13}$$

where $\mu' := GM'$ is the standard gravitational parameter of the body of mass M', $\hat{\boldsymbol{r}}' := \boldsymbol{r}'/r'$ is the versor of the position vector $\boldsymbol{r}'$ from M' to M, and δ_{ij} is the Kronecker delta (Olver et al., 2010).

Its average with respect to T'_{K}, to the zero order in J'_2, is (Iorio, 2014b)

$$\langle\mathsf{T}_{11}\rangle'_{\mathrm{N}} = -\frac{\mu'\left(1 + 3\cos 2I' + 6\sin^2 I'\cos 2\Omega'\right)}{8a'^3\left(1 - e'^2\right)^{3/2}}, \tag{9.14}$$

$$\langle\mathsf{T}_{22}\rangle'_{\mathrm{N}} = -\frac{\mu'\left(1 + 3\cos 2I' - 6\sin^2 I'\cos 2\Omega'\right)}{8a'^3\left(1 - e'^2\right)^{3/2}}, \tag{9.15}$$

$$\langle\mathsf{T}_{33}\rangle'_{\mathrm{N}} = \frac{\mu'\left(1 + 3\cos 2I'\right)}{4a'^3\left(1 - e'^2\right)^{3/2}}, \tag{9.16}$$

$$\langle\mathsf{T}_{12}\rangle'_{\mathrm{N}} = -\frac{3\mu'\sin^2 I'\sin 2\Omega'}{4a'^3\left(1 - e'^2\right)^{3/2}}, \tag{9.17}$$

$$\langle\mathsf{T}_{13}\rangle'_{\mathrm{N}} = \frac{3\mu'\sin 2I'\sin\Omega'}{4a'^3\left(1 - e'^2\right)^{3/2}}, \tag{9.18}$$

$$\langle\mathsf{T}_{23}\rangle'_{\mathrm{N}} = -\frac{3\mu'\sin 2I'\cos\Omega'}{4a'^3\left(1 - e'^2\right)^{3/2}}. \tag{9.19}$$

9.1.2 The 1pN Gravitoelectric Effects

The 1pN gravitoelectric tidal matrix is (Mashhoon et al., 1989)

$$\begin{aligned}\mathsf{T}_{ij}^{1\mathrm{pN}} = &-\frac{\mu'^2}{c^2 r'^4}\left(3\delta_{ij} - 9\hat{r}'_i\hat{r}'_j\right)\\ &+\frac{\mu'}{c^2 r'^3}\left\{3\left[v'^2\delta_{ij} - v'_i v'_j + 3\left(\boldsymbol{v}'\cdot\hat{\boldsymbol{r}}'\right)\hat{r}'_{(i}v'_{j)}\right]\right.\\ &\left.-3\left(\boldsymbol{v}'\cdot\hat{\boldsymbol{r}}'\right)^2\delta_{ij} - 6\hat{r}'_i\hat{r}'_j v'^2\right\},\ i,j = 1,2,3.\end{aligned} \tag{9.20}$$

In Equation (9.20), $\boldsymbol{v}'$ is the velocity vector of M with respect to M', and parentheses around indices denote symmetrization.

Its average with respect to T'_{K} is (Iorio, 2014b)

$$\langle \mathsf{T}_{11}\rangle'_{1\mathrm{pN}} = -\frac{3\mu'^2 e'^2}{32c^2 a'^4\left(1-e'^2\right)^{5/2}}\mathcal{T}_{11}^{1\mathrm{pN}}, \tag{9.21}$$

$$\langle \mathsf{T}_{22}\rangle'_{1\mathrm{pN}} = \frac{3\mu'^2 e'^2}{32c^2 a'^4\left(1-e'^2\right)^{5/2}}\mathcal{T}_{22}^{1\mathrm{pN}}, \tag{9.22}$$

$$\langle \mathsf{T}_{33}\rangle'_{1\mathrm{pN}} = \frac{3\mu'^2 e'^2}{4c^2 a'^4\left(1-e'^2\right)^{5/2}}\mathcal{T}_{33}^{1\mathrm{pN}}, \tag{9.23}$$

$$\langle \mathsf{T}_{12}\rangle'_{1\mathrm{pN}} = -\frac{3\mu'^2 e'^2}{16c^2 a'^4\left(1-e'^2\right)^{5/2}}\mathcal{T}_{12}^{1\mathrm{pN}}, \tag{9.24}$$

$$\langle \mathsf{T}_{13}\rangle'_{1\mathrm{pN}} = -\frac{3\mu'^2 e'^2}{4c^2 a'^4\left(1-e'^2\right)^{5/2}}\mathcal{T}_{13}^{1\mathrm{pN}}, \tag{9.25}$$

$$\langle \mathsf{T}_{23}\rangle'_{1\mathrm{pN}} = \frac{3\mu'^2 e'^2}{4c^2 a'^4\left(1-e'^2\right)^{5/2}}\mathcal{T}_{23}^{1\mathrm{pN}}, \tag{9.26}$$

where the explicit expressions of the coefficients $\mathcal{T}_{ij}^{1\mathrm{pN}},\ i,j = 1,2,3$ are displayed in Equations (H.9)–(H.14) of Appendix H. It should be noted that, if the motion of $\mathcal{S}$ about M' is circular, that is, for $e' = 0$, Equations (9.21)–(9.26) vanish.

9.1.3 The 1pN Gravitomagnetic Effects

The gravitomagnetic tidal matrix is (Mashhoon et al., 1989)

$$\begin{aligned}
\mathsf{T}_{ij}^{\mathrm{gvm}} = &-\frac{6GJ'}{c^2 r'^4}\left[3\left(\boldsymbol{v}'\times\hat{\boldsymbol{J}}'\right)_{(i}\hat{r}'_{j)} + \left(\hat{\boldsymbol{r}}'\times\hat{\boldsymbol{J}}'\right)_{(i}v'_{j)}\right.\\
&+\hat{\boldsymbol{r}}'\cdot\left(\boldsymbol{v}'\times\hat{\boldsymbol{J}}'\right)\left(\delta_{ij}-5\hat{r}'_i\hat{r}'_j\right)\\
&\left.-5\left(\hat{\boldsymbol{r}}'\cdot\boldsymbol{v}'\right)\left(\hat{\boldsymbol{r}}'\times\hat{\boldsymbol{J}}'\right)_{(i}\hat{r}'_{j)}\right],\ i,j = 1,2,3.
\end{aligned} \tag{9.27}$$

In Equation (9.27), $\hat{\boldsymbol{J}}'$ is the spin unit vector of the body of mass M', and parentheses around indices denote symmetrization.

Its average with respect to T'_{K} is (Iorio, 2014b)

$$\langle \mathsf{T}_{11}\rangle'_{\mathrm{gvm}} = -\frac{3GJ'n'_{\mathrm{K}}}{64c^2a'^3\left(1-e'^2\right)^3}\mathcal{T}_{11}^{\mathrm{gvm}}, \tag{9.28}$$

$$\langle \mathsf{T}_{22}\rangle'_{\mathrm{gvm}} = -\frac{3GJ'n'_{\mathrm{K}}}{64c^2a'^3\left(1-e'^2\right)^3}\mathcal{T}_{22}^{\mathrm{gvm}}, \tag{9.29}$$

$$\langle \mathsf{T}_{33}\rangle'_{\mathrm{gvm}} = -\frac{3GJ'n'_{\mathrm{K}}}{16c^2a'^3\left(1-e'^2\right)^3}\mathcal{T}_{33}^{\mathrm{gvm}}, \tag{9.30}$$

$$\langle \mathsf{T}_{12}\rangle'_{\mathrm{gvm}} = -\frac{3GJ'n'_{\mathrm{K}}}{64c^2a'^3\left(1-e'^2\right)^3}\mathcal{T}_{12}^{\mathrm{gvm}}, \tag{9.31}$$

$$\langle \mathsf{T}_{13}\rangle'_{\mathrm{gvm}} = \frac{3GJ'n'_{\mathrm{K}}}{32c^2a'^3\left(1-e'^2\right)^3}\mathcal{T}_{13}^{\mathrm{gvm}}, \tag{9.32}$$

$$\langle \mathsf{T}_{23}\rangle'_{\mathrm{gvm}} = -\frac{3GJ'n'_{\mathrm{K}}}{64c^2a'^3\left(1-e'^2\right)^3}\mathcal{T}_{23}^{\mathrm{gvm}}, \tag{9.33}$$

where the explicit expressions of the coefficients $\mathcal{T}_{ij}^{\mathrm{gvm}}$, $i,j = 1,2,3$ are displayed in Equations (H.15)–(H.20) of Appendix H.

9.2 Tidal Orbital Effects in a Kinematically Rotating Frame

Here, the reference frame $\mathcal{K}$ attached to $\mathcal{S}$ and moving geodesically in the deformed spacetime of M' is considered dynamically non-rotating, but kinematically rotating. It implies that the de Sitter–Fokker and Pugh–Schiff precessions of its axes are not corrected for, thus impacting the orbital dynamics within $\mathcal{S}$.

9.2.1 The Impact of the de Sitter–Fokker Precession

The averaged rates of change of the Keplerian orbital elements of the test particle in its motion about M due to the de Sitter–Fokker precession (Renzetti, 2012b) can be computed with Equations (2.27)–(2.32) starting from the disturbing function

$$\mathfrak{R}_{\mathrm{dS}} = -\frac{3\mu'\boldsymbol{h}'\boldsymbol{\cdot}\boldsymbol{h}}{2c^2r'^3}, \tag{9.34}$$

where $\boldsymbol{h}' = n'_{\mathrm{K}}a'^2\sqrt{1-e'^2}\,\hat{\boldsymbol{h}}'$ and $\boldsymbol{h} = n_{\mathrm{K}}a^2\sqrt{1-e^2}\,\hat{\boldsymbol{h}}$ are the orbital angular momenta per unit mass characterizing the motions of $\mathcal{S}$ about M' and of the test particle around M, respectively. Equation (9.34), up to the minus sign and the test particle's mass,[4] can be inferred, for example, from Barker and O'Connell (1979, Equation (2.17), p. 155) for the potential energy

$$\mathcal{V}_{S_1} = \frac{G}{c^2 r^3}\left(2 + \frac{3m_2}{2m_1}\right)\boldsymbol{S}^{(1)} \cdot (\boldsymbol{r} \times \boldsymbol{P}) \tag{9.35}$$

of two bodies of masses m_2 and m_1, one of which carries the angular momentum $\boldsymbol{S}^{(1)}$, separated by a distance r and moving with relative velocity $\boldsymbol{v}$ in the limit $m_2 \gg m_1$, with the identification $m_2 \rightarrow M^{'}$, $m_1 \rightarrow M$. Furthermore, it is assumed that $\boldsymbol{S}^{(1)}$ is the orbital angular momentum of the test particle's motion about M. Thus, r in Barker and O'Connell (1979, Equation (2.17), p. 155) has to be identified with $r^{'}$, and $\boldsymbol{r} \times \boldsymbol{P}$ is the orbital angular momentum $\boldsymbol{r}^{'} \times M\boldsymbol{v}^{'}$ of the motion of $\mathcal{S}$ around $M^{'}$. See also Ohanian and Ruffini (2013). It can be noted that

$$\boldsymbol{v} \cdot \nabla_{v}\mathfrak{R}_{\mathrm{dS}} = \mathfrak{R}_{\mathrm{dS}}. \tag{9.36}$$

The average of Equation (9.34) over $T_{\mathrm{K}}^{'}$ returns (Renzetti, 2012b)

$$\langle\mathfrak{R}\rangle_{\mathrm{dS}}^{'} = -\frac{3\mu^{'} n_{\mathrm{K}}^{'} a^2 n_{\mathrm{K}}\sqrt{1-e^2}\,\hat{\boldsymbol{h}}^{'} \cdot \hat{\boldsymbol{h}}}{2c^2 a^{'}\left(1-e^{'2}\right)}. \tag{9.37}$$

By calculating Equations (2.27)–(2.32) with Equation (9.37) yields[5] (Renzetti, 2012b)

$$\left\langle\frac{da}{dt}\right\rangle_{\mathrm{dS}}^{'} = 0, \tag{9.38}$$

$$\left\langle\frac{de}{dt}\right\rangle_{\mathrm{dS}}^{'} = 0, \tag{9.39}$$

$$\left\langle\frac{dI}{dt}\right\rangle_{\mathrm{dS}}^{'} = \frac{3\mu^{'} n_{\mathrm{K}}^{'}\hat{\boldsymbol{h}}^{'} \cdot \hat{\boldsymbol{l}}}{2c^2 a^{'}\left(1-e^{'2}\right)}, \tag{9.40}$$

$$\left\langle\frac{d\Omega}{dt}\right\rangle_{\mathrm{dS}}^{'} = \frac{3\mu^{'} n_{\mathrm{K}}^{'}\csc I\,\hat{\boldsymbol{h}}^{'} \cdot \hat{\boldsymbol{m}}}{2c^2 a^{'}\left(1-e^{'2}\right)}, \tag{9.41}$$

$$\left\langle\frac{d\omega}{dt}\right\rangle_{\mathrm{dS}}^{'} = \frac{3\mu^{'} n_{\mathrm{K}}^{'}\hat{\boldsymbol{h}}^{'} \cdot \left(\hat{\boldsymbol{h}} - \cot I\,\hat{\boldsymbol{m}}\right)}{2c^2 a^{'}\left(1-e^{'2}\right)}, \tag{9.42}$$

[4] Recall that the disturbing function is the opposite of the potential energy per unit mass of the pK effect of interest.

[5] Equation (9.43) comes from Equation (9.36) and the last term in Equation (2.32), being the sum of the first two equal to zero.

$$\left\langle \frac{d\eta}{dt} \right\rangle'_{\rm dS} = \frac{9\mu' n'_{\rm K}\sqrt{1-e^2}\,\hat{\boldsymbol{h}}' \cdot \hat{\boldsymbol{h}}}{2c^2 a' \left(1-e'^2\right)}. \tag{9.43}$$

It should be noted that Equations (9.38)–(9.43) are completely general since they hold for any orbital configurations for the motions around both M and M'.

The de Sitter precession of the perigee of the lunar orbit in the external field of the Sun was accurately measured with increasing accuracy over the years by means of the LLR technique (Bertotti et al., 1987; Shapiro et al., 1988; Williams et al., 1996; Nordtvedt, 1999; Williams et al., 2004; Merkowitz, 2010; Hofmann and Müller, 2018).

9.2.2 The Impact of the Spin of the Distant Body

The disturbing function $\mathfrak{R}_{\rm gvm}$ arising from the gravitomagnetic three-body potential induced by the angular momentum $\boldsymbol{J}'$ of the distant body of mass M' on the internal dynamics of $\mathcal{S}$ is

$$\mathfrak{R}_{\rm gvm} = -\frac{GJ'}{c^2 r'^3}\left[-\left(\hat{\boldsymbol{J}}' \cdot \boldsymbol{h}\right) + 3\left(\boldsymbol{h}\cdot\hat{\boldsymbol{r}}'\right)\left(\hat{\boldsymbol{J}}'\cdot\hat{\boldsymbol{r}}'\right)\right]. \tag{9.44}$$

It turns out that

$$\boldsymbol{v}\cdot\nabla_v \mathfrak{R}_{\rm gvm} = \mathfrak{R}_{\rm gvm}. \tag{9.45}$$

Equation (9.44), up to the minus sign and the test particle's mass,[6] can be inferred, for example, from Barker and O'Connell (1979, Equation (2.19), p. 155) for the interaction potential energy

$$\mathcal{V}_{S_1,S_2} = \frac{G}{c^2 r^3}\left[\frac{3\left(\boldsymbol{S}^{(1)}\cdot\boldsymbol{r}\right)\left(\boldsymbol{S}^{(2)}\cdot\boldsymbol{r}\right)}{r^2} - \boldsymbol{S}^{(1)}\cdot\boldsymbol{S}^{(2)}\right] \tag{9.46}$$

of two spins $\boldsymbol{S}^{(1)}$, $\boldsymbol{S}^{(2)}$ separated by a distance r by assuming that the spin $\boldsymbol{S}^{(1)}$ is the orbital angular momentum of the test particle's motion about M, while $\boldsymbol{S}^{(2)}$ is the spin angular momentum $\boldsymbol{J}'$ of the distant third body of mass M'. Thus, r in Barker and O'Connell (1979, Equation (2.19), p. 155) has to be identified with r'.

The orbit average of Equation (9.44) over $T'_{\rm K}$ yields (Iorio, 2019b)

$$\begin{aligned}\langle\mathfrak{R}\rangle'_{\rm gvm} = {} & \frac{GJ' n_{\rm K} a^2\sqrt{1-e^2}}{8c^2 a'^3\left(1-e'^2\right)^{3/2}}\left(2\cos I\left[\hat{J}'_z\left(1+3\cos 2I'\right)\right.\right.\\ & \left.\left. + 3\sin 2I'\left(-\hat{J}'_y\cos\Omega' + \hat{J}'_x\sin\Omega'\right)\right]\right.\end{aligned}$$

[6] Recall that the disturbing function is the opposite of the potential energy per unit mass of the pK effect of interest.

$$+ \sin I \left\{\hat{J}_y^{'}\left[\cos\Omega - 3\cos\left(\Omega - 2\Omega^{'}\right)\right] - \hat{J}_x^{'}\left[\sin\Omega + 3\sin\left(\Omega - 2\Omega^{'}\right)\right] + 6\cos\left(\Omega - \Omega^{'}\right)\left[\hat{J}_z^{'}\sin 2I^{'} + \cos 2I^{'}\left(\hat{J}_y^{'}\cos\Omega^{'} - \hat{J}_x^{'}\sin\Omega^{'}\right)\right]\right\}\Big). \tag{9.47}$$

Inserting Equation (9.47) in Equations (2.27)–(2.32) allows one to calculate the mean orbital precessions which are[7] (Iorio, 2019b)

$$\left\langle\frac{da}{dt}\right\rangle_{\rm gvm}^{'} = 0, \tag{9.48}$$

$$\left\langle\frac{de}{dt}\right\rangle_{\rm gvm}^{'} = 0, \tag{9.49}$$

$$\left\langle\frac{dI}{dt}\right\rangle_{\rm gvm}^{'} = -\frac{GJ^{'}}{8c^2a^{'3}\left(1-e^{'2}\right)^{3/2}}\left\{-\hat{J}_x^{'}\left[\cos\Omega + 3\cos\left(\Omega - 2\Omega^{'}\right)\right] - \hat{J}_y^{'}\left[\sin\Omega - 3\sin\left(\Omega - 2\Omega^{'}\right)\right] - 6\sin\left(\Omega - \Omega^{'}\right)\left[\hat{J}_z^{'}\sin 2I^{'} + \cos 2I^{'}\left(\hat{J}_y^{'}\cos\Omega^{'} - \hat{J}_x^{'}\sin\Omega^{'}\right)\right]\right\}, \tag{9.50}$$

$$\left\langle\frac{d\Omega}{dt}\right\rangle_{\rm gvm}^{'} = \frac{GJ^{'}\csc I}{8c^2a^{'3}\left(1-e^{'2}\right)^{3/2}}\Big(-2\sin I\left[\hat{J}_z^{'}\left(1 + 3\cos 2I^{'}\right) + 3\sin 2I^{'}\left(-\hat{J}_y^{'}\cos\Omega^{'} + \hat{J}_x^{'}\sin\Omega^{'}\right)\right] + \cos I\left\{\hat{J}_y^{'}\left[\cos\Omega - 3\cos\left(\Omega - 2\Omega^{'}\right)\right] - \hat{J}_x^{'}\left[\sin\Omega + 3\sin\left(\Omega - 2\Omega^{'}\right)\right] + 6\cos\left(\Omega - \Omega^{'}\right)\left[\hat{J}_z^{'}\sin 2I^{'} + \cos 2I^{'}\left(\hat{J}_y^{'}\cos\Omega^{'} - \hat{J}_x^{'}\sin\Omega^{'}\right)\right]\right\}\Big), \tag{9.51}$$

$$\left\langle\frac{d\omega}{dt}\right\rangle_{\rm gvm}^{'} = -\frac{GJ^{'}\csc I}{8c^2a^{'3}\left(1-e^{'2}\right)^{3/2}}\left\{\hat{J}_y^{'}\left[\cos\Omega - 3\cos\left(\Omega - 2\Omega^{'}\right)\right] - \hat{J}_x^{'}\left[\sin\Omega + 3\sin\left(\Omega - 2\Omega^{'}\right)\right] + 6\cos\left(\Omega - \Omega^{'}\right)\left[\hat{J}_z^{'}\sin 2I^{'} + \cos 2I^{'}\left(\hat{J}_y^{'}\cos\Omega^{'} - \hat{J}_x^{'}\sin\Omega^{'}\right)\right]\right\}, \tag{9.52}$$

[7] Equation (9.53) comes from Equation (9.45) and the last term in Equation (2.32), being the sum of the first two equal to zero.

$$
\begin{aligned}
\left\langle\frac{d\eta}{dt}\right\rangle'_{\rm gvm} = &-\frac{3GJ'\sqrt{1-e^2}}{8c^2a'^3\left(1-e'^2\right)^{3/2}}\left(2\cos I\left[\hat{J}'_z\left(1+3\cos 2I'\right)\right.\right.\\
&\left.+3\sin 2I'\left(-\hat{J}'_y\cos\Omega'+\hat{J}'_x\sin\Omega'\right)\right]\\
&+\sin I\left\{\hat{J}'_y\left[\cos\Omega-3\cos\left(\Omega-2\Omega'\right)\right]\right.\\
&-\hat{J}'_x\left[\sin\Omega+3\sin\left(\Omega-2\Omega'\right)\right]+6\cos\left(\Omega-\Omega'\right)\left[\hat{J}'_z\sin 2I'\right.\\
&\left.\left.\left.+\cos 2I'\left(\hat{J}'_y\cos\Omega'-\hat{J}'_x\sin\Omega'\right)\right]\right\}\right). \qquad (9.53)
\end{aligned}
$$

For a previous, approximate calculation restricted to the orbital angular momentum of the Moon orbiting the Earth in the gravitomagnetic field of the rotating Sun, see Gill et al. (1992, Section (3.3.3)).

10

Modified Models of Gravity: Orbital Precessions

In the following sections, some modifications of the Newtonian inverse-square law are taken into account; those that arise from power-law and Yukawa-like additional potentials, logarithmic-type potential, some dark matter density profiles, once-per-revolution and constant accelerations, and gravitomagnetic violations of the Lorentz invariance in the gravitational sector.

The resulting secular orbital precessions are analytically worked out in their full generality, extending and generalizing an earlier calculation by Adkins and McDonnell (2007) for some of the aforementioned extra-potentials.

10.1 Power-Law Modified Potentials

Power-law modifications of the usual inverse-square law (Fischbach et al., 2001; Adelberger et al., 2003) arise in the most disparate theoretical frameworks; they are far too numerous to be exhaustively mentioned here. For r^{-2} and r^{-3}-type extra potentials, see, for example, Iorio (2012b,d); Iorio and Ruggiero (2018), and references therein. It should be noted that, in fact, not all the power-law modified potentials are necessarily to be ascribed to alternative models of gravity; suffice it to say that the general relativistic Reissner-Nordström spacetime metric (Reissner, 1916; Weyl, 1917; Nordström, 1918) encompasses just a $1/r^2$ correction to the standard $1/r$ Newtonian potential. The quadrupolar term of the Newtonian multipolar expansion of the potential of a matter ring[1] (Ciftja et al., 2009), calculated in a point in the plane of the annulus at great distance r from it, goes just as $1/r^3$ (Demetrian, 2006). Furthermore, the cosmological constant[2] Λ (O'Raifeartaigh et al., 2018), fully accommodated within GTR (Ashtekar, 2017), gives rise to an

[1] It is an adequate model for a fast revolving, tight inner binary in a hierarchical triple system with a distant companion.

[2] It is the simplest possible explanation for dark energy driving the observed cosmological accelerated expansion (Peebles and Ratra, 2003; Harvey, 2009). But see also Lahav (2020).

additional potential quadratic in r (Rindler, 2001). Also the quadrupolar term of the Newtonian multipolar expansion of the potential of a matter ring (Ciftja et al., 2009), calculated in a point in the plane of the annulus at distance r smaller than its radius, is directly proportional to r^2 (Demetrian, 2006).

Here, the disturbing function $\mathfrak{R}$ of such kinds of modified models gravity is generally written as

$$\mathfrak{R}^{\mathrm{pl}} = \frac{\mathrm{K}}{r^n}, \tag{10.1}$$

where K is a dimensional quantity containing some numerical factors, and the free parameter(s) characterizing the model under consideration, to be determined or constrained by observations, and, possibly, the specific binary system at hand as well as its masses, and so on. The latter case does not necessarily occur; suffice it to say that, for $n = -2$, one has (Rindler, 2001; Kerr et al., 2003)

$$\mathrm{K} = \frac{\Lambda c^2}{6}, \tag{10.2}$$

arising from the Schwarzschild–de Sitter (Stuchlík and Hledík, 1999) or Kottler (Kottler, 1918) spacetime. In Equation (10.2), Λ is dimensionally the reciprocal of an area, and plays the role of the cosmological constant (Rindler, 2001).

The average of Equation (10.1) over one orbital revolution, performed by using the eccentric anomaly E as a fast variable of integration to simplify the calculation, turns out to be

$$\begin{aligned}\langle\mathfrak{R}\rangle^{\mathrm{pl}} = &\frac{\mathrm{K}}{4} a^{-2n}\left(1-e^2\right)^n \Bigg[3a^n\left(1-e\right)\left(1+e\right)^n {}_2F_1\left(-\frac{1}{2}, n; 2; \frac{2e}{e-1}\right) \\ &+ 3a^n\left(1+e\right)\left(1-e\right)^n {}_2F_1\left(-\frac{1}{2}, n; 2; \frac{2e}{1+e}\right) \\ &- a^n\left(1+e\right)^n\left(1-3e+4e\,n\right) {}_2F_1\left(\frac{1}{2}, n; 2; \frac{2e}{e-1}\right) \\ &+ a^n\left(1-e\right)^n\left(-1-3e+4e\,n\right) {}_2F_1\left(\frac{1}{2}, n; 2; \frac{2e}{1+e}\right)\Bigg],\end{aligned} \tag{10.3}$$

where ${}_2F_1\left(\mathfrak{a}, \mathfrak{b}; \mathfrak{c}; \mathfrak{x}\right)$ is the Gauss hypergeometric function (Barnes, 1908; Olver et al., 2010). It should be remarked that Equation (10.3) retains its validity for any integer n, either positive or negative; furthermore, it holds for arbitrarily eccentric orbits, being an exact result in e.

From Equation (10.3) and Equations (2.27)–(2.32), it can be straightforwardly inferred that the only generally nonvanishing orbital precessions are those of ω and η. They are

$$\left\langle\frac{d\omega}{dt}\right\rangle^{\mathrm{pl}} = -\frac{(-1+n)\,\mathrm{K}\sqrt{1-e^2}\left[a^2\left(1-e^2\right)\right]^{-1-n}}{4en_{\mathrm{K}}}$$
$$\left\{-3\,(-1+e)\,a^n\,(1+e)^n\,{}_2F_1\left(-\frac{1}{2},n;2;\frac{2e}{e-1}\right)\right.$$
$$-3\,(1+e)^2\,a^n\,(1-e)^n\,{}_2F_1\left(-\frac{1}{2},n;2;\frac{2e}{1-e}\right)$$
$$+\,(-1+e)\,a^n\,(1+e)^n\,(3-3e+4en)\,{}_2F_1\left(\frac{1}{2},n;2;\frac{2e}{e-1}\right)$$
$$\left.-\,(1+e)\,a^n\,(1-e)^n\,(-3-3e+4en)\,{}_2F_1\left(\frac{1}{2},n;2;\frac{2e}{1-e}\right)\right\}, \quad (10.4)$$

$$\left\langle\frac{d\eta}{dt}\right\rangle^{\mathrm{pl}} = \frac{\mathrm{K}\left[a^2\left(1-e^2\right)\right]^{-n}}{4a^2n_{\mathrm{K}}e}$$
$$\left(-3\,(-1+e)\,a^n\,(1+e)^n\,(-1+e+n+en)\,{}_2F_1\left(-\frac{1}{2},n;2;\frac{2e}{e-1}\right)\right.$$
$$+3\,(1+e)\,a^n\,(1-e)^n\,[1+e+(-1+e)\,n]\,{}_2F_1\left(-\frac{1}{2},n;2;\frac{2e}{1-e}\right)$$
$$+a^n\,(1+e)^n\left\{-3\,(-1+e)^2+[3+(-8+e)\,e]\,n\right.$$
$$\left.+4e\,(1+e)\,n^2\right\}{}_2F_1\left(\frac{1}{2},n;2;\frac{2e}{e-1}\right)$$
$$+a^n\,(1-e)^n\left\{-3\,(1+e)^2+[3+e\,(8+e)]\,n\right.$$
$$\left.\left.+4\,(-1+e)\,en^2\right\}{}_2F_1\left(\frac{1}{2},n;2;\frac{2e}{1-e}\right)\right). \quad (10.5)$$

For $n=2$, corresponding to a $1/r^2$ extra-potential (Adelberger et al., 2007; Maeda and Dadhich, 2007; Iorio and Saridakis, 2012; Ruggiero and Radicella, 2015; Ali and Khalil, 2016; Bhattacharya and Chakraborty, 2017), Equations (10.4)–(10.5) yield

$$\left\langle\frac{d\omega}{dt}\right\rangle^{\mathrm{pl}} = \frac{\mathrm{K}}{n_{\mathrm{K}}a^4\left(1-e^2\right)}, \quad (10.6)$$

$$\left\langle\frac{d\eta}{dt}\right\rangle^{\mathrm{pl}} = \frac{3\mathrm{K}}{n_{\mathrm{K}}a^4\sqrt{1-e^2}}. \quad (10.7)$$

For $n=3$, corresponding to a $1/r^3$ extra-potential[3] (Bonanno and Reuter, 2000; Sotiriou and Zhou, 2014; Chakraborty and SenGupta, 2017), Equations (10.4)–(10.5) yield

[3] As pointed out by Iorio and Ruggiero (2018), some of these models may not be constrained by observations since they do not contain free parameters. Furthermore, some of them may describe only black holes or wormholes, while some others are valid for material bodies such as a star. There are also other $1/r^3$ models which, however, are valid only at particle physics scales; thus, they cannot be constrained by orbital motions.

$$\left\langle \frac{d\omega}{dt} \right\rangle^{\text{pl}} = \frac{3\text{K}}{n_{\text{K}} a^5 \left(1 - e^2\right)^2}, \tag{10.8}$$

$$\left\langle \frac{d\eta}{dt} \right\rangle^{\text{pl}} = \frac{3\text{K}}{n_{\text{K}} a^5 \left(1 - e^2\right)^{3/2}}. \tag{10.9}$$

For $n = -2$, corresponding to a cosmological constant–type extra-potential quadratic in r (Rindler, 2001; Kerr et al., 2003; Iorio and Saridakis, 2012), Equations (10.4)–(10.5) yield

$$\left\langle \frac{d\omega}{dt} \right\rangle^{\text{pl}} = \frac{3\text{K}\sqrt{1 - e^2}}{n_{\text{K}}}, \tag{10.10}$$

$$\left\langle \frac{d\eta}{dt} \right\rangle^{\text{pl}} = -\frac{\left(7 + 3e^2\text{K}\right)}{n_{\text{K}}}. \tag{10.11}$$

10.2 Yukawa Modified Potential

Another very popular modification of the Newtonian inverse-square law is expressed as a Yukawa-like additional potential (Yukawa, 1935); see, for example, Nieto and Goldman (1991), Bertolami and Páramos (2005), Reynaud and Jaekel (2005), Moffat (2006), Bertolami et al. (2007), and references therein.

The resulting disturbing function $\mathfrak{R}^{\text{Y}}$ is customarily modelled as

$$\mathfrak{R}^{\text{Y}} = \frac{\text{K}_{\text{Y}}}{r} \exp\left(-\frac{r}{\lambda_{\text{Y}}}\right), \tag{10.12}$$

where λ_{Y} is a characteristic scale distance, and K_{Y} is usually proportional to the product of the mass M of the primary by a dimensionless parameter α_{Y} characterizing the intensity of the putative Yukawa interaction.

The average of Equation (10.12) over one orbital period, performed by adopting the eccentric anomaly E as a fast variable of integration to facilitate the calculation, is

$$\langle\mathfrak{R}\rangle^{\text{Y}} = \frac{\text{K}_{\text{Y}}}{a} \exp\left(-\frac{a}{\lambda_{\text{Y}}}\right) \mathcal{I}_0\left(\frac{ae}{\lambda_{\text{Y}}}\right), \tag{10.13}$$

where $\mathcal{I}_0(\ldots)$ is the modified Bessel function of the first kind of order $s = 0$ (Olver et al., 2010). It should be noted that Equation (10.13) holds for any value of λ_{Y}; furthermore, it is an exact result in e being, thus, valid for highly eccentric orbits as well.

From Equation (10.13) and Equations (2.27)–(2.32), it can be inferred that only ω and η experience nonvanishing secular precessions. They are

$$\left\langle \frac{d\omega}{dt} \right\rangle^{\text{Y}} = \frac{\text{K}_{\text{Y}}\sqrt{1 - e^2}}{a^2 e n_{\text{K}} \lambda_{\text{Y}}} \exp\left(-\frac{a}{\lambda_{\text{Y}}}\right) \mathcal{I}_1\left(\frac{ae}{\lambda_{\text{Y}}}\right), \tag{10.14}$$

$$\left\langle \frac{d\eta}{dt} \right\rangle^{\mathrm{Y}} = \frac{\mathrm{K_Y}}{a^3 n_{\mathrm{K}} \lambda_{\mathrm{Y}}} \exp\left(-\frac{a}{\lambda_{\mathrm{Y}}}\right) \left[2\left(a + \lambda_{\mathrm{Y}}\right) \mathcal{I}_0\left(\frac{ae}{\lambda_{\mathrm{Y}}}\right) - \frac{a\left(1+e^2\right)}{e} \mathcal{I}_1\left(\frac{ae}{\lambda_{\mathrm{Y}}}\right) \right], \tag{10.15}$$

where $\mathcal{I}_1(\ldots)$ is the modified Bessel function of the first kind of order $s = 1$ (Olver et al., 2010). For previous derivations of Equation (10.14), see Burgess and Cloutier (1988); Iorio (2012c).

10.3 Logarithmic Potential

The logarithmic potential, giving rise to the disturbing function

$$\mathfrak{R}^{\log} = \mathrm{K} \ln\left(\frac{r}{\mathcal{L}}\right), \tag{10.16}$$

where $\mathcal{L}$ is some characteristic length scale of the scenario at hand, is often used for explaining the action of dark matter at galactic scales in terms of alternative models of gravity (Das and Sur, 2022; Nash, 2023). It yields a $1/r$ pK acceleration.

The average over one orbital period of Equation (10.16), calculated with the eccentric anomaly E as a fast variable of integration, turns out to be

$$\langle\mathfrak{R}\rangle^{\log} = \mathrm{K}\left\{1 - \sqrt{1-e^2} + \ln\left[\frac{a}{2\mathcal{L}}\left(1 + \sqrt{1-e^2}\right)\right]\right\}. \tag{10.17}$$

From Equation (10.17) and Equations (2.27)–(2.32), it can be inferred that only ω and η experience nonvanishing secular precessions. They are

$$\left\langle \frac{d\omega}{dt} \right\rangle^{\log} = \frac{\mathrm{K}\left(-1 + e^2 + \sqrt{1-e^2}\right)}{n_{\mathrm{K}} a^2 e^2}, \tag{10.18}$$

$$\left\langle \frac{d\eta}{dt} \right\rangle^{\log} = \frac{\mathrm{K}\left(-3 + e^2 - 2\sqrt{1-e^2}\right)}{n_{\mathrm{K}} a^2 \left(1 + \sqrt{1-e^2}\right)}. \tag{10.19}$$

It can be noted that the limits of Equations (10.18)–(10.19) for $e \to 0$ are finite.

10.4 Once-Per-Revolution (1-cpr) Accelerations

Once-per-revolution empirical accelerations (1-cpr) are often modeled in satellite data reductions in order to account for any possible unknown, time-dependent accelerations, of whatever physical origin, affecting the orbital motions just at the orbital frequency (Tapley et al., 2004a).

When precise orbit determination (POD) is the desired outcome of data reductions to achieve some specific tasks,[4] the only goal is to obtain the smallest possible post-fit residuals in any way; in such circumstances, empirical 1-cpr accelerations are modeled and estimated. Instead, when a data reduction is specifically aimed to test, say, some pN features of motion, the 1-cpr accelerations should *not* be estimated because, otherwise, the effect(s) one is interested in would be partly or totally removed from the signal, being likely absorbed in the estimated values of the former ones.

In the following, the orbital precessions due to a generic 1-cpr acceleration are explicitly worked out.

In the most general case, the radial, transverse, and normal components can be written as

$$A_r^{1\mathrm{cpr}} = C_r^{1\mathrm{cpr}} \cos f + S_r^{1\mathrm{cpr}} \sin f, \tag{10.20}$$

$$A_\tau^{1\mathrm{cpr}} = C_\tau^{1\mathrm{cpr}} \cos f + S_\tau^{1\mathrm{cpr}} \sin f, \tag{10.21}$$

$$A_h^{1\mathrm{cpr}} = C_h^{1\mathrm{cpr}} \cos f + S_h^{1\mathrm{cpr}} \sin f, \tag{10.22}$$

where the coefficients $C_r^{1\mathrm{cpr}} \ldots S_h^{1\mathrm{cpr}}$ are constants.

By inserting Equations (10.20)–(10.22) in Equations (2.13)–(2.18), one gets

$$\left\langle \frac{da}{dt} \right\rangle^{1\mathrm{cpr}} = \frac{2\left(-1+e^2+\sqrt{1-e^2}\right)\left(S_r^{1\mathrm{cpr}} - C_\tau^{1\mathrm{cpr}}\right)}{n_{\mathrm{K}} e}, \tag{10.23}$$

$$\begin{aligned}\left\langle \frac{de}{dt} \right\rangle^{1\mathrm{cpr}} &= \frac{1}{2n_{\mathrm{K}} a e^2}\left\{2\left(-1+e^2\right)^2\left(-1+1/\sqrt{1-e^2}\right)S_r^{1\mathrm{cpr}}\right.\\ &\left.+\left[2+2e^4-2\sqrt{1-e^2}+e^2\left(-4+5\sqrt{1-e^2}\right)\right]C_\tau^{1\mathrm{cpr}}\right\},\end{aligned} \tag{10.24}$$

$$\left\langle \frac{dI}{dt} \right\rangle^{1\mathrm{cpr}} = \frac{\left(1+2e^2\right)C_h^{1\mathrm{cpr}}\cos\omega + \left(-1+e^2\right)S_h^{1\mathrm{cpr}}\sin\omega}{2n_{\mathrm{K}} a\sqrt{1-e^2}}, \tag{10.25}$$

$$\left\langle \frac{d\Omega}{dt} \right\rangle^{1\mathrm{cpr}} = \frac{\csc I\left[-\left(-1+e^2\right)S_h^{1\mathrm{cpr}}\cos\omega + \left(1+2e^2\right)C_h^{1\mathrm{cpr}}\sin\omega\right]}{2n_{\mathrm{K}} a\sqrt{1-e^2}}, \tag{10.26}$$

$$\begin{aligned}\left\langle \frac{d\omega}{dt} \right\rangle^{1\mathrm{cpr}} &= \frac{1}{2n_{\mathrm{K}} a e^3\sqrt{1-e^2}}\left(-\left(-1+e^2\right)\left\{2\left[1-\sqrt{1-e^2}\right.\right.\right.\\ &\left.+e^2\left(-2+\sqrt{1-e^2}\right)\right]C_r^{1\mathrm{cpr}} + \left[2-2\sqrt{1-e^2}\right.\\ &\left.\left.+e^2\left(-1+2\sqrt{1-e^2}\right)\right]S_\tau^{1\mathrm{cpr}}\right\}\\ &\left.+e^3\cot I\left[\left(-1+e^2\right)S_h^{1\mathrm{cpr}}\cos\omega - \left(1+2e^2\right)C_h^{1\mathrm{cpr}}\sin\omega\right]\right),\end{aligned} \tag{10.27}$$

[4] It is the case of, for example, remote sensing, altimetry and gravity field mapping.

$$\left\langle \frac{d\eta}{dt} \right\rangle^{1\text{cpr}} = \frac{1}{2n_{\text{K}}ae^3} \left\{ 2\left[-1 + \sqrt{1-e^2} + e^2\left(3 - 2\sqrt{1-e^2}\right) \right.\right.$$
$$\left. + e^4\left(1 + \sqrt{1-e^2}\right)\right] C_r^{1\text{cpr}} + (-1 + e^2)$$
$$\left.\left[2 - 2\sqrt{1-e^2} + e^2\left(-1 + 2\sqrt{1-e^2}\right)\right] S_\tau^{1\text{cpr}} \right\} \tag{10.28}$$

10.5 Constant and Uniform Acceleration

Here, a constant pK acceleration is treated, irrespectively of its possible physical origin. In the most general case, it can be decomposed as

$$A_r^{\text{cst}} = C_r^{\text{cst}}, \tag{10.29}$$
$$A_\tau^{\text{cst}} = C_\tau^{\text{cst}}, \tag{10.30}$$
$$A_h^{\text{cst}} = C_h^{\text{cst}}, \tag{10.31}$$

where the coefficients C_r^{cst}, C_τ^{cst}, C_h^{cst} are constants.

It may be recalled that a constant, radial acceleration directed towards the Sun was adopted as one of the most popular models of the so-called Pioneer anomaly (Turyshev and Toth, 2010) for as long as it was believed to be an anomalous gravitational effect before it was ultimately explained in terms of standard non-gravitational effects (Scheffer, 2003; Bertolami et al., 2008; Rievers et al., 2009, 2010; Rievers and Lämmerzahl, 2011; Francisco et al., 2012; Turyshev et al., 2012; Modenini and Tortora, 2014). Furthermore, a radial and constant acceleration enters the equations of motion in the framework of the Modified General Relativity (MGR) (Nash, 2023) aimed to explain the dark matter phenomenology at galactic scales.

By inserting Equations (10.29)–(10.31) in Equations (2.13)–(2.18), one gets the following orbital precessions

$$\left\langle \frac{da}{dt} \right\rangle^{\text{cst}} = \frac{2\sqrt{1-e^2}C_\tau^{\text{cst}}}{n_{\text{K}}}, \tag{10.32}$$

$$\left\langle \frac{de}{dt} \right\rangle^{\text{cst}} = -\frac{3e\sqrt{1-e^2}C_\tau^{\text{cst}}}{2n_{\text{K}}a}, \tag{10.33}$$

$$\left\langle \frac{dI}{dt} \right\rangle^{\text{cst}} = -\frac{3eC_h^{\text{cst}}\cos\omega}{2n_{\text{K}}a\sqrt{1-e^2}}, \tag{10.34}$$

$$\left\langle \frac{d\Omega}{dt} \right\rangle^{\text{cst}} = -\frac{3eC_h^{\text{cst}}\csc I\sin\omega}{2n_{\text{K}}a\sqrt{1-e^2}}, \tag{10.35}$$

$$\left\langle \frac{d\omega}{dt} \right\rangle^{\text{cst}} = \frac{-2\left(-1+e^2\right) C_r^{\text{cst}} + 3eC_h^{\text{cst}} \cot I \sin\omega}{2n_{\text{K}} a\sqrt{1-e^2}}, \tag{10.36}$$

$$\left\langle \frac{d\eta}{dt} \right\rangle^{\text{cst}} = -\frac{3C_r^{\text{cst}}}{n_{\text{K}} a}. \tag{10.37}$$

10.6 Tidal-Type Matrix Acceleration

For any extra-acceleration of the form

$$A_i = \sum_{j=1}^{3} \mathsf{M}_{ij} x_j,\ i,j = 1,2,3, \tag{10.38}$$

where $\mathsf{M}_{ij},\ i,j = 1,2,3$ is a traceless and symmetric matrix, Equations (9.7)–(9.12) hold provided that the matrix elements can be considered constant over a full orbital revolution of the test particle around its primary.

10.7 Dark Matter Distributions

So far, several authors have put dynamical constraints on the dark matter distribution within our solar system and in the GC from orbital motions (Anderson et al., 1989, 1995; Grøn and Soleng, 1996; Iorio, 2006c; Sereno and Jetzer, 2006; Khriplovich and Pitjeva, 2006; Khriplovich, 2007; Zakharov et al., 2007; Frère et al., 2008; Adler, 2009; Iorio, 2010a; Saadat et al., 2010; Zakharov et al., 2010; De Risi et al., 2012; Pitjev and Pitjeva, 2013; Iorio, 2013b).

In many cases, more or less approximate expressions for the anomalous perihelion precession induced by certain spherically symmetric dark matter distributions were used, in particular by considering nearly circular orbits. Here, such a restriction is overcome by calculating *exact* expressions, which can, thus, yield more accurate constraints – with the caveat exposed in Chapter 1 – in view of the increasing level of accuracy in determining the orbits of some of the major bodies of the solar system. Moreover, the present results can be used also with systems characterized by highly eccentric orbits such as, for example, the S stars in the GC.

10.7.1 Exponential Mass Density Profile

By adopting an exponentially decaying mass density profile[5] (Pitjev and Pitjeva, 2013),

$$\rho_{\text{exp}}^{\text{dm}} = \rho_0 \exp\left(-\frac{r}{\lambda_{\text{dm}}}\right), \tag{10.39}$$

[5] The density profile of Equation (10.39) is a particular case of the Einasto profile (Einasto, 1965), often adopted to describe dark matter halos in galaxies (Merritt et al., 2006).

where $\lambda_{\rm dm}$ is a characteristic scale length, the Poisson equation yields the following disturbing function:

$$\mathfrak{R}^{\rm dm}_{\rm exp} = \frac{4\pi G\rho_0}{r}\lambda^3_{\rm dm}\left[2 - \left(2 + \frac{r}{\lambda_{\rm dm}}\right)\exp\left(-\frac{r}{\lambda_{\rm dm}}\right)\right]. \tag{10.40}$$

By averaging Equation (10.40) over one orbital period with the eccentric anomaly E as a fast variable of integration, one gets (Iorio, 2013b)

$$\begin{aligned}\langle\mathfrak{R}\rangle^{\rm dm}_{\rm exp} = {} & \frac{4\pi G\rho_0\lambda^2_{\rm dm}}{a}\left\{2\lambda_{\rm dm} + \exp\left(-\frac{r}{\lambda_{\rm dm}}\right)\left[ae\mathcal{I}_1\left(\frac{ae}{\lambda_{\rm dm}}\right)\right.\right.\\ & \left.\left. - (a + 2\lambda_{\rm dm})\,\mathcal{I}_0\left(\frac{ae}{\lambda_{\rm dm}}\right)\right]\right\}.\end{aligned} \tag{10.41}$$

The resulting nonvanishing orbital precessions, calculated with Equations (2.27)–(2.32), turn out to be (Iorio, 2013b)

$$\begin{aligned}\left\langle\frac{d\omega}{dt}\right\rangle^{\rm dm}_{\rm exp} = {} & \frac{4\pi\sqrt{1-e^2}G\rho_0\lambda_{\rm dm}}{n_{\rm K}ae}\exp\left(-\frac{r}{\lambda_{\rm dm}}\right)\left\{e\mathcal{I}_0\left(\frac{ae}{\lambda_{\rm dm}}\right)\right.\\ & \left. - \left(1 + \frac{2\lambda_{\rm dm}}{a}\right)\mathcal{I}_1\left(\frac{ae}{\lambda_{\rm dm}}\right)\right\},\end{aligned} \tag{10.42}$$

$$\begin{aligned}\left\langle\frac{d\eta}{dt}\right\rangle^{\rm dm}_{\rm exp} = {} & \frac{4\pi G\rho_0\lambda_{\rm dm}}{n_{\rm K}a^3 e}\exp\left(-\frac{r}{\lambda_{\rm dm}}\right)\left\{4e\lambda^2_{\rm dm}\exp\left(\frac{r}{\lambda_{\rm dm}}\right)\right.\\ & - e\left[a^2\left(3+e^2\right) + 4a\lambda + 4\lambda^2\right]\mathcal{I}_0\left(\frac{ae}{\lambda_{\rm dm}}\right)\\ & \left. + a\left[a\left(1+3e^2\right) + 2\lambda\left(1+2e^2\right)\right]\mathcal{I}_1\left(\frac{ae}{\lambda_{\rm dm}}\right)\right\}.\end{aligned} \tag{10.43}$$

10.7.2 Power-Law Mass Density Profile

By adopting a power-law mass density profile

$$\rho^{\rm dm}_{\rm pl} = \rho_0\left(\frac{r}{\lambda_{\rm dm}}\right)^{-\gamma},\ \gamma > 0, \tag{10.44}$$

generally adopted for the galactic halos and in several dark matter–related studies (Merritt et al., 2006; Gillessen et al., 2009), the Poisson equation yields the following disturbing function:

$$\mathfrak{R}^{\rm dm}_{\rm pl} = \frac{4\pi G\rho_0}{(3-\gamma)(2-\gamma)\lambda^{-\gamma}_{\rm dm}}r^{2-\gamma}. \tag{10.45}$$

By averaging Equation (10.45) over one orbital period with the eccentric anomaly E as a fast variable of integration, one gets

$$
\begin{aligned}
\left\langle\mathfrak{R}\right\rangle_{\mathrm{pl}}^{\mathrm{dm}} = & \frac{\pi G a^{2-\gamma}\left(1-e^{2}\right)^{-\gamma}\lambda_{\mathrm{dm}}^{\gamma}\rho_{0}}{\left(-1+\gamma\right)\left(6-5\gamma+\gamma^{2}\right)^{2}}\left[\mathtt{U}_{1}^{\mathrm{dm}}{}_{2}F_{1}\left(-\frac{1}{2},\gamma;1;\frac{2e}{-1+e}\right)\right. \\
& +\mathtt{U}_{2}^{\mathrm{dm}}{}_{2}F_{1}\left(-\frac{1}{2},\gamma;1;\frac{2e}{1+e}\right) \\
& +\mathtt{U}_{3}^{\mathrm{dm}}{}_{2}F_{1}\left(\frac{1}{2},\gamma;1;\frac{2e}{-1+e}\right) \\
& \left.+\mathtt{U}_{4}^{\mathrm{dm}}{}_{2}F_{1}\left(\frac{1}{2},\gamma;1;\frac{2e}{1+e}\right)\right],
\end{aligned}
\tag{10.46}
$$

where ${}_{2}F_{1}\left(\mathfrak{a},\mathfrak{b};\mathfrak{c};\mathfrak{x}\right)$ is the generalized hypergeometric function ${}_{2}F_{1}$ (Olver et al., 2010), and the coefficients $\mathtt{U}_{j}^{\mathrm{dm}}, j = 1,2,3,4$ are explicitly displayed in Equations (I.1)–(I.4) of Appendix I.

The resulting nonvanishing orbital precessions, calculated with Equations (2.27)–(2.32), turn out to be[6] (Iorio, 2013b)

$$
\begin{aligned}
\left\langle\frac{d\omega}{dt}\right\rangle_{\mathrm{pl}}^{\mathrm{dm}} = & \frac{\pi G a^{-\gamma}\left(1-e^{2}\right)^{-1/2-\gamma}\lambda_{\mathrm{dm}}^{\gamma}\rho_{0}}{2e^{2}n_{\mathrm{K}}\left(-3+\gamma\right)^{2}\left(-2+\gamma\right)^{2}\left(-1+\gamma\right)} \\
& \left[\mathtt{P}_{1}^{\mathrm{dm}}{}_{2}F_{1}\left(-\frac{1}{2},\gamma;1;\frac{2e}{-1+e}\right)\right. \\
& +\mathtt{P}_{2}^{\mathrm{dm}}{}_{2}F_{1}\left(-\frac{1}{2},\gamma;1;\frac{2e}{1+e}\right) \\
& +\mathtt{P}_{3}{}_{2}^{\mathrm{dm}}F_{1}\left(\frac{1}{2},\gamma;1;\frac{2e}{-1+e}\right) \\
& \left.+\mathtt{P}_{4}^{\mathrm{dm}}{}_{2}F_{1}\left(\frac{1}{2},\gamma;1;\frac{2e}{1+e}\right)\right],
\end{aligned}
\tag{10.47}
$$

$$
\begin{aligned}
\left\langle\frac{d\eta}{dt}\right\rangle_{\mathrm{pl}}^{\mathrm{dm}} = & \frac{\pi G a^{-\gamma}\left(1-e^{2}\right)^{-\gamma}\lambda_{\mathrm{dm}}^{\gamma}\rho_{0}}{2e^{2}n_{\mathrm{K}}\left(-3+\gamma\right)^{2}\left(-2+\gamma\right)^{2}\left(-1+\gamma\right)} \\
& \left[\mathtt{H}_{1}^{\mathrm{dm}}{}_{2}F_{1}\left(-\frac{1}{2},\gamma;1;\frac{2e}{-1+e}\right)\right. \\
& +\mathtt{H}_{2}^{\mathrm{dm}}{}_{2}F_{1}\left(-\frac{1}{2},\gamma;1;\frac{2e}{1+e}\right) \\
& +\mathtt{H}_{3}^{\mathrm{dm}}{}_{2}F_{1}\left(\frac{1}{2},\gamma;1;\frac{2e}{-1+e}\right) \\
& \left.+\mathtt{H}_{4}^{\mathrm{dm}}{}_{2}F_{1}\left(\frac{1}{2},\gamma;1;\frac{2e}{1+e}\right)\right].
\end{aligned}
\tag{10.48}
$$

[6] The orbital effects of Equation (10.46) have been computed more or less explicitly and at various levels of approximations in Khriplovich (2007), Frère et al. (2008), Saadat et al. (2010), De Risi et al. (2012), and Zakharov et al. (2007, 2010).

The coefficients $\mathtt{P}_j^{\mathrm{dm}}, j = 1,2,3,4$ and $\mathtt{H}_j^{\mathrm{dm}}, j = 1,2,3,4$ are explicitly displayed in Equations (I.5)–(I.12) of Appendix I.

10.8 Lorentz-Violating Gravitomagnetic Acceleration

The Standard Model Extension (SME) is a theoretical framework encompassing generic violations of the Lorentz symmetry for both gravity and electromagnetism (Kostelecký and Potting, 1995; Colladay and Kostelecký, 1997, 1998; Kostelecký, 2004). In general, there are 20 coefficients for Lorentz violation in the gravitational sector; by assuming spontaneous Lorentz-symmetry breaking, the main effects in the weak-field approximation are accounted for by the traceless coefficients $\bar{s}^{\sigma\lambda}, \sigma, \lambda = 0,1,2,3$ (Bailey and Kostelecký, 2006) containing nine independent quantities

According to Bailey (2010), in the weak-field and slow-motion approximation, a test particle moving with velocity $\boldsymbol{v}$ at distance r from a central, static body of mass M experiences a Lorentz-violating gravitomagnetic acceleration

$$\boldsymbol{A}^{\mathrm{Lgvm}} = \frac{\boldsymbol{v}}{c} \times \boldsymbol{B}^{\mathrm{Lgvm}}, \tag{10.49}$$

with

$$\boldsymbol{B}^{\mathrm{Lgvm}} := \frac{2\mu}{cr^3}\left(\boldsymbol{s} \times \boldsymbol{r}\right), \tag{10.50}$$

where

$$\boldsymbol{s} := -\bar{s}^{0j}, j = 1,2,3 \tag{10.51}$$

is a vector made of the off-diagonal SME Lorentz-violating coefficients.

The radial, transverse, and normal components of Equation (10.49) are

$$A_r^{\mathrm{Lgvm}} = -\frac{2n_{\mathrm{K}}\mu\left(1 + e\cos f\right)^3\left(\mathtt{sm}\cos u - \mathtt{sl}\sin u\right)}{ca\left(1 - e^2\right)^{5/2}}, \tag{10.52}$$

$$A_\tau^{\mathrm{Lgvm}} = \frac{2en_{\mathrm{K}}\mu\left(1 + e\cos f\right)^2\sin f\left(\mathtt{shl}\cos u + \mathtt{shm}\sin u\right)}{ca\left(1 - e^2\right)^{5/2}}, \tag{10.53}$$

$$A_h^{\mathrm{Lgvm}} = \frac{2en_{\mathrm{K}}\mu\left(1 + e\cos f\right)^2\sin f\,\mathtt{sh}}{ca\left(1 - e^2\right)^{5/2}}. \tag{10.54}$$

The coefficients sl, sm, sh, shl, and shm entering Equations (10.52)–(10.54) are defined in Equations (D.44)–(D.48) in Appendix D.

Inserting Equations (10.52)–(10.54) in Equations (2.13)–(2.18) yields (Iorio, 2012e)

$$\left\langle \frac{da}{dt} \right\rangle^{\mathrm{Lgvm}} = 0, \tag{10.55}$$

$$\left\langle \frac{de}{dt} \right\rangle^{\mathrm{Lgvm}} = \frac{2\mu\left(-1+e^2+\sqrt{1-e^2}\right)\left(\mathtt{sl}\cos\omega+\mathtt{sm}\sin\omega\right)}{ca^2e^2}, \tag{10.56}$$

$$\left\langle \frac{dI}{dt} \right\rangle^{\mathrm{Lgvm}} = \frac{2\mu\left(\sqrt{1-e^2}-1\right)\mathtt{sh}\sin\omega}{ca^2e\sqrt{1-e^2}}, \tag{10.57}$$

$$\left\langle \frac{d\Omega}{dt} \right\rangle^{\mathrm{Lgvm}} = -\frac{2\mu\left(\sqrt{1-e^2}-1\right)\csc I\,\mathtt{sh}\cos\omega}{ca^2e\sqrt{1-e^2}}, \tag{10.58}$$

$$\begin{aligned}\left\langle \frac{d\omega}{dt} \right\rangle^{\mathrm{Lgvm}} &= \frac{\mu}{ca^2e^3\left(1-e^2\right)}\left\{-2\left(-1+e^2\right)\left(-1+\sqrt{1-e^2}\right)\mathtt{sl}\sin\omega\right.\\ &+\left[-2e^2\left(-1+e^2+\sqrt{1-e^2}\right)\cot I\,\mathtt{sh}\right.\\ &\left.\left.+2\left(-1+e^2\right)\left(-1+\sqrt{1-e^2}\right)\mathtt{sm}\right]\cos\omega\right\},\end{aligned} \tag{10.59}$$

$$\left\langle \frac{d\eta}{dt} \right\rangle^{\mathrm{Lgvm}} = -\frac{2\mu\left(-1+e^2+\sqrt{1-e^2}\right)\left(\mathtt{sm}\cos\omega-\mathtt{sl}\sin\omega\right)}{ca^2e^3}. \tag{10.60}$$

It can be noted that, while Equations (10.56)–(10.58) are well-defined in the limit for $e \rightarrow 0$, it is not so for Equations (10.59)–(10.60).

Appendix A

List of Acronyms and Abbreviations

ano	anomalistic
″	arcsecond
BH	black hole
CCF	cross-correlation function
cpr	cycle-per-revolution
cst	constant
cty	century
d	day
DD	Damour–Deruelle
dec.	declination
dm	dark matter
dra	draconitic
dS	de Sitter
e	equatorial
ECOs	exotic compact objects
EHT	Event Horizon Telescope
EOS	equation of state
EP	equivalence principle
EPM	Ephemeris of Planets and the Moon
EPRV	Extreme Precision Radial Velocity
ESA	European Space Agency
ESO	European Southern Observatory
ESPRESSO	Echelle SPectrograph for Rocky Exoplanets and Stable Spectroscopic Observations
EURO	Elliptical Uranian Relativity Orbiter
exp	exponential
EXPRES	EXtreme PREcision Spectrograph
FWHM	full-width at half-maximum

GC	Galactic Centre
GP-B	Gravity Probe B
GRACE	Gravity Recovery and Climate Experiment
GRACE-FO	GRACE Follow-On
GRAIL	Gravity Recovery and Interior Laboratory
GTR	general theory of relativity
gvm	gravitomagnetic
hr	hour
HERO	Highly Elliptical Relativity Orbiter
IAU	International Astronomical Union
ICRF	International Celestial Reference Frame
IORIO	In-Orbit Relativity Iuppiter Observatory, or IOvis Relativity In–orbit Observatory
JPL	Jet Propulsion Laboratory
K	Keplerian
kg	kilogram
km	kilometre
LAGEOS	LAser GEOdynamic Satellite
LARES	LAser RElativity Satellite
LETSGO	LEnse-Thirring Sun-Geo Orbiter
LIGO	Laser Interferometer Gravitational-wave Observatory
Lgvm	Lorentz-violating gravitomagnetic
LISA	Laser Interferometer Space Antenna
LLR	Lunar Laser Ranging
log	logarithmic
LT	Lense–Thirring
m	metre
MAROON-X	M dwarf Advanced Radial velocity Observer Of Neighbouring eXoplanets
mas	milliarcosecond
MESSENGER	MErcury Surface, Space ENvironment, GEochemistry, and Ranging
MGR	Modified General Relativity
MGS	Mars Global Surveyor
MicroSCOPE	Micro-Satellite à traînée Compensée pour l'Observation du Principe d'Equivalence
μas	microarcsecond
mm	millimetre
MOG	MOdified Gravity
MOND	MOdified Newtonian Dynamics

N	Newtonian
NASA	National Aeronautics and Space Administration
NMoI	normalized moment of inertia
ns	neutron star
p	planet
pK	post-Keplerian
pl	power-law
PLR	Planetary Laser Ranging
pN	post-Newtonian
po	polar
POD	Precise Orbit Determination
PPN	parameterized post-Newtonian
PS	Panhans–Soffel
psr	pulsar
RA	right ascension
s	second
°	sexagesimal degree
SED	spectral energy distribution
Sgr A*	Sagittarius A*
sid	sidereal
SINFONI	SINgle Faint Object Near-IR Investigation
SLR	Satellite Laser Ranging
SMBH	supermassive black hole
SME	Standard Model Extension
STR	Special Theory of Relativity
STVG	Scalar Tensor Vector Gravity
tid	tidal
TOAs	times of arrival
VLT	Very Large Telescope
wd	white dwarf
XMM	X-ray Multi-Mirror
Y	Yukawa
yr	year

Appendix B
Notations and Definitions

Here, some basic notations and definitions used throughout the text are presented (Soffel, 1989; Brumberg, 1991; Bertotti et al., 2003; Kopeikin et al., 2011; Poisson and Will, 2014; Soffel and Han, 2019). Indexes denoted with Greek letters run from 0 to 3, while those dubbed with Latin ones run from 1 to 3.

B.1 Some Constants of Nature and Astronomical Quantities

c	Speed of light in vacuum
G	Newtonian constant of gravitation
Υ	Vernal Equinox at some reference epoch
$M_\odot$	Sun's mass
$\mu_\odot := GM_\odot$	Sun's standard gravitational parameter
$R_{\mathrm{e}}^\odot$	Sun's equatorial radius
$R_{\mathrm{po}}^\odot$	Sun's polar radius
$J_2^\odot$	Sun's dimensionless quadrupole mass moment
$\boldsymbol{J}_\odot$	Sun's angular momentum
$J_\odot$	Magnitude of the Sun's angular momentum
$\hat{\boldsymbol{J}}_\odot$	Sun's spin axis
$\alpha_{J_\odot}$	RA of the Sun's north pole of rotation
$\delta_{J_\odot}$	dec. of the Sun's north pole of rotation
$M_\oplus$	Earth's mass
$\mu_\oplus := GM_\oplus$	Earth's standard gravitational parameter
$R_{\mathrm{e}}^\oplus$	Earth's equatorial radius
$R_{\mathrm{po}}^\oplus$	Earth's polar radius
$J_2^\oplus$	Earth's dimensionless quadrupole mass moment
$\boldsymbol{J}_\oplus$	Earth's angular momentum
$J_\oplus$	Magnitude of the Earth's angular momentum

$\hat{\boldsymbol{J}}_{\oplus}$	Earth's spin axis
$M_{♃}$	Jupiter's mass
$\mu_{♃} := GM_{♃}$	Jupiter's standard gravitational parameter
$R_e^{♃}$	Jupiter's equatorial radius
$R_{\mathrm{po}}^{♃}$	Jupiter's polar radius
$J_2^{♃}$	Jupiter's dimensionless quadrupole mass moment
$\boldsymbol{J}_{♃}$	Jupiter's angular momentum
$J_{♃}$	Magnitude of the Jupiter's angular momentum
$\hat{\boldsymbol{J}}_{♃}$	Jupiter's spin axis
$\alpha_{J_{♃}}$	RA of the Jupiter's north pole of rotation
$\delta_{J_{♃}}$	dec. of the Jupiter's north pole of rotation
$M_{\bullet}$	Black hole's mass
$\mu_{\bullet} := GM_{\bullet}$	Black hole's standard gravitational parameter
$\mathcal{R}_{\bullet} := 2\mu_{\bullet}/c^2$	Black hole's Schwarzschild radius
$\mathbb{M}_{\bullet}^{\ell}$	Mass moment of degree ℓ of a Kerr black hole
$\mathbb{J}_{\bullet}^{\ell}$	Spin moment of degree ℓ of a Kerr black hole
$\boldsymbol{J}_{\bullet}$	Angular momentum of a Kerr black hole
$\zeta_{\bullet}$	Azimuth of the spin axis of a Kerr black hole in the plane of the sky
$i_{\bullet}$	Inclination of the spin axis of a Kerr black hole to the line of sight
$\hat{\boldsymbol{J}}_{\bullet} = \{\sin i_{\bullet} \cos\zeta_{\bullet}, \sin i_{\bullet} \sin\zeta_{\bullet}, \cos i_{\bullet}\}$	Spin axis of a Kerr black hole
χ_g	Dimensionless spin parameter of a Kerr black hole; $\lvert\chi_g\rvert \leq 1$
$J_{\bullet} = \chi_g M_{\bullet}^2 G/c$	Magnitude of the angular momentum of a Kerr black hole
$\mathcal{Q}_2^{\bullet} = -J_{\bullet}^2/\left(c^2 M_{\bullet}\right)$	Quadrupole mass moment of a Kerr black hole
ξ_{ns}	Dimensionless quadrupole parameter of a neutron star

B.2 Parameters of a Mass-Energy Source and Spacetime Variables

n_{d}	Number of dimensions of a Lorentzian manifold
κ_g	Einstein's gravitational constant
$\mathsf{T}^{\sigma\lambda}, \sigma, \lambda = 0, 1, 2, 3$	Stress-energy tensor
$\mathcal{U}$	Exterior gravitational potential of an arbitrary mass distribution
$\mathcal{V}$	Interaction potential energy due to some pK effect
ρ	Density of an arbitrary mass distribution
ρ_0	Scaling parameter of the mass density profile
λ_{dm}	Characteristic length scale of a dark matter density profile

γ	Exponent of dark matter power-law density profile
$\mathsf{g}_{\sigma\lambda},\ \sigma,\lambda = 0,1,2,3$	Spacetime metric tensor
$\mathsf{g}^{\sigma\lambda},\ \sigma,\lambda = 0,1,2,3$	Inverse of the spacetime metric tensor
$\Gamma^{\sigma}{}_{\upsilon\iota},\ \sigma,\upsilon,\iota = 0,1,2,3$	Christoffel symbols of the second kind
λ	Affine parameter of geodesics
τ	Proper time of a moving material object
t	Coordinate time
$x^0 := ct$	Temporal coordinate
$\mathsf{R}^{\epsilon}{}_{\sigma\psi\lambda},\ \epsilon,\sigma,\psi,\lambda = 0,1,2,3$	Riemann curvature tensor of the spacetime
$\mathsf{R}_{\sigma\lambda},\ \sigma,\lambda = 0,1,2,3$	Ricci curvature tensor of the spacetime
R	Trace of the Ricci tensor
$\mathsf{h}_{\sigma\lambda},\ \sigma,\lambda = 0,1,2,3$	Post-Newtonian corrections to the Minkowskian components of the spacetime metric tensor
β_{PPN}	A parameter of the PPN formalism
γ_{PPN}	A parameter of the PPN formalism

B.3 Relevant Mathematical Functions and Notations

$\epsilon_{ijw},\ i,j,w = 1,2,3$	3-dimensional Levi-Civita symbol
$\delta_{ij},\ i,j = 1,2,3$	Kronecker delta
$\mathcal{P}_{\ell}\,(\cdots)$	Legendre polynomial of degree ℓ
$\mathcal{J}_s\,(\ldots)$	Bessel function of the first kind of order s
$\mathcal{I}_s\,(\ldots)$	Modified Bessel function of the first kind of order s
${}_2F_1\,(\mathfrak{a},\mathfrak{b};\mathfrak{c};\mathfrak{x})$	Generalized hypergeometric function
$\boldsymbol{\nabla}_{\boldsymbol{q}} := \{\partial/\partial q_1,\ \partial/\partial q_2,\ \partial/\partial q_3,\ \}$	Gradient with respect to the components of the generic vector $\boldsymbol{q} = \{q1,\ q2,\ q3\}$

B.4 Relevant Physical Parameters of the Central Body

D	Characteristic size
M	Mass
$\mu := GM$	Standard gravitational parameter
$\mathcal{U}$	Newtonian exterior gravitational potential (arbitrarily shaped body)
L	Lagrangian per unit mass
U_{N}	Newtonian exterior gravitational potential (spherically symmetric body)
$\mathfrak{R} := -\left(\mathcal{U} - U_{\mathrm{N}}\right)$	Disturbing function
R_{e}	Equatorial radius

R_{po}	Polar radius
$\varepsilon := \sqrt{1 - \left(\frac{R_{po}}{R_e}\right)^2}$	Ellipticity
$\mathfrak{I}$	Moment of inertia
$\mathfrak{i}$	Normalized moment of inertia (NMoI)
k_2	Love number
$k_2' := k_2/2$	Apsidal constant
ω	Angular speed
P	Rotational period
$q_c := \omega^2 R_e^3/\mu$	Centrifugal quadrupole parameter
M'	Mass of a nearby tide-raising body
a'	Semimajor axis of the relative orbit of M with respect to M'
$q_t := -3\left(R_e/a'\right)^3\left(M'/M\right)$	Tidal quadrupole parameter
$J_2 := (k_2/3)\left[q_c - (q_t/2)\right]$	Zonal harmonic coefficient of degree $\ell = 2$ of a body in hydrostatic equilibrium
$\widetilde{f} : \left[(1 + k_2)/2\right] q_c$	Dimensionless parameter entering the Darwin–Radau approximate relation between NMoI and J_2 for a body in hydrostatic equilibrium
$Q_2 := -J_2 M R_e^2$	Dimensional quadrupole mass moment
ΔU_ℓ	Term of degree $\ell = 2, 3, 4, \ldots$ of the Newtonian gravitational potential of an axisymmetric body
$\boldsymbol{J}$	Angular momentum
J	Magnitude of the angular momentum
α_J	RA of the north pole of rotation
δ_J	Dec. of the north pole of rotation
$\hat{\boldsymbol{J}} = \{\cos\alpha_J \cos\delta_J, \sin\alpha_J \cos\delta_J, \sin\delta_J\}$	Spin unit vector
$R_\star$	Equatorial radius of an exoplanet's host star
$b := a\left(1 - e^2\right)\cos I / R_\star\left(1 + e\sin\omega\right)$	Impact parameter of an exoplanet
$i_\star$	Inclination of the spin axis of an exoplanet's host star to the line of sight
$\zeta_\star$	Azimuthal angle of the spin axis of an exoplanet's host star in the plane of the sky
λ	sky-projected spin-orbit angle of a transiting exoplanet
$v_e^\star$	Equatorial rotational velocity of an exoplanet's host star
$u_\star := v_e^\star \sin i_\star$	Projected rotational velocity of an exoplanet's host star

λ_{Y}	Yukawa potential scale distance
K_{Y}	Yukawa potential dimensional strength intensity
α_{Y}	Yukawa potential dimensionless strength intensity
K	Dimensional strength parameter of power-law extra-potentials
$\mathcal{L}$	Characteristic length scale of a logarithmic potential
W^{oct}	Spin-octupole gravitomagnetic potential
$\boldsymbol{B}^{\mathrm{oct}} := -\nabla W^{\mathrm{oct}}$	Spin octupole gravitomagnetic field
$\boldsymbol{s} := \left\{-\bar{s}^{01}, -\bar{s}^{02}, -\bar{s}^{03}\right\}$	Vector of the off-diagonal SME Lorentz-violating coefficients $-\bar{s}^{0j}, j = 1, 2, 3$
$\boldsymbol{B}^{\mathrm{Lgvm}}$	Lorentz-violating gravitomagnetic field

B.5 Relevant Physical Parameters of a Binary System

M_{A}	Mass of the body A
M_{B}	Mass of the body B
$M_{\mathrm{b}} := M_{\mathrm{A}} + M_{\mathrm{B}}$	Total mass
$x_{\mathrm{A}} := M_{\mathrm{A}}/M_{\mathrm{b}}$	Normalized mass of the body A to the system's total mass
$x_{\mathrm{B}} := M_{\mathrm{B}}/M_{\mathrm{b}} = 1 - x_{\mathrm{A}}$	Normalized mass of the body B to the system's total mass
$\mu_{\mathrm{b}} := GM_{\mathrm{b}}$	Standard gravitational parameter
$M_{\mathrm{red}} := M_{\mathrm{A}}M_{\mathrm{B}}/M_{\mathrm{b}}$	Reduced mass
$\boldsymbol{H}_{\mathrm{b}}$	Orbital angular momentum
$H_{\mathrm{b}} = M_{\mathrm{red}}\sqrt{\mu_{\mathrm{b}}a\left(1-e^2\right)}$	Magnitude of the orbital angular momentum
$\nu := M_{\mathrm{red}}/M_{\mathrm{b}} = M_{\mathrm{A}}M_{\mathrm{B}}/M_{\mathrm{b}}^2$	Symmetric mass ratio ($0 \leq \nu \leq 1/4$)
m_{c}	Mass of the unseen companion of the emitting body in a binary
μ_{c}	Standard gravitational parameter of the unseen companion of the emitting body in a binary
$\gamma_{\mathrm{A,B}} := M_{\mathrm{A\,or\,B}}/M_{\mathrm{b}}$	Ratio of the visible to the total mass
$\boldsymbol{J}_{\mathrm{A}}$	Angular momentum of the body A
$\boldsymbol{J}_{\mathrm{B}}$	Angular momentum of the body B
J_{A}	Magnitude of the angular momentum of the body A
J_{B}	magnitude of the angular momentum of the body B
$\hat{\boldsymbol{J}}_{\mathrm{A}}$	Spin axis unit vector of the body A
$\hat{\boldsymbol{J}}_{\mathrm{B}}$	Spin axis unit vector of the body B
$\boldsymbol{S} := \left(1 + 3M_{\mathrm{B}}/4M_{\mathrm{A}}\right)\boldsymbol{J}_{\mathrm{A}} + \left(1 + 3M_{\mathrm{A}}/4M_{\mathrm{B}}\right)\boldsymbol{J}_{\mathrm{B}}$	Weighted sum of the spins of both bodies

Q_2^{A}	Dimensional quadrupole mass moment of the body A
Q_2^{B}	Dimensional quadrupole mass moment of the body B
J_2^{A}	Dimensionless quadrupole mass moment of the body A
J_2^{B}	Dimensionless quadrupole mass moment of the body B
R_{A}	Equatorial radius of the body A
R_{B}	Equatorial radius of the body B
$\vartheta := R_{\mathrm{p}}/R_{\star}$	Ratio of the radius of an exoplanet to that of its parent star
δt_D	Total transit duration of transiting exoplanets
$\delta t_{\mathrm{i/e}}$	Ingress/egress transit duration of transiting exoplanets
δt_H	Full width at half maximum primary transit duration of transiting exoplanets
t_{cj}	Time of inferior conjunction of transiting exoplanets
$\rho := \sqrt{(\boldsymbol{r}_{\mathrm{A}} - \boldsymbol{r}_{\mathrm{B}}) \cdot (\boldsymbol{r}_{\mathrm{A}} - \boldsymbol{r}_{\mathrm{B}})}$	Two-body range
$\hat{\boldsymbol{\rho}} := (\boldsymbol{r}_{\mathrm{A}} - \boldsymbol{r}_{\mathrm{B}}) / \rho$	Range unit vector
$\Delta\rho = (\boldsymbol{\Delta r}_{\mathrm{A}} - \boldsymbol{\Delta r}_{\mathrm{B}}) \cdot \hat{\boldsymbol{\rho}}$	Two-body range shift
$\dot{\rho} := (\boldsymbol{v}_{\mathrm{A}} - \boldsymbol{v}_{\mathrm{B}}) \cdot \hat{\boldsymbol{\rho}}$	Two-body range rate
$\hat{\boldsymbol{\rho}}_v := \left[(\boldsymbol{v}_{\mathrm{A}} - \boldsymbol{v}_{\mathrm{B}}) - \dot{\rho}\hat{\boldsymbol{\rho}}\right] / \rho$	Unit vector needed to construct the range rate shift
$\Delta\dot{\rho} = (\boldsymbol{\Delta v}_{\mathrm{A}} - \boldsymbol{\Delta v}_{\mathrm{B}}) \cdot \hat{\boldsymbol{\rho}} + (\boldsymbol{\Delta r}_{\mathrm{A}} - \boldsymbol{\Delta r}_{\mathrm{B}}) \cdot \hat{\boldsymbol{\rho}}_{\mathrm{v}}$	Two-body range rate shift
$\widetilde{t}_{\mathrm{psr}}$	Time of arrivals (TOAs) of the radio pulses emitted by a pulsar
$\delta\widetilde{t}_{\mathrm{psr}} = m_{\mathrm{c}} \sin I \sin u / M_{\mathrm{b}} c\,(1 + e\cos f)$	Variation of the TOAs due to the barycentric orbital motion of the pulsar

B.6 Relevant Physical and Orbital Parameters of a Test Particle

$\boldsymbol{A}$	pK Perturbing acceleration
$\boldsymbol{A}_{\mathrm{N}}$	Newtonian inverse-square acceleration
$\alpha\,(t)$	RA
$\delta\,(t)$	Dec.
a	Semimajor axis
$n_{\mathrm{K}} := \sqrt{\mu/a^3}$	Keplerian mean motion
$T_{\mathrm{K}} := 2\pi/n_{\mathrm{K}}$	Keplerian orbital period

T_j^{pK}	pK j-type (anomalistic, draconitic, sidereal) orbital period
$\Delta T_j^{\mathrm{pK}} := T_j^{\mathrm{pK}} - T_{\mathrm{K}}$	pK j-type correction to the Keplerian orbital period
e	eccentricity
$\beta := \left(1 - \sqrt{1-e^2}\right)/e$	Auxiliary quantity depending on e
$p := a\left(1 - e^2\right)$	Semilatus rectum
I	Inclination of the orbital plane to the reference plane $\{x, y\}$
☊	Ascending node
☋	Descending node
Ω	Longitude of the ascending node
ω	Argument of pericentre
$k := e\sin\omega$	First Laplace–Lagrange parameter
$q := e\cos\omega$	Second Laplace–Lagrange parameter
$\varpi := \Omega + \omega$	Longitude of the pericentre
$\widetilde{q} := e\cos\varpi$	Nonsingular equinoctial element
$\widetilde{k} := e\sin\varpi$	Nonsingular equinoctial element
$f\,(t)$	True anomaly
t_0	Arbitrary moment of time chosen as initial instant
f_0	True anomaly at some arbitrary moment of time t_0
$\overline{\Delta\kappa} := \int_{f_0}^{f_0+2\pi}(d\kappa/df)\,df$	Net shift per orbit of the (perturbed) orbital element κf
$w_\kappa := \overline{\Delta\kappa}/2\pi$	Fractional net shift per orbit of the orbital element κ
$\langle d\kappa/dt\rangle := \overline{\Delta\kappa}/T_{\mathrm{K}} = (n_{\mathrm{K}}/2\pi)\int_{f_0}^{f_0+2\pi}(d\kappa/df)\,df$	Average rate of change of the orbital element $\kappa\ (f)$
$\Delta\Theta\,(f) := \arctan\left[\sqrt{\frac{1-e}{1+e}}\tan\left(\frac{f_0}{2}\right)\right] - \arctan\left[\sqrt{\frac{1-e}{1+e}}\tan\left(\frac{f}{2}\right)\right]$	Function of the true anomaly used in some calculation
$u\,(t) := \omega + f\,(t)$	Argument of latitude
$u_0 := \omega + f_0$	Argument of latitude at t_0
$l\,(t) := \varpi + f\,(t)$	True longitude
$l_0 := \varpi + f_0$	True longitude at t_0
$\mathcal{M}\,(t)$	Mean anomaly
η	Mean anomaly at epoch

$\Phi(t) := \int_{t_0}^{t} \Delta n_{\mathrm{K}}(t')\,\mathrm{d}t' = -\frac{3}{2}\frac{n_{\mathrm{K}}}{a}\int_{f_0}^{f} \Delta a(f')\,\frac{dt}{df'}\,\mathrm{d}f'$	One of the two terms entering the instantaneous shift of the mean anomaly
$E(t)$	Eccentric anomaly
$\boldsymbol{r}(t)$	Position vector with respect to the central body
$\boldsymbol{\Delta r}(t)$	Instantaneous shift of the position vector
$r(t)$	Time-dependent distance from the central body
r_0	Constant radius of a circular orbit
$\hat{\boldsymbol{r}} := \boldsymbol{r}/r = \{\cos\Omega\cos u(t) - \cos I\sin\Omega\sin u(t),\ \sin\Omega\cos u(t) + \cos I\cos\Omega\sin u(t),\ \sin I\sin u(t)\}$	Radial unit vector
$r_J := \hat{\boldsymbol{J}}\cdot\hat{\boldsymbol{r}}$	Cosine of the angle between $\hat{\boldsymbol{J}}$ and $\boldsymbol{r}$
$x(t) = r(t)\left[\cos\Omega\cos u(t) - \cos I\sin\Omega\sin u(t)\right]$	Keplerian x coordinate
$y(t) = r(t)\left[\sin\Omega\cos u(t) + \cos I\cos\Omega\sin u(t)\right]$	Keplerian y coordinate
$z(t) = r(t)\left[\sin I\sin u(t)\right]$	Keplerian z coordinate
$\phi(t) := \arctan\left[y(t)/x(t)\right]$	Azimuthal angle in the reference plane $\{x, y\}$
$\varphi(t)$	Azimuthal angle in the equatorial plane of the primary
$\boldsymbol{v}(t)$	Velocity vector
$\boldsymbol{\Delta v}(t)$	Instantaneous shift of the velocity vector
$v(t)$	Magnitude of the velocity vector
$v_{r}, := \boldsymbol{v}\cdot\hat{\boldsymbol{r}}$	Projection of $\boldsymbol{v}$ onto the direction of $\boldsymbol{r}$
$v_J := \boldsymbol{v}\cdot\hat{\boldsymbol{J}}$	Projection of $\boldsymbol{v}$ onto the direction of $\hat{\boldsymbol{J}}$
V	Radial velocity of a spectroscopic binary
V^0	Radial velocity of the barycentre of a spectroscopic binary
$K := m_{\mathrm{c}} n_{\mathrm{K}} a\sin I / M_{\mathrm{b}}\sqrt{1-e^2}$	Semiamplitude of the radial velocity curve of a spectroscopic binary
$\hat{\boldsymbol{l}} := \{\cos\Omega,\ \sin\Omega,\ 0\}$	Unit vector directed along the line of the nodes towards the ascending node
$\hat{\boldsymbol{m}} := \{-\cos I\sin\Omega,\ \cos I\cos\Omega,\ \sin I$	Unit vector directed transversely to the line of the nodes in the orbital plane
$\boldsymbol{h} = \boldsymbol{r}\times\boldsymbol{v}$	Orbital angular momentum per unit mass
$h = n_{\mathrm{K}} a^2\sqrt{1-e^2}$	Magnitude of the orbital angular momentum per unit mass

$\hat{\boldsymbol{h}} := \{\sin I \sin\Omega,\ -\sin I \cos\Omega,\ \cos I\}$	Unit vector of the orbital angular momentum such that $\hat{\boldsymbol{l}} \times \hat{\boldsymbol{m}} = \hat{\boldsymbol{h}}$
$\boldsymbol{C}$	Laplace–Runge–Lenz vector per unit mass
$\hat{\boldsymbol{C}} =$	Laplace–Runge–Lenz unit vector per unit mass
$\hat{\boldsymbol{\tau}} := \hat{\boldsymbol{h}} \times \hat{\boldsymbol{r}} = \{-\sin u\,(t) \cos\Omega - \cos I \sin\Omega \cos u\,(t),\ -\sin\Omega \sin u\,(t) + \cos I \cos\Omega \cos u\,(t),\ \sin I \cos u\,(t)\}$	Transverse unit vector
$A_r := \boldsymbol{A} \cdot \hat{\boldsymbol{r}}$	Radial component of $\boldsymbol{A}$
$A_\tau := \boldsymbol{A} \cdot \hat{\boldsymbol{\tau}}$	Transverse component of $\boldsymbol{A}$
$A_h := \boldsymbol{A} \cdot \hat{\boldsymbol{h}}$	Normal component of $\boldsymbol{A}$
C_r^{1cpr}	Cosine coefficient of the radial component of a 1-cpr acceleration
S_r^{1cpr}	Sine coefficient of the radial component of a 1-cpr acceleration
C_τ^{1cpr}	Cosine coefficient of the transverse component of a 1-cpr acceleration
S_τ^{1cpr}	Sine coefficient of the transverse component of a 1-cpr acceleration
C_h^{1cpr}	Cosine coefficient of the normal component of a 1-cpr acceleration
S_h^{1cpr}	Sine coefficient of the normal component of a 1-cpr acceleration
$\mathcal{R}_r\,(f) := \boldsymbol{\Delta r} \cdot \hat{\boldsymbol{r}}$	Radial component of the instantaneous shift of the position vector
$\mathcal{R}_\tau\,(f) := \boldsymbol{\Delta r} \cdot \hat{\boldsymbol{\tau}}$	Transverse component of the instantaneous shift of the position vector
$\mathcal{R}_h\,(f) := \boldsymbol{\Delta r} \cdot \hat{\boldsymbol{h}}$	Normal component of the instantaneous shift of the position vector
$\mathcal{V}_r\,(f) := \boldsymbol{\Delta v} \cdot \hat{\boldsymbol{r}}$	Radial component of the instantaneous shift of the velocity vector
$\mathcal{V}_\tau\,(f) := \boldsymbol{\Delta v} \cdot \hat{\boldsymbol{\tau}}$	Transverse component of the instantaneous shift of the velocity vector
$\mathcal{V}_h\,(f) := \boldsymbol{\Delta v} \cdot \hat{\boldsymbol{h}}$	Normal component of the instantaneous shift of the velocity vector
n_{DD}	Damour–Deruelle mean motion (Damour and Deruelle, 1985, Equation (3.7))

e_{T} Damour–Deruelle proper time eccentricity (Damour and Deruelle, 1986, p. 272)
e_t A member of the Damour–Deruelle parametrization (Damour and Deruelle, 1985, Equation (3.8b))
e_θ A member of the Damour–Deruelle parametrization (Damour and Deruelle, 1985, Equation (4.13))
e_r A member of the Damour–Deruelle parametrization (Damour and Deruelle, 1985, Equation (6.3b))
$\tilde{\delta}$ A member of the Damour–Deruelle parametrization (Damour and Deruelle, 1986, Equation (20))
da_0 A function of the osculating Keplerian semimajor axis, eccentricity and true anomaly at epoch (Klioner and Kopeikin, 1994, Equation (14))
de_0 A function of the osculating Keplerian semimajor axis, eccentricity and true anomaly at epoch (Klioner and Kopeikin, 1994, Equation (14))
a_R A function of the osculating Keplerian semimajor axis and eccentricity (Klioner and Kopeikin, 1994, Equation (28))
e_R A function of the osculating Keplerian semimajor axis and eccentricity (Klioner and Kopeikin, 1994, Equation (29))
ξ_1 A constant of integration (Kopeikin and Potapov, 1994, Equation (4.5))
ξ_2 A constant of integration (Kopeikin and Potapov, 1994, Equation (4.6))

B.7 Relevant Parameters of a Hierarchical Triple System

B.7.1 Physical and Orbital Parameters of the Third Body

M' Mass
$\boldsymbol{J}'$ Angular momentum
$\boldsymbol{J}' - \left\{\hat{J}'_x, \hat{J}'_y, \hat{J}'_z\right\}$ Unit vector of the angular momentum
r' Distance
a' Semimajor axis
n'_{K} Keplerian mean motion
T'_{K} Orbital period
e' Eccentricity
$\boldsymbol{h}'$ Orbital angular momentum per unit mass
$\hat{\boldsymbol{h}}'$ Unit vector of the orbital angular momentum
I' Inclination
Ω' Longitude of the ascending node
ω' Argument of pericentre
f' True anomaly

Appendix C

pK Disturbing Functions from the Spacetime Metric

In GTR, the disturbing function $\mathfrak{R}$ for a given mass-energy distribution to be used in Equations (2.27)–(2.32) can be obtained from the pK part of the Lagrangian per unit mass L (Brumberg, 1991; Soffel and Brumberg, 1991) in the following way.

For a stationary distribution of mass-energy, the spacetime metric tensor components are, in spatially isotropic or harmonic coordinates to the pN order[1] (Brumberg, 1991),

$$\mathsf{g}_{00} \simeq 1 + \mathsf{h}_{00} = 1 + \frac{2\mathcal{U}}{c^2} + \frac{2\mathcal{U}^2}{c^4} + \mathcal{O}\left(1/c^6\right), \tag{C.1}$$

$$\mathsf{g}_{0i} \simeq \mathsf{h}_{0i} = \mathcal{O}\left(1/c^3\right), \; i = 1, 2, 3, \tag{C.2}$$

$$\mathsf{g}_{ij} \simeq -1 + \mathsf{h}_{ij} = -\left(1 - \frac{2\mathcal{U}}{c^2}\right)\delta_{ij} + \mathcal{O}\left(1/c^4\right), \; i,j = 1, 2, 3, \tag{C.3}$$

where $\mathsf{h}_{\alpha\beta}$, $\alpha, \beta = 0, 1, 2, 3$ are the pN corrections to the Minkowskian values of the spacetime metric tensor, h_{0i}, $i = 1, 2, 3$ are connected to the mass-energy currents of the source, $\mathcal{U}$ is the exterior gravitational potential, and

$$\delta_{ij} := \begin{cases} 1 \text{ for } i = j \\ 0 \text{ for } i \neq j, \end{cases} \quad i,j = 1, 2, 3, \tag{C.4}$$

is the Kronecker delta (Olver et al., 2010). For an isolated, rotating body of mass M and angular momentum J, assumed spherically symmetric, one has

$$\mathcal{U} = U_{\mathrm{N}}(r) = -\frac{\mu}{r}, \tag{C.5}$$

$$\mathsf{h}_{0i} = \frac{2G\epsilon_{ijw}J^j x^w}{c^3 r^3}, \; i = 1, 2, 3, \tag{C.6}$$

[1] The terms $\mathsf{h}_{\mu\nu}$, $\mu, \nu = 0, 1, 2, 3$ are the pN corrections to the Minkowskian components of the spacetime metric tensor.

where

$$\epsilon_{ijw} :== \begin{cases} +1 & \text{if } (i, j, w) \text{ is } (1, 2, 3), (2, 3, 1), \text{ or } (3, 1, 2) \\ -1 & \text{if } (i, j, w) \text{ is } (3, 2, 1), (1, 3, 2), \text{ or } (2, 1, 3) \\ 0 & \text{if } i = j, \text{ or } j = w, \text{ or } w = i \end{cases} \tag{C.7}$$

is the 3–dimensional Levi-Civita symbol (Olver et al., 2010), and x^k, $k = 1, 2, 3$ are the Cartesian coordinates x, y, z; in Equation (C.6), the Einstein summation convention (Olver et al., 2010) is applied to the dummy summation indexes j and w.

To the 1pN level, the Lagrangian per unit mass is given by (Brumberg, 1991, p. 56, Equation (2.2.53))

$$L = L_{\mathrm{N}} + L^{\mathrm{1pN}}, \tag{C.8}$$

where[2]

$$L_{\mathrm{N}} = \frac{1}{2}v^2 - \frac{1}{2}c^2 \mathsf{h}_{00}^{(1/c^2)}, \tag{C.9}$$

$$L^{\mathrm{1pN}} = -\frac{1}{2}c^2 \mathsf{h}_{00}^{(1/c^4)} + \frac{v^4}{8c^2} - \frac{1}{4}\mathsf{h}_{00}v^2 + \frac{c^2}{8}\mathsf{h}_{00}^2 - \frac{1}{2}\mathsf{h}_{ij}v^i v^j - c\mathsf{h}_{0j}v^j, \tag{C.10}$$

where $\mathsf{h}_{00}^{(1/c^2)}$ and $\mathsf{h}_{00}^{(1/c^4)}$ denote the 1pN and 2pN parts of h_{00}, respectively; both of them are needed to keep the Lagrangian to the 1pN level. To this aim, it is intended that only $\mathsf{h}_{00}^{(1/c^2)}$ enters the third and fourth terms of Equation (C.10).

C.1 The 1pN Gravitoelectric Lagrangian per Unit Mass

As an example, from Equation (C.5) and by assuming $\mathsf{h}_{0i} = 0$, $i = 1, 2, 3$, Equation (C.10) yields the 1pN gravitoelectric Lagrangian per unit mass of a static, spherically symmetric body. It turns out to be

$$L^{\mathrm{1pN}} = \frac{r^2 v^4 + 12 r v^2 \mu - 4\mu^2}{8c^2 r^2}. \tag{C.11}$$

Calculated onto the unperturbed Keplerian ellipse with Equations (2.7)–(2.8) and Equation (2.11), Equation (C.11) becomes

$$L^{\mathrm{1pN}}(f) = \frac{\mu^2 \left\{9 + 26e^2 + e^4 + 4e\left[4\left(2 + e^2\right)\cos f + 3e\cos 2f\right]\right\}}{8a^2c^2\left(1 - e^2\right)^2}. \tag{C.12}$$

[2] Here, the velocity components v^i, $i = 1, 2, 3$ are calculated with respect to the coordinate time t (Brumberg, 1991).

Furthermore, it is

$$\boldsymbol{v} \cdot \nabla_{v} L^{\mathrm{1pN}} = \frac{r v^4 + 6 v^2 \mu}{2 c^2 r}, \tag{C.13}$$

which, evaluated onto the Keplerian ellipse, yields

$$\boldsymbol{v} \cdot \nabla_{v} L^{\mathrm{1pN}} (f) = \frac{\mu^2 \left(1 + e^2 + 2 e \cos f\right) \left(7 + e^2 + 8 e \cos f\right)}{2 c^2 a^2 \left(1 - e^2\right)^2}. \tag{C.14}$$

The averages of Equation (C.12) and Equation (C.14) are calculated in Equations (3.19)–(3.20) to obtain Equations (3.13)–(3.17) to obtain Equations (3.13)–(3.18).

C.2 The 1pN Lense–Thirring Lagrangian per Unit Mass

From the last term in Equation (C.10) containing the off-diagonal metric tensor components related to the body's angular momentum, one has

$$L^{\mathrm{LT}} = -\frac{2GJ}{c^2 r^3} \left(\hat{\boldsymbol{J}} \times \boldsymbol{r}\right) \cdot \boldsymbol{v}. \tag{C.15}$$

Equation (C.15), up to the minus sign and the test particle's mass,[3] can be inferred, for example, from Barker and O'Connell (1979, Equation (2.18), p. 155) for the potential energy

$$\mathcal{V}_{S_2} = \frac{G}{c^2 r^3} \left(2 + \frac{3 m_1}{2 m_2}\right) \boldsymbol{S}^{(2)} \cdot (\boldsymbol{r} \times \boldsymbol{P}) \tag{C.16}$$

of two bodies of masses m_2 and m_1, one of which carries the angular momentum $\boldsymbol{S}^{(2)}$, separated by a distance r and moving with relative velocity $\boldsymbol{v}$ in the limit $m_2 \gg m_1$, with the identification $m_2 \to M$, $m_1 \to m$, where m is the test particle's mass. Furthermore, it is assumed that $\boldsymbol{S}^{(2)}$ is the spin angular momentum of the massive body M. Thus, $\boldsymbol{r} \times \boldsymbol{P}$ in Barker and O'Connell (1979, Equation (2.18), p. 155) is the orbital angular momentum $\boldsymbol{r} \times m\boldsymbol{v}$ of the motion of the test particle around M. Calculated onto the unperturbed Keplerian ellipse with Equations (2.7)–(2.8) and Equation (2.11), Equation (C.15) becomes

$$L^{\mathrm{LT}} (f) = \frac{2 n_{\mathrm{K}} G J \left(1 + e \cos f\right)^2}{c^2 a \left(1 - e^2\right)^{5/2}} \left\{ \left[\left(\hat{\boldsymbol{J}} \times \hat{\boldsymbol{m}}\right) \cdot \hat{\boldsymbol{l}} \right] \sin u \left(e \sin \omega + \sin u\right) \right.$$
$$\left. - \left[\left(\hat{\boldsymbol{J}} \times \hat{\boldsymbol{l}}\right) \cdot \hat{\boldsymbol{m}} \right] \cos u \left(e \cos \omega + \cos u\right) \right\}. \tag{C.17}$$

[3] Recall that the disturbing function is the opposite of the potential energy per unit mass of the pK effect of interest.

Note that, since Equation (C.15) is linear in $\boldsymbol{v}$, it is

$$\boldsymbol{v} \cdot \nabla_{\boldsymbol{v}} L^{\mathrm{LT}} = L^{\mathrm{LT}}. \tag{C.18}$$

The average of Equation (C.17) is calculated in Equation (5.36) to obtain Equations (5.30)–(5.35).

C.3 The Newtonian J_2 Lagrangian per Unit Mass

It should be noted that Equation (C.9) is able to provide a pK Lagrangian of Newtonian origin if deviations from Equation (C.5) in the external potential of the source are present. It is just the case when the oblateness J_2 of the central body is taken into account. Indeed, from

$$\mathcal{U} = U_{\mathrm{N}} + \Delta U_2 = -\frac{\mu}{r}\left[1 - \left(\frac{R_{\mathrm{e}}}{r}\right)^2 J_2 \mathcal{P}_2\left(r_J\right)\right] \tag{C.19}$$

and the second term of Equation (C.9), it turns out that the pK Lagrangian due to J_2 is

$$L^{J_2} = -\frac{\mu J_2 R_{\mathrm{e}}^2 \left(-1 + 3 r_J^2\right)}{2 r^3}. \tag{C.20}$$

By evaluating Equation (C.20) onto the Keplerian ellipse, one gets

$$L^{J_2}\left(f\right) = -\frac{\mu J_2 R_{\mathrm{e}}^2 \left(1 + e\cos f\right)^3}{2a^3 \left(1 - e^2\right)^3}\left[-1 + 3\left(\mathtt{Jl}\cos u + \mathtt{Jm}\sin u\right)^2\right], \tag{C.21}$$

where $\mathtt{Jl}$ and $\mathtt{Jm}$ are given by Equations (D.1)–(D.2).

The average of Equation (C.21) is calculated in Equation (7.34) to obtain Equations (7.28)–(7.33).

Appendix D

Spin-Orbit Orientation and Lorentz-Violating Coefficients

Let the cosines of the angles between $\hat{\boldsymbol{J}}$ and the unit vectors $\hat{\boldsymbol{l}}, \hat{\boldsymbol{m}}, \hat{\boldsymbol{h}}$ be defined as follows:

$$\texttt{Jl} := \hat{\boldsymbol{J}} \cdot \hat{\boldsymbol{l}} \tag{D.1}$$

$$\texttt{Jm} := \hat{\boldsymbol{J}} \cdot \hat{\boldsymbol{m}} \tag{D.2}$$

$$\texttt{Jh} := \hat{\boldsymbol{J}} \cdot \hat{\boldsymbol{h}} \tag{D.3}$$

Here, the coefficients $\widehat{T}_j, j = 1, 2, \ldots 6$, characterizing the mutual spin-orbit orientation to the second order in $\hat{\boldsymbol{J}}$, are displayed. They are

$$\widehat{T}_1 := 1, \tag{D.4}$$

$$\widehat{T}_2 := \texttt{Jl}^2 + \texttt{Jm}^2, \tag{D.5}$$

$$\widehat{T}_3 := \texttt{Jl}^2 - \texttt{Jm}^2, \tag{D.6}$$

$$\widehat{T}_4 := \texttt{Jh}\,\texttt{Jl}, \tag{D.7}$$

$$\widehat{T}_5 := \texttt{Jh}\,\texttt{Jm}, \tag{D.8}$$

$$\widehat{T}_6 := \texttt{Jl}\,\texttt{Jm}. \tag{D.9}$$

The introduction of $\widehat{T}_1$ is motivated just by consistency reasons in writing down expressions like, for example, Equations (7.22)–(7.27).

By parameterizing $\hat{\boldsymbol{J}}$ as, for example,

$$\hat{J}_x = \cos\delta_J \cos\alpha_J, \tag{D.10}$$

$$\hat{J}_y = \cos\delta_J \sin\alpha_J, \tag{D.11}$$

$$\hat{J}_z = \sin\delta_J, \tag{D.12}$$

the building blocks of $\widehat{T}_j, j = 1, 2, \ldots 6$, namely Equations (D.1)–(D.3), can be written as

$$\texttt{Jl} = \cos\delta_J \cos\left(\alpha_J - \Omega\right), \tag{D.13}$$

$$\mathtt{Jm} = \sin I \sin \delta_J + \cos I \cos \delta_J \sin(\alpha_J - \Omega), \tag{D.14}$$

$$\mathtt{Jh} = \cos I \sin \delta_J - \cos \delta_J \sin I \sin(\alpha_J - \Omega). \tag{D.15}$$

Furthermore, it is

$$\left(\hat{\boldsymbol{J}} \times \hat{\boldsymbol{m}}\right) \cdot \hat{\boldsymbol{l}} - \left(\hat{\boldsymbol{J}} \times \hat{\boldsymbol{l}}\right) \cdot \hat{\boldsymbol{m}} = 2\cos I \sin \delta_J - 2\cos \delta_J \sin I \sin(\alpha_J - \Omega). \tag{D.16}$$

D.1 Equatorial Orbits: General Spin Orientation

Equatorial orbits are characterized by the common direction of the orbital angular momentum $\boldsymbol{h}$ and the spin angular momentum $\boldsymbol{J}$ of the primary. Furthermore, since both $\hat{\boldsymbol{l}}$ and $\hat{\boldsymbol{m}}$ lie in the equatorial plane, they are orthogonal to $\hat{\boldsymbol{J}}$. Thus,

$$\mathtt{Jl} = 0, \tag{D.17}$$

$$\mathtt{Jm} = 0, \tag{D.18}$$

$$\mathtt{Jh} = \pm 1, \tag{D.19}$$

and Equations (D.4)–(D.9) reduce to

$$\widehat{T}_1 = 1, \tag{D.20}$$

$$\widehat{T}_2 = 0, \tag{D.21}$$

$$\widehat{T}_3 = 0, \tag{D.22}$$

$$\widehat{T}_4 = 0, \tag{D.23}$$

$$\widehat{T}_5 = 0, \tag{D.24}$$

$$\widehat{T}_6 = 0. \tag{D.25}$$

D.2 Polar Orbits: General Spin Orientation

In polar orbits, the spin angular momentum $\boldsymbol{J}$ of the central body, assumed arbitrarily oriented in space, lies somewhere in the orbital plane between $\hat{\boldsymbol{l}}$ and $\hat{\boldsymbol{m}}$, so that it is orthogonal to the orbital angular momentum $\boldsymbol{h}$; this condition is accomplished with $I = 90°$ and $\Omega = \alpha$. According to Equations (D.13)–(D.15), one has

$$\mathtt{Jl} = \cos \delta_J, \tag{D.26}$$

$$\mathtt{Jm} = \sin \delta_J, \tag{D.27}$$

$$\mathtt{Jh} = 0, \tag{D.28}$$

so that Equations (D.4)–(D.9) become

$$\widehat{T}_1 = 1, \tag{D.29}$$

$$\widehat{T}_2 = 1, \tag{D.30}$$

$$\widehat{T}_3 = \cos 2\delta_J, \tag{D.31}$$

$$\widehat{T}_4 = 0, \tag{D.32}$$

$$\widehat{T}_5 = 0, \tag{D.33}$$

$$\widehat{T}_6 = \cos \delta_J \sin \delta_J. \tag{D.34}$$

D.3 Known Spin Orientation

If, as in the case of the Earth, the direction of $\boldsymbol{J}$ is known with sufficient accuracy, it can be assumed as, say, the z direction of an equatorial reference frame. Thus, for a generic orbital configuration referred to the latter and $\delta_J = 90^\circ$, it is

$$\mathtt{Jl} = 0, \tag{D.35}$$

$$\mathtt{Jm} = \sin I, \tag{D.36}$$

$$\mathtt{Jh} = \cos I, \tag{D.37}$$

as per Equations (D.13)–(D.15); Equations (D.4)–(D.9) become

$$\widehat{T}_1 = 1, \tag{D.38}$$

$$\widehat{T}_2 = \sin^2 I, \tag{D.39}$$

$$\widehat{T}_3 = -\sin^2 I, \tag{D.40}$$

$$\widehat{T}_4 = 0, \tag{D.41}$$

$$\widehat{T}_5 = \cos I \sin I, \tag{D.42}$$

$$\widehat{T}_6 = 0. \tag{D.43}$$

D.4 Lorentz-Violating Gravitomagnetic Coefficients

The coefficients entering Equations (10.52)–(10.54) are defined as

$$\mathtt{sl} := \boldsymbol{s} \cdot \hat{\boldsymbol{l}}, \tag{D.44}$$

$$\mathtt{sm} := \boldsymbol{s} \cdot \hat{\boldsymbol{m}}, \tag{D.45}$$

$$\mathtt{sh} := \boldsymbol{s} \cdot \hat{\boldsymbol{h}}, \tag{D.46}$$

$$\mathtt{shl} := \boldsymbol{s} \cdot \left(\hat{\boldsymbol{h}} \times \hat{\boldsymbol{l}}\right), \tag{D.47}$$

$$\mathtt{shm} := \boldsymbol{s} \cdot \left(\hat{\boldsymbol{h}} \times \hat{\boldsymbol{m}}\right). \tag{D.48}$$

Appendix E

Coefficients of the Lense–Thirring Orbital Shifts

The coefficients entering the instantaneous LT shifts of the Keplerian orbital elements of Equations (5.26)–(5.28) are listed here. They are

E.1 The Inclination

$$\begin{aligned}\mathcal{I}_1^{\mathrm{LT}} &= -4\left(f - f_0\right) + 2\left(\sin 2u_0 - \sin 2u\right) + e\left[\sin\left(3f_0 + 2\omega\right)\right.\\&\quad - \sin\left(3f + 2\omega\right) - 4\left(\sin f - \sin f_0\right) + \sin\left(f_0 + 2\omega\right)\\&\quad \left. - \sin\left(f + 2\omega\right)\right], \end{aligned} \tag{E.1}$$

$$\begin{aligned}\mathcal{I}_2^{\mathrm{LT}} &= -2\left(\cos 2u_0 + \cos 2u\right) + e\left\{-3\left[\cos\left(3f_0 + 2\omega\right)\right.\right.\\&\quad \left. - \cos\left(3f + 2\omega\right)\right] + 2\left(\cos f - \cos f_0\right) - \cos\left(f_0 + 2\omega\right)\\&\quad \left. + \cos\left(f + 2\omega\right)\right\}. \end{aligned} \tag{E.2}$$

E.2 The Longitude of the Ascending Node

$$\begin{aligned}\mathcal{N}_1^{\mathrm{LT}} &= -2\left(\cos 2u_0 - \cos 2u\right) + e\left\{-3\left[\cos\left(3f_0 + 2\omega\right)\right.\right.\\&\quad \left. - \cos\left(3f + 2\omega\right)\right] - 2\cos f + 2\cos f_0 - \cos\left(f_0 + 2\omega\right)\\&\quad \left. + \cos\left(f + 2\omega\right)\right\}, \end{aligned} \tag{E.3}$$

$$\begin{aligned}\mathcal{N}_2^{\mathrm{LT}} &= 2\left[-2\left(f - f_0\right) - \sin 2u_0 + \sin 2u\right] + e\left[-\sin\left(3f_0 + 2\omega\right)\right.\\&\quad + \sin\left(3f + 2\omega\right) - 4\left(\sin f - \sin f_0\right) - \sin\left(f_0 + 2\omega\right)\\&\quad \left. + \sin\left(f + 2\omega\right)\right]. \end{aligned} \tag{E.4}$$

E.3 The Argument of Pericentre

$$\begin{aligned}
\mathcal{G}_1^{\mathrm{LT}} &= e\cot I\,\{2\,(\cos 2u_0 - \cos 2u) + e\,[\cos(3f_0 + 2\omega) - \cos(3f + 2\omega) \\
&\quad + 2\,(\cos f - \cos f_0) + \cos(f_0 + 2\omega) - \cos(f + 2\omega)]\}\,, \qquad \text{(E.5)}\\
\mathcal{G}_2^{\mathrm{LT}} &= e\cot I\,\{2\,[2\,(f - f_0) + \sin 2u_0 - \sin 2u] + e\,[\sin(3f_0 + 2\omega) \\
&\quad - \sin(3f + 2\omega) + 4\,(\sin f - \sin f_0) + \sin(f_0 + 2\omega) \\
&\quad - \sin(f + 2\omega)]\}\,, \qquad \text{(E.6)}\\
\mathcal{G}_3^{\mathrm{LT}} &= 4\,\{\sin f - \sin f_0 + e\,[2\,(f - f_0) + e\,(\sin f - \sin f_0)]\}\,. \qquad \text{(E.7)}
\end{aligned}$$

Appendix F

Coefficients of the Newtonian J_2 Orbital Shifts

Here, the coefficients $\mathcal{A}_1^{J_2}, \ldots \mathcal{H}_6^{J_2}$ entering Equations (7.22)–(7.27) of the Newtonian J_2-driven instantaneous variations of the Keplerian orbital elements are explicitly displayed. In the following, $[f \rightarrow f_0]$ denotes the same preceding expression in which f is replaced with f_0.

F.1 The Semimajor Axis

$$\mathcal{A}_1^{J_2} := 4\,e\left[-3\left(4+e^2\right)\cos f + e\left(-6\cos 2f - e\cos 3f\right)\right] - [f \rightarrow f_0], \tag{F.1}$$

$$\mathcal{A}_2^{J_2} := 6\,e\left[3\left(4+e^2\right)\cos f + e\left(6\cos 2f + e\cos 3f\right)\right] - [f \rightarrow f_0], \tag{F.2}$$

$$\mathcal{A}_3^{J_2} := 3\left(e^3\cos(f-2\omega) + 6\,e\left\{\left[2\,e + \left(4+e^2\right)\cos f\right]\cos 2u + e\cos(4f+2\omega)\right\} + e^3\cos(5f+2\omega) - 16\sin f\sin(f+2\omega)\right) - [f \rightarrow f_0], \tag{F.3}$$

$$\mathcal{A}_4^{J_2} := 0, \tag{F.4}$$

$$\mathcal{A}_5^{J_2} := 0, \tag{F.5}$$

$$\mathcal{A}_6^{J_2} := 6\left(16\cos(f+2\omega)\sin f + e\left\{-e^2\sin(f-2\omega) + 6\left[2\,e + \left(4+e^2\right)\cos f\right]\sin 2u + 6\,e\sin(4f+2\omega) + e^2\sin(5f+2\omega)\right\}\right) - [f \rightarrow f_0]. \tag{F.6}$$

F.2 The Eccentricity

$$\mathcal{E}_1^{J_2} := 4\left[3\left(4+e^2\right)\cos f + e\left(6\cos 2f + e\cos 3f\right)\right] - [f \rightarrow f_0], \tag{F.7}$$

$$\mathcal{E}_2^{J_2} := -6\left[3\left(4+e^2\right)\cos f + e\left(6\cos 2f + e\cos 3f\right)\right] - [f \rightarrow f_0], \tag{F.8}$$

$$\begin{aligned}\mathcal{E}_3^{J_2} := & -4\,[3\,\cos(f+2\omega)+7\,\cos(3f+2\omega)]+e\{-e\,[3\,\cos(f-2\omega)\\ & +33\,\cos(f+2\omega)+17\,\cos(3f+2\omega)+3\,\cos(5f+2\omega)]\\ & +36\,\sin 2f\,\sin 2u+120\,\sin f\,\sin(f+2\omega)\}-[f\to f_0]\,,\end{aligned} \tag{F.9}$$

$$\mathcal{E}_4^{J_2} := 0, \tag{F.10}$$

$$\mathcal{E}_5^{J_2} := 0, \tag{F.11}$$

$$\begin{aligned}\mathcal{E}_6^{J_2} := & \,6\,e^2\,\sin(f-2\omega)-8\,[3\,\sin(f+2\omega)+7\,\sin(3f+2\omega)]\\ & -2\,e\,\{24\,[3\,\cos f\,\cos 2u+5\,\cos(f+2\omega)]\,\sin f\\ & +e\,[33\,\sin(f+2\omega)+17\,\sin(3f+2\omega)+3\,\sin(5f+2\omega)]\}\\ & -[f\to f_0]\,.\end{aligned} \tag{F.12}$$

F.3 The Inclination

$$\mathcal{I}_1^{J_2} := 0, \tag{F.13}$$

$$\mathcal{I}_2^{J_2} := 0, \tag{F.14}$$

$$\mathcal{I}_3^{J_2} := 0, \tag{F.15}$$

$$\begin{aligned}\mathcal{I}_4^{J_2} := & \,6f+6\,e\,\sin f+3\,\sin 2u+3\,e\,\sin(f+2\omega)\\ & +e\,\sin(3f+2\omega)-[f\to f_0]\,,\end{aligned} \tag{F.16}$$

$$\begin{aligned}\mathcal{I}_5^{J_2} := & -\{3\,\cos 2u+e\,[3\,\cos(f+2\omega)+\cos(3f+2\omega)]\}\\ & -[f\to f_0]\,,\end{aligned} \tag{F.17}$$

$$\mathcal{I}_6^{J_2} := 0. \tag{F.18}$$

F.4 The Longitude of the Ascending Node

$$\mathcal{N}_1^{J_2} := 0, \tag{F.19}$$

$$\mathcal{N}_2^{J_2} := 0, \tag{F.20}$$

$$\mathcal{N}_3^{J_2} := 0, \tag{F.21}$$

$$\begin{aligned}\mathcal{N}_4^{J_2} := & -\{3\,\cos 2u+e\,[3\,\cos(f+2\omega)+\cos(3f+2\omega)]\}\\ & -[f\to f_0]\,,\end{aligned} \tag{F.22}$$

$$\begin{aligned}\mathcal{N}_5^{J_2} := & \,6f+6\,e\,\sin f-3\,\sin 2u\\ & -e\,[3\,\sin(f+2\omega)+\sin(3f+2\omega)]-[f\to f_0]\,,\end{aligned} \tag{F.23}$$

$$\mathcal{N}_6^{J_2} := 0. \tag{F.24}$$

F.5 The Argument of Pericentre

$$\mathcal{G}_1^{J_2} := 48\,ef + 8\,\left(6 + 5\,e^2 + 6\,e\,\cos f + e^2\,\cos 2f\right)\,\sin f - [f \to f_0]\,, \tag{F.25}$$

$$\mathcal{G}_2^{J_2} := 6\,\left[-12\,ef - 2\,\left(6 + 5\,e^2 + 6\,e\,\cos f + e^2\,\cos 2f\right)\,\sin f\right] - [f \to f_0]\,, \tag{F.26}$$

$$\begin{aligned}\mathcal{G}_3^{J_2} := {} & 4\,[3\,\sin(f+2\omega) - 7\,\sin(3f+2\omega)] \\ & - e\,\{36\,[3\,\cos(f+2\omega) + \cos(3f+2\omega)]\,\sin f \\ & + e\,[3\,\sin(f-2\omega) + 21\,\sin(f+2\omega) \\ & + 11\,\sin(3f+2\omega) + 3\,\sin(5f+2\omega)]\} - [f \to f_0]\,,\end{aligned} \tag{F.27}$$

$$\begin{aligned}\mathcal{G}_4^{J_2} := {} & -8\,e\,\{3\,\cos 2u + e\,[3\,\cos(f+2\omega) + \cos(3f+2\omega)]\}\,\cot I \\ & - [f \to f_0]\,,\end{aligned} \tag{F.28}$$

$$\begin{aligned}\mathcal{G}_5^{J_2} := {} & -8\,e\,\cot I\,\{-6f + 3\,\sin 2u \\ & + e\,[-6\,\sin f + 3\,\sin(f+2\omega) + \sin(3f+2\omega)]\} - [f \to f_0]\,,\end{aligned} \tag{F.29}$$

$$\begin{aligned}\mathcal{G}_6^{J_2} := {} & -6\,e^2\,\cos(f-2\omega) + 6(-4 + 7\,e^2)\,\cos(f+2\omega) \\ & + 56\,\cos(3f+2\omega) + 2\,e\,\{11\,e\,\cos(3f+2\omega) \\ & + 3\,e\,\cos(5f+2\omega) - 36\,\sin f\,[3\,\sin(f+2\omega) \\ & + \sin(3f+2\omega)]\} - [f \to f_0]\,.\end{aligned} \tag{F.30}$$

F.6 The Mean Anomaly at Epoch

$$\begin{aligned}\mathcal{H}_1^{J_2} := {} & -4\left[3\left(4 - 5e^2\right)\sin f - 12ef + e\,(6\sin 2f + e\sin 3f)\right] \\ & - [f \to f_0]\,,\end{aligned} \tag{F.31}$$

$$\begin{aligned}\mathcal{H}_2^{J_2} := {} & 12\left[-6ef + \left(6 - 7e^2 + 6e\cos f + e^2\cos 2f\right)\sin f\right] \\ & - [f \to f_0]\,,\end{aligned} \tag{F.32}$$

$$\begin{aligned}\mathcal{H}_3^{J_2} := {} & -4\,[3\sin(f+2\omega) - 7\sin(3f+2\omega)] \\ & - e\,\{36\sin 2u - 18\sin(4f+2\omega) + e\,[-3\sin(f-2\omega) \\ & + 51\sin(f+2\omega) + 13\sin(3f+2\omega) - 3\sin(5f+2\omega)]\} \\ & - [f \to f_0]\,,\end{aligned} \tag{F.33}$$

$$\mathcal{H}_4^{J_2} := 0, \tag{F.34}$$

$$\mathcal{H}_5^{J_2} := 0, \tag{F.35}$$

$$\begin{aligned}\mathcal{H}_6^{J_2} := {} & 8\,[3\cos(f+2\omega) - 7\cos(3f+2\omega)] \\ & + 2e\,\{36\cos 2u - 18\cos(4f+2\omega) + e\,[3\cos(f-2\omega) \\ & + 51\cos(f+2\omega) + 13\cos(3f+2\omega) - 3\cos(5f+2\omega)]\} \\ & - [f \to f_0]\,.\end{aligned} \tag{F.36}$$

F.7 $\Phi\left(t\right)$

Here, the coefficients $\mathcal{F}_j^{J_2}, j = 1, 2, \ldots 6$ entering Equation (7.48) for the Newtonian J_2-induced contribution to the instantaneous variation of the mean anomaly due to $\Phi\left(f\right)$ are displayed.

$$\begin{aligned}
\mathcal{F}_1^{J_2} &= -2\Delta\Theta\left(f\right)\left(1+e\cos f_0\right)^3 \\
&\quad - \frac{\sqrt{1-e^2}}{2\left(1+e\cos f\right)\left(1+e\cos f_0\right)}\left[-2\left(-1+e^2\right)\left(f-f_0\right)\right. \\
&\quad + e\left(\cos f_0\left[-2\left(-1+e^2\right)\left(f-f_0\right)\right.\right. \\
&\quad \left. + e\left(4+e^2\right)\sin f\right] + \left(4+e^2\right)\left(\sin f - \sin f_0\right) \\
&\quad + \cos f\left\{-2\left(-1+e^2\right)\left(1+e\cos f_0\right)\left(f-f_0+e\sin f\right)\right. \\
&\quad \left. + e\left[-4-e^2+2e\left(-1+e^2\right)\cos f_0\right]\sin f_0\right\} \\
&\quad \left.\left. + e\left(-1+e^2\right)\sin 2f_0\right)\right],
\end{aligned} \tag{F.37}$$

$$\begin{aligned}
\mathcal{F}_2^{J_2} &= 3\Delta\Theta\left(f\right)\left(1+e\cos f_0\right)^3 \\
&\quad - \frac{3\sqrt{1-e^2}}{4\left(1+e\cos f\right)\left(1+e\cos f_0\right)}\left(\left(1+e\cos f_0\right)\left\{2\left(-1+e^2\right)\left(f\right.\right.\right. \\
&\quad \left. - f_0\right)\left(1+e\cos f\right) + e\left[-4-e^2+2e\left(-1+e^2\right)\cos f\right]\sin f\right\} \\
&\quad \left. - e\left(1+e\cos f\right)\left[-4-e^2+2e\left(-1+e^2\right)\cos f_0\right]\sin f_0\right),
\end{aligned} \tag{F.38}$$

$$\begin{aligned}
\mathcal{F}_3^{J_2} &= 3\Delta\Theta\left(f\right)\left(1+e\cos f_0\right)^3\sin 2u_0 \\
&\quad + \frac{\sqrt{1-e^2}}{16\left(1+e\cos f\right)\left(1+e\cos f_0\right)}\left[2\left(1+e\cos f\right)\cos 2\omega\left(-21e\sin f_0\right.\right. \\
&\quad \left. + \left(-1+e^2\right)\left[\left(6+4e^2\right)\sin 2f_0 + e\left(5\sin 3f_0 + e\sin 4f_0\right)\right]\right) \\
&\quad + 2\left(1+e\cos f_0\right)\left(\cos 2\omega\left\{21e\sin f - \left(-1+e^2\right)\left[\left(6+4e^2\right)\sin 2f\right.\right.\right. \\
&\quad \left.\left. + e\left(5\sin 3f + e\sin 4f\right)\right]\right\} - 4\left(-1+e^2\right)\left(1+e\cos f\right)\left(\cos f\right. \\
&\quad \left. - \cos f_0\right)\left[e\left(2+\cos 2f\right) + 3\cos f_0 + \cos f\left(3+2e\cos f_0\right)\right. \\
&\quad \left.\left.\left. + e\cos 2f_0\right]\sin 2\omega\right)\right],
\end{aligned} \tag{F.39}$$

$$\mathcal{F}_4^{J_2} = 0, \tag{F.40}$$

$$\mathcal{F}_5^{J_2} = 0, \tag{F.41}$$

$$\begin{aligned}
\mathcal{F}_6^{J_2} &= 6\Delta\Theta\left(f\right)\left(1+e\cos f_0\right)^3\sin 2u_0 \\
&\quad - \frac{\left(1-e^2\right)^{3/2}\left(4e\cos^3 f + 3\cos 2f - 4e\cos^3 f_0 - 3\cos 2f_0\right)\cos 2\omega}{2}.
\end{aligned} \tag{F.42}$$

Appendix G

Coefficients of the Total Mixed Net Orbital Shifts of the Order of J_2/c^2

Here, the total mixed net shifts per orbit of the order of $\mathcal{O}\left(J_2/c^2\right)$, calculated in Section 8.2.1, are treated. The coefficients $\widetilde{\mathcal{A}}_1^{J_2/c^2}, \ldots \widetilde{\mathcal{H}}_6^{J_2/c^2}$ entering Equations (8.31)–(8.36) are as follows.

G.1 The Semimajor Axis

$$\widetilde{\mathcal{A}}_1^{J_2/c^2} := 0, \tag{G.1}$$

$$\widetilde{\mathcal{A}}_2^{J_2/c^2} := 0, \tag{G.2}$$

$$\begin{aligned}\widetilde{\mathcal{A}}_3^{J_2/c^2} := {} & 8\left(1+e\cos f_0\right)^3 \cos 2\omega \sin 2f_0 + \left\{4e\left(3+e^2\right)\cos f_0\right.\\ & + 4\left(2+3e^2\right)\cos 2f_0 + e\left[3\left(4+e^2\right)\cos 3f_0\right.\\ & \left.\left.+ e\left(12+e^2+6\cos 4f_0 + e\cos 5f_0\right)\right]\right\}\sin 2\omega,\end{aligned} \tag{G.3}$$

$$\widetilde{\mathcal{A}}_4^{J_2/c^2} := 0, \tag{G.4}$$

$$\widetilde{\mathcal{A}}_5^{J_2/c^2} := 0, \tag{G.5}$$

$$\begin{aligned}\widetilde{\mathcal{A}}_6^{J_2/c^2} := {} & -2\left\{4e\left(3+e^2\right)\cos f_0 + 4\left(2+3e^2\right)\cos 2f_0\right.\\ & + e\left[3\left(4+e^2\right)\cos 3f_0 + e\left(12+e^2+6\cos 4f_0\right.\right.\\ & \left.\left.\left. + e\cos 5f_0\right)\right]\right\}\cos 2\omega + 16\left(1+e\cos f_0\right)^3 \sin 2f_0 \sin 2\omega.\end{aligned} \tag{G.6}$$

G.2 The Eccentricity

$$\widetilde{\mathcal{E}}_1^{J_2/c^2} := 0, \tag{G.7}$$

$$\widetilde{\mathcal{E}}_2^{J_2/c^2} := 0, \tag{G.8}$$

$$\begin{aligned}\widetilde{\mathcal{E}}_3^{J_2/c^2} :=& -\{4\left[3\sin\left(f_0+2\omega\right)+7\sin\left(3f_0+2\omega\right)\right]\\&+e\left[-3e\sin\left(f_0-2\omega\right)+\left(20+19e^2\right)\sin 2\omega\right.\\&+60\sin u_0+18\sin\left(4f_0+2\omega\right)\\&+33e\sin\left(f_0+2\omega\right)+17e\sin\left(3f_0+2\omega\right)\\&\left.+3e\sin\left(5f_0+2\omega\right)\right]\},\end{aligned} \tag{G.9}$$

$$\widetilde{\mathcal{E}}_4^{J_2/c^2} := 0, \tag{G.10}$$

$$\widetilde{\mathcal{E}}_5^{J_2/c^2} := 0, \tag{G.11}$$

$$\begin{aligned}\widetilde{\mathcal{E}}_6^{J_2/c^2} :=& 8\left[3\cos\left(f_0+2\omega\right)+7\cos\left(3f_0+2\omega\right)\right]\\&+2e\left[3e\cos\left(f_0-2\omega\right)+\left(20+19e^2\right)\cos 2\omega\right.\\&+60\cos u_0+18\cos\left(4f_0+2\omega\right)\\&+33e\cos\left(f_0+2\omega\right)+17e\cos\left(3f_0+2\omega\right)\\&\left.+3e\cos\left(5f_0+2\omega\right)\right].\end{aligned} \tag{G.12}$$

G.3 The Inclination

$$\widetilde{\mathcal{I}}_1^{J_2/c^2} := 0, \tag{G.13}$$

$$\widetilde{\mathcal{I}}_2^{J_2/c^2} := 0, \tag{G.14}$$

$$\widetilde{\mathcal{I}}_3^{J_2/c^2} := 0, \tag{G.15}$$

$$\begin{aligned}\widetilde{\mathcal{I}}_4^{J_2/c^2} :=& 5e^2+3\cos u_0+e\left[-16\cos f_0+2e\cos 2\omega\right.\\&\left.+3\cos\left(f_0+2\omega\right)+\cos\left(3f_0+2\omega\right)\right],\end{aligned} \tag{G.16}$$

$$\widetilde{\mathcal{I}}_5^{J_2/c^2} := 3\sin u_0+e\left[2e\sin 2\omega+3\sin\left(f_0+2\omega\right)+\sin\left(3f_0+2\omega\right)\right], \tag{G.17}$$

$$\widetilde{\mathcal{I}}_6^{J_2/c^2} := 0. \tag{G.18}$$

G.4 The Longitude of the Ascending Node

$$\widetilde{\mathcal{N}}_1^{J_2/c^2} := 0, \tag{G.19}$$

$$\widetilde{\mathcal{N}}_2^{J_2/c^2} := 0, \tag{G.20}$$

$$\widetilde{\mathcal{N}}_3^{J_2/c^2} := 0, \tag{G.21}$$

$$\widetilde{\mathcal{N}}_4^{J_2/c^2} := 3\sin u_0+e\left[2e\sin 2\omega+3\sin\left(f_0+2\omega\right)+\sin\left(3f_0+2\omega\right)\right], \tag{G.22}$$

$$\begin{aligned}\widetilde{\mathcal{N}}_5^{J_2/c^2} &:= 5e^2 - 3\cos u_0 - e\,[16\cos f_0 + 2e\cos 2\omega \\ &\quad + 3\cos(f_0 + 2\omega) + \cos(3f_0 + 2\omega)]\,,\end{aligned} \tag{G.23}$$

$$\widetilde{\mathcal{N}}_6^{J_2/c^2} := 0. \tag{G.24}$$

G.5 The Argument of Pericentre

$$\widetilde{\mathcal{G}}_1^{J_2/c^2} := -4e\left(44 + 17e^2 - 64e\cos f_0\right)\sin I, \tag{G.25}$$

$$\widetilde{\mathcal{G}}_2^{J_2/c^2} := 6e\left(44 + 17e^2 - 64e\cos f_0\right)\sin I, \tag{G.26}$$

$$\begin{aligned}\widetilde{\mathcal{G}}_3^{J_2/c^2} &:= 2\,\{4\,[-3\cos(f_0 + 2\omega) + 7\cos(3f_0 + 2\omega)] \\ &\quad + e\left[-3e\cos(f_0 - 2\omega) + 2\left(-10 + 9e^2\right)\cos 2\omega + 60\cos u_0\right. \\ &\quad + 18\cos(4f_0 + 2\omega) + 45e\cos(f_0 + 2\omega) \\ &\quad + 19e\cos(3f_0 + 2\omega) + 3e\cos(5f_0 + 2\omega)]\}\sin I,\end{aligned} \tag{G.27}$$

$$\begin{aligned}\widetilde{\mathcal{G}}_4^{J_2/c^2} &:= -16e\cos I\,\{3\sin u_0 + e\,[2e\sin 2\omega + 3\sin(f_0 + 2\omega) \\ &\quad + \sin(3f_0 + 2\omega)]\}\,,\end{aligned} \tag{G.28}$$

$$\begin{aligned}\widetilde{\mathcal{G}}_5^{J_2/c^2} &:= 16e\cos I\left\{-5e^2 + 3\cos u_0 + e\,[16\cos f_0 + 2e\cos 2\omega\right. \\ &\quad + 3\cos(f_0 + 2\omega) + \cos(3f_0 + 2\omega)]\}\,,\end{aligned} \tag{G.29}$$

$$\begin{aligned}\widetilde{\mathcal{G}}_6^{J_2/c^2} &:= 4\sin I\,\{4\,[-3\sin(f_0 + 2\omega) + 7\sin(3f_0 + 2\omega)] \\ &\quad + e\left[3e\sin(f_0 - 2\omega) + 2\left(-10 + 9e^2\right)\sin 2\omega + 60\sin u_0\right. \\ &\quad + 18\sin(4f_0 + 2\omega) + 45e\sin(f_0 + 2\omega) \\ &\quad + 19e\sin(3f_0 + 2\omega) + 3e\sin(5f_0 + 2\omega)]\}\,.\end{aligned} \tag{G.30}$$

G.6 The Mean Anomaly at Epoch

$$\begin{aligned}\widetilde{\mathcal{H}}_1^{J_2/c^2} &:= 4e\left\{88 + 5e^4 - 16\sqrt{1-e^2} - 3e^2\left(21 + 8\sqrt{1-e^2}\right)\right. \\ &\quad - e\left[3e^2\left(7 + 4\sqrt{1-e^2}\right) + 8\left(-17 + 6\sqrt{1-e^2}\right)\right]\cos f_0 \\ &\quad \left. + e^2\left[8\left(5 - 3\sqrt{1-e^2}\right)\cos 2f_0 + e\left(5 - 4\sqrt{1-e^2}\right)\cos 3f_0\right]\right\},\end{aligned} \tag{G.31}$$

$$\begin{aligned}\widetilde{\mathcal{H}}_2^{J_2/c^2} &:= 6e\left\{-88 - 5e^4 + 16\sqrt{1-e^2} + 3e^2\left(21 + 8\sqrt{1-e^2}\right)\right. \\ &\quad + e\left[3e^2\left(7 + 4\sqrt{1-e^2}\right) + 8\left(-17 + 6\sqrt{1-e^2}\right)\right]\cos f_0 \\ &\quad \left. + 4e^2\sqrt{1-e^2}\,(6\cos 2f_0 + e\cos 3f_0) - 5e^2\,(8\cos 2f_0 + e\cos 3f_0)\right\},\end{aligned} \tag{G.32}$$

$$\begin{aligned}
\widetilde{\mathcal{H}}_3^{J_2/c^2} :=\ & 3e^2\left(2-7e^2\right)\cos\left(f_0-2\omega\right) \\
& + 96e\sqrt{1-e^2}\left(1+e\cos f_0\right)^3\cos u_0 \\
& + 8\left[3\cos\left(f_0+2\omega\right)-7\cos\left(3f_0+2\omega\right)\right] \\
& + e\left[-2\left(-20+7e^2+13e^4\right)\cos 2\omega\right. \\
& - 12\left(14+11e^2\right)\cos u_0 - 18\left(2+3e^2\right)\cos\left(4f_0+2\omega\right) \\
& - 3e\left(74+9e^2\right)\cos\left(f_0+2\omega\right) - e\left(138+31e^2\right)\cos\left(3f_0+2\omega\right) \\
& \left.- 3e\left(2+3e^2\right)\cos\left(5f_0+2\omega\right)\right],
\end{aligned} \tag{G.33}$$

$$\widetilde{\mathcal{H}}_4^{J_2/c^2} := 0, \tag{G.34}$$

$$\widetilde{\mathcal{H}}_5^{J_2/c^2} := 0, \tag{G.35}$$

$$\begin{aligned}
\widetilde{\mathcal{H}}_6^{J_2/c^2} :=\ & -2\left[3e^2\left[2+e^2\left(-7+4\sqrt{1-e^2}\right)\right]\sin\left(f_0-2\omega\right)\right. \\
& + 2e\left[-20+13e^4+e^2\left(7-36\sqrt{1-e^2}\right)\right]\sin 2\omega \\
& + 8\left[-3\sin\left(f_0+2\omega\right)+7\sin\left(3f_0+2\omega\right)\right] \\
& + e\left(12\left[14-8\sqrt{1-e^2}+e^2\left(11-12\sqrt{1-e^2}\right)\right]\sin u_0\right. \\
& + 18\left[2+e^2\left(3-4\sqrt{1-e^2}\right)\right]\sin\left(4f_0+2\omega\right) \\
& + e\left\{3\left[74-48\sqrt{1-e^2}+e^2\left(9-12\sqrt{1-e^2}\right)\right]\sin\left(f_0+2\omega\right)\right. \\
& + \left[138-144\sqrt{1-e^2}+e^2\left(31-36\sqrt{1-e^2}\right)\right]\sin\left(3f_0+2\omega\right) \\
& \left.\left.\left.+ 3\left[2+e^2\left(3-4\sqrt{1-e^2}\right)\right]\sin\left(5f_0+2\omega\right)\right\}\right)\right].
\end{aligned} \tag{G.36}$$

Appendix H
The Coefficients of the Tidal Orbital Precessions

H.1 The General Expressions of the Net Orbital Precessions

Here, the coefficients entering Equations (9.8)–(9.12) are explicitly displayed.

H.1.1 The Eccentricity

The coefficient $\mathcal{E}^{\text{tid}}$ entering Equation (9.8) is

$$
\begin{aligned}
\mathcal{E}^{\text{tid}} = & -2 \sin 2\omega \sin \Omega \left(2\mathsf{T}_{21} \cos \Omega + \mathsf{T}_{33} \sin \Omega\right) \cos^2 I \\
& + 2 \{2 \sin I \sin 2\omega \left(\mathsf{T}_{32} \cos \Omega - \mathsf{T}_{31} \sin \Omega\right) \\
& + \cos 2\omega \left[2\mathsf{T}_{21} \cos 2\Omega + \left(2\mathsf{T}_{22} + \mathsf{T}_{33}\right) \sin 2\Omega\right]\} \cos I \\
& + 4 \cos 2\omega \sin I \left(\mathsf{T}_{31} \cos \Omega + \mathsf{T}_{32} \sin \Omega\right) \\
& + \sin 2\omega \left[\mathsf{T}_{22} \left(\cos 2I + 3\right) \cos 2\Omega\right. \\
& \left. + \mathsf{T}_{33} \left(-\cos 2I + \cos 2\Omega + 2\right) - 2\mathsf{T}_{21} \sin 2\Omega\right].
\end{aligned} \tag{H.1}
$$

H.1.2 The Inclination

The coefficient $\mathcal{I}^{\text{tid}}$ entering Equation (9.9) is

$$
\begin{aligned}
\mathcal{I}^{\text{tid}} = & \, 10e^2 \cos 2I \sin 2\omega \left(\mathsf{T}_{32} \cos \Omega - \mathsf{T}_{31} \sin \Omega\right) \\
& - \frac{5}{2} e^2 \sin 2I \sin 2\omega \left[-3\mathsf{T}_{33} + \left(2\mathsf{T}_{22} + \mathsf{T}_{33}\right) \cos 2\Omega - 2\mathsf{T}_{21} \sin 2\Omega\right] \\
& + 2 \cos I \left(5e^2 \cos 2\omega + 3e^2 + 2\right) \left(\mathsf{T}_{31} \cos \Omega + \mathsf{T}_{32} \sin \Omega\right) \\
& - \left(5e^2 \cos 2\omega + 3e^2 + 2\right) \sin I \left[2\mathsf{T}_{21} \cos 2\Omega\right. \\
& \left. + \left(2\mathsf{T}_{22} + \mathsf{T}_{33}\right) \sin 2\Omega\right].
\end{aligned} \tag{H.2}
$$

H.1.3 The Longitude of the Ascending Node

The coefficient $\mathcal{N}^{\rm tid}$ entering Equation (9.10) is

$$\begin{aligned}\mathcal{N}^{\rm tid} &= 20e^2 \sin 2\omega \left\{\cos I \left(\mathsf{T}_{31} \cos \Omega + \mathsf{T}_{32} \sin \Omega\right) \right.\\ &\quad \left. - \sin I \left[\mathsf{T}_{21} \cos 2\Omega + \left(2\mathsf{T}_{22} + \mathsf{T}_{33}\right) \cos \Omega \sin \Omega\right]\right\} \\ &\quad + 4 \cos 2I \left(5e^2 \cos 2\omega - 3e^2 - 2\right) \left(\mathsf{T}_{31} \sin \Omega - \mathsf{T}_{32} \cos \Omega\right) \\ &\quad + \left(5e^2 \cos 2\omega - 3e^2 - 2\right) \sin 2I \left[-3\mathsf{T}_{33}\right. \\ &\quad \left. + \left(2\mathsf{T}_{22} + \mathsf{T}_{33}\right) \cos 2\Omega - 2\mathsf{T}_{21} \sin 2\Omega\right] \end{aligned} \tag{H.3}$$

H.1.4 The Argument of Pericentre

The numerical coefficients $k_i^{\rm tid}$, $i = 1, 2$ entering Equation (9.11) are

$$k_1^{\rm tid} = 1, \tag{H.4}$$

$$k_2^{\rm tid} = \frac{1}{2}, \tag{H.5}$$

while $\mathcal{P}_i^{\rm tid}$, $i = 1, 2$ are

$$\begin{aligned}\mathcal{P}_1^{\rm tid} &= -2\left(-1 + e^2\right) \cos^2 \omega \left[3\mathsf{T}_{33}\right. \\ &\quad + 2\left(2\mathsf{T}_{22} + \mathsf{T}_{33}\right) \cos 2\Omega + \mathsf{T}_{32} \cos \Omega \cot I \\ &\quad \left. - \left(8\mathsf{T}_{21} \cos \Omega + \mathsf{T}_{31} \cot I\right) \sin \Omega\right], \end{aligned} \tag{H.6}$$

$$\begin{aligned}\mathcal{P}_2^{\rm tid} &= \sin^2 \omega \left[3\mathsf{T}_{33} \left(-1 + 6e^2 + 5 \cos 2I\right)\right. \\ &\quad + \left(2\mathsf{T}_{22} + \mathsf{T}_{33}\right) \left(-7 + 2e^2 - 5 \cos 2I\right) \cos 2\Omega \\ &\quad + 4\mathsf{T}_{32} \left(-4 + 4e^2 + 5 \cos 2I\right) \cos \Omega \cot I \\ &\quad - 4\mathsf{T}_{31} \left(-4 + 4e^2 + 5 \cos 2I\right) \cot I \sin \Omega \\ &\quad \left. + 2\mathsf{T}_{21} \left(7 - 2e^2 + 5 \cos 2I\right) \sin 2\Omega\right] - 5\left[2 - 3e^2\right. \\ &\quad \left. + \left(-2 + e^2\right) \cos 2I\right] \sin 2\omega \csc I \left(\mathsf{T}_{31} \cos \Omega + \mathsf{T}_{32} \sin \Omega\right) \\ &\quad + 2 \sin 2\omega \cos I \left\{5\left(-2 + e^2\right) \mathsf{T}_{21} \cos 2\Omega\right. \\ &\quad \left. + \left[\left(-8 + 3e^2\right) \mathsf{T}_{22} - 3\mathsf{T}_{33}\right] \sin 2\Omega\right\} \\ &\quad + \left[-4\left(-1 + e^2\right) \mathsf{T}_{11} + e^2 \mathsf{T}_{33}\right] \cos I \sin 2\omega \sin 2\Omega. \end{aligned} \tag{H.7}$$

H.1.5 The Mean Anomaly at Epoch

The coefficient $\mathcal{H}^{\rm tid}$ entering Equation (9.12) is

$$\begin{aligned}\mathcal{H}^{\rm tid} &= -5\left(1 + e^2\right) \cos 2\omega \left[\left(2\mathsf{T}_{22} + \mathsf{T}_{33}\right) \left(3 + \cos 2I\right) \cos 2\Omega\right. \\ &\quad + 6\mathsf{T}_{33} \sin^2 I + 4 \sin 2I \left(\mathsf{T}_{32} \cos \Omega - \mathsf{T}_{31} \sin \Omega\right) \end{aligned}$$

$$
\begin{aligned}
&-2\mathrm{T}_{21}\left(3+\cos 2I\right)\sin 2\Omega] - \left(7+3e^2\right)\{\mathrm{T}_{33}\left(1+3\cos 2I\right) \\
&+4\sin 2I\left(-\mathrm{T}_{32}\cos\Omega + \mathrm{T}_{31}\sin\Omega\right) \\
&+2\sin^2 I\left[\left(2\mathrm{T}_{22}+\mathrm{T}_{33}\right)\cos 2\Omega - 2\mathrm{T}_{21}\sin 2\Omega\right]\} \\
&+20\left(1+e^2\right)\sin 2\omega\,\{2\sin I\left(\mathrm{T}_{31}\cos\Omega + \mathrm{T}_{32}\sin\Omega\right) \\
&+\cos I\left[2\mathrm{T}_{21}\cos 2\Omega + \left(2\mathrm{T}_{22}+\mathrm{T}_{33}\right)\sin 2\Omega\right]\}\,. \qquad \text{(H.8)}
\end{aligned}
$$

H.2 The Net 1pN Gravitoelectric Tidal Coefficients

The 1pN gravitoelectric tidal coefficients of Equations (9.21)–(9.26), averaged over $T_{\mathrm{K}}^{'}$, turn out to be

$$
\begin{aligned}
\mathcal{T}_{11}^{\mathrm{1pN}} &= 12\cos 2I^{'} + 2\cos 2\omega^{'}\left(1-\cos 2I^{'}\right) \\
&+ \cos 2\Omega^{'}\left[24\sin^2 I^{'} + 2\left(\cos 2I^{'}+3\right)\cos 2\omega^{'}\right] \\
&- 8\cos I^{'}\sin 2\Omega^{'}\sin 2\omega^{'} + 4, \qquad \text{(H.9)} \\
\mathcal{T}_{22}^{\mathrm{1pN}} &= -12\cos 2I^{'} - 2\cos 2\omega^{'}\left(1-\cos 2I^{'}\right) \\
&+ \cos 2\Omega^{'}\left[24\sin^2 I^{'} + 2\left(\cos 2I^{'}+3\right)\cos 2\omega^{'}\right] \\
&- 8\cos I^{'}\sin 2\Omega^{'}\sin 2\omega^{'} - 4, \qquad \text{(H.10)} \\
\mathcal{T}_{33}^{\mathrm{1pN}} &= \sin^2 I^{'}\cos 2\omega^{'} + 3\cos 2I^{'} + 1, \qquad \text{(H.11)} \\
\mathcal{T}_{12}^{\mathrm{1pN}} &= 4\cos I^{'}\cos 2\Omega^{'}\sin 2\omega^{'} \\
&+ \left[12\sin^2 I^{'} + \left(\cos 2I^{'}+3\right)\cos 2\omega^{'}\right]\sin 2\Omega^{'}, \qquad \text{(H.12)} \\
\mathcal{T}_{13}^{\mathrm{1pN}} &= \sin I^{'}\left[\cos\Omega^{'}\sin 2\omega^{'} + \cos I^{'}\left(\cos 2\omega^{'}-6\right)\sin\Omega^{'}\right], \qquad \text{(H.13)} \\
\mathcal{T}_{23}^{\mathrm{1pN}} &= \sin I^{'}\left[\cos I^{'}\left(\cos 2\omega^{'}-6\right)\cos\Omega^{'} - \sin 2\omega^{'}\sin\Omega^{'}\right]. \qquad \text{(H.14)}
\end{aligned}
$$

H.3 The Net Gravitomagnetic Tidal Coefficients

The gravitomagnetic tidal coefficients of Equations (9.28)–(9.33), averaged over $T_{\mathrm{K}}^{'}$, turn out to be

$$
\begin{aligned}
\mathcal{T}_{11}^{\mathrm{gvm}} &= 40e^{'2}\sin 2\omega^{'}\left(2\hat{J}_z^{'}\cos 2I^{'} + 3\hat{J}_x^{'}\sin 2I^{'}\sin\Omega^{'}\right)\sin 2\Omega^{'} \\
&+ 5\cos 2\omega^{'}\left\{12\sin 3I^{'}\left(\hat{J}_x^{'}\sin\Omega^{'} - \hat{J}_y^{'}\cos\Omega^{'}\right)\sin^2\Omega^{'}\right. \\
&\left.+ \sin I^{'}\left[\hat{J}_y^{'}\left(\cos\Omega^{'} + 15\cos 3\Omega^{'}\right) - 3\hat{J}_x^{'}\left(\sin\Omega^{'} + 5\sin 3\Omega^{'}\right)\right]\right\}e^{'2}
\end{aligned}
$$

$$
\begin{aligned}
&- 20\left(3e'^2 + 2\right)\hat{J}_z' \cos 3I' - 4\cos I' \left\{5\hat{J}_z' \cos 2\omega' \left[6\sin^2 I'\right.\right. \\
&+ \left.\left(3\cos 2I' + 1\right)\cos 2\Omega'\right] e'^2 - 10\hat{J}_y' \sin I' \sin 2\omega' \left(\sin\Omega' - 3\sin 3\Omega'\right) e'^2 \\
&+ \left.\left(3e'^2 + 2\right)\hat{J}_z' \left(20\cos 2\Omega' \sin^2 I' + 3\right)\right\} \\
&+ 2\left(3e'^2 + 2\right)\left\{20\sin 3I' \left(\hat{J}_y' \cos\Omega' - \hat{J}_x' \sin\Omega'\right)\sin^2\Omega'\right. \\
&+ \left.\sin I' \left[\hat{J}_y' \left(\cos\Omega' + 15\cos 3\Omega'\right) - 3\hat{J}_x' \left(\sin\Omega' + 5\sin 3\Omega'\right)\right]\right\},
\end{aligned}
\tag{H.15}
$$

$$
\begin{aligned}
\mathcal{T}_{22}^{\mathrm{gvm}} &= 20e'^2\hat{J}_x' \left(\cos\Omega' + 3\cos 3\Omega'\right)\sin 2I' \sin 2\omega' \\
&- 80e'^2\hat{J}_z' \cos 2I' \sin 2\omega' \sin 2\Omega' - 20\left(3e'^2 + 2\right)\hat{J}_z' \cos 3I' \\
&+ 20\left(3e'^2\cos 2\omega' - 6e'^2 - 4\right)\cos^2\Omega' \sin 3I' \left(\hat{J}_x' \sin\Omega' - \hat{J}_y' \cos\Omega'\right) \\
&+ 4e'^2\cos I' \left\{5\hat{J}_z' \cos 2\omega' \left[\left(3\cos 2I' + 1\right)\cos 2\Omega' - 6\sin^2 I'\right]\right. \\
&+ 120e'^2\hat{J}_y' \cos^2\Omega' \sin I' \sin 2\omega' \sin\Omega' \\
&+ \left.\left(3e'^2 + 2\right)\hat{J}_z' \left(20\cos 2\Omega' \sin^2 I' - 3\right)\right\} \\
&+ \left(5e'^2\cos 2\omega' + 6e'^2 + 4\right)\sin I' \left[3\hat{J}_y' \left(\cos\Omega'\right.\right. \\
&- \left.\left.5\cos 3\Omega'\right) - \hat{J}_x' \left(\sin\Omega' + 15\sin 3\Omega'\right)\right],
\end{aligned}
\tag{H.16}
$$

$$
\begin{aligned}
\mathcal{T}_{33}^{\mathrm{gvm}} &= -20e'^2\sin 2I' \sin 2\Omega' \left(\hat{J}_x' \cos\Omega' + \hat{J}_y' \sin\Omega'\right) \\
&+ 5e'^2\cos 2\Omega' \left[12\hat{J}_z' \cos I' \sin^2 I' + \hat{J}_y' \cos\Omega' \left(3\sin 3I' - \sin I'\right)\right. \\
&+ \left.\hat{J}_x' \left(\sin I' - 3\sin 3I'\right)\sin\Omega'\right] + 2\left(3e'^2 + 2\right)\left[3\hat{J}_z' \cos I'\right. \\
&+ \left.5\hat{J}_z' \cos 3I' - \left(\sin I' + 5\sin 3I'\right)\left(\hat{J}_y' \cos\Omega' - \hat{J}_x' \sin\Omega'\right)\right],
\end{aligned}
\tag{H.17}
$$

$$
\begin{aligned}
\mathcal{T}_{12}^{\mathrm{gvm}} &= -80e'^2\hat{J}_z' \cos 2I' \cos 2\Omega' \sin 2\omega' \\
&+ 20e'^2\sin 2I' \sin 2\omega' \left[\hat{J}_y' \left(\cos\Omega' + 3\cos 3\Omega'\right)\right. \\
&+ \left.\hat{J}_x' \left(\sin\Omega' - 3\sin 3\Omega'\right)\right] \\
&- 10\hat{J}_z' \cos 3I' \left(3e'^2\cos 2\omega' - 6e'^2 - 4\right)\sin 2\Omega' \\
&- 10\hat{J}_z' \cos I' \left(5e'^2\cos 2\omega' + 6e'^2 + 4\right)\sin 2\Omega'
\end{aligned}
$$

$$
\begin{aligned}
&+ 10\left(3e'^2 \cos 2\omega' - 6e'^2 - 4\right)\sin 3I'\left(\hat{J}'_y \cos\Omega' - \hat{J}'_x \sin\Omega'\right)\sin 2\Omega' \\
&+ \left(5e'^2 \cos 2\omega' + 6e'^2 + 4\right)\sin I'\left[\hat{J}'_x\left(\cos\Omega' + 15\cos 3\Omega'\right)\right. \\
&\left. - \hat{J}'_y\left(\sin\Omega' - 15\sin 3\Omega'\right)\right],
\end{aligned} \tag{H.18}
$$

$$
\begin{aligned}
\mathcal{T}^{\mathrm{gvm}}_{13} &= -10e'^2\hat{J}'_z \cos 2\omega'\left(\sin I' - 3\sin 3I'\right)\sin\Omega' \\
&+ 40e'^2 \sin 2\omega'\left[\hat{J}'_z \cos\Omega' \sin 2I' + \cos 2I'\left(\hat{J}'_y \cos 2\Omega' - \hat{J}'_x \sin 2\Omega'\right)\right] \\
&- 4\left(3e'^2 + 2\right)\hat{J}'_z\left(\sin I' + 5\sin 3I'\right)\sin\Omega' \\
&+ 5e'^2 \cos 3I'\left[3\cos 2\omega'\left(\hat{J}'_x \cos 2\Omega' + \hat{J}'_y \sin 2\Omega'\right) + 2\left(3e'^2 + 2\right)\hat{J}'_x\right] \\
&+ e'^2 \cos I'\left\{5\cos 2\omega'\left[\hat{J}'_x\left(12\sin^2 I' + 5\cos 2\Omega'\right) + 5\hat{J}'_y \sin 2\Omega'\right]\right. \\
&\left. + 2\left(3e'^2 + 2\right)\left[20\left(\hat{J}'_x \cos 2\Omega' + \hat{J}'_y \sin 2\Omega'\right)\sin^2 I' + 3\hat{J}'_x\right]\right\},
\end{aligned} \tag{H.19}
$$

$$
\begin{aligned}
\mathcal{T}^{\mathrm{gvm}}_{23} &= 10e'^2 \cos 2\omega'\left\{\hat{J}'_y\left[-12\cos I' \sin^2 I' + \left(5\cos I' + 3\cos 3I'\right)\cos 2\Omega'\right]\right. \\
&\left. - 2\hat{J}'_z \cos\Omega'\left(\sin I' - 3\sin 3I'\right) - \hat{J}'_x\left(5\cos I' + 3\cos 3I'\right)\sin 2\Omega'\right\} \\
&- 80e'^2 \sin 2\omega'\left[\hat{J}'_z \sin 2I' \sin\Omega' + \cos 2I'\left(\hat{J}'_x \cos 2\Omega' + \hat{J}'_y \sin 2\Omega'\right)\right] \\
&+ 4\left(3e'^2 + 2\right)\left\{-5\hat{J}'_y \cos 3I' - 2\hat{J}'_z \cos\Omega'\left(\sin I' + 5\sin 3I'\right)\right. \\
&\left. + \cos I'\left[20\sin^2 I'\left(\hat{J}'_y \cos 2\Omega' - \hat{J}'_x \sin 2\Omega'\right) - 3\hat{J}'_y\right]\right\}.
\end{aligned} \tag{H.20}
$$

Appendix I

Coefficients of the Power-Law Mass Density Profile Orbital Precessions

Here, the coefficients of the averaged disturbing function and of the resulting precessions of the pericentre and the mean anomaly at epoch of Section 10.7.2 are explicitly displayed.

I.1 The Disturbing Function

The coefficients $\mathtt{U}_j^{\mathrm{dm}}, j = 1,2,3,4$ of the averaged disturbing function of Equation (10.46) turn out to be

$$\mathtt{U}_1^{\mathrm{dm}} = (-1+e)(1+e)^{\gamma}\left[11+e^2(-2+\gamma)^2+3(-4+\gamma)\gamma\right], \tag{I.1}$$

$$\mathtt{U}_2^{\mathrm{dm}} = -(1+e)(1-e)^{\gamma}\left[11+e^2(-2+\gamma)^2+3(-4+\gamma)\gamma\right], \tag{I.2}$$

$$\begin{aligned}\mathtt{U}_3^{\mathrm{dm}} = {} & (1+e)^{1+\gamma}\left[-1-2e(5+2e)+10\gamma+12e(2+e)\gamma\right. \\ & \left.-9(1+e)^2\gamma^2+2(1+e)^2\gamma^3\right],\end{aligned} \tag{I.3}$$

$$\begin{aligned}\mathtt{U}_4^{\mathrm{dm}} = {} & -(-1+e)(1-e)^{\gamma}\left[-1-2e(-1+\gamma)^2(-5+2\gamma)\right. \\ & \left.+(-2+\gamma)\gamma(-5+2\gamma)+e^2(-2+\gamma)^2(-1+2\gamma)\right].\end{aligned} \tag{I.4}$$

I.2 The Argument of Pericentre

The coefficients $\mathtt{P}_j^{\mathrm{dm}}, j = 1,2,3,4$ of the precession of the pericentre of Equation (10.47) turn out to be

$$\begin{aligned}\mathtt{P}_1^{\mathrm{dm}} = {} & \left(-1+e^2\right)(1+e)^{\gamma}\left\{1+10e+4e^2-2\left[5+6e(2+e)\right]\gamma\right. \\ & +9(1+e)^2\gamma^2-2(1+e)^2\gamma^3+11+5e^2(-2+\gamma)^2 \\ & +2e^3(-3+\gamma)(-2+\gamma)^2+3(-4+\gamma)\gamma \\ & \left.+2e(-1+\gamma)\left[11+3(-4+\gamma)\gamma\right]\right\},\end{aligned} \tag{I.5}$$

$$
\begin{aligned}
\mathrm{P}_2^{\mathrm{dm}} = & -(-1+e)(1+e)\{2(1+e)(1-e)^{\gamma}(-3+\gamma)(-2+\gamma) \\
& \left[-1+e^2(-2+\gamma)+\gamma\right]\},
\end{aligned} \tag{I.6}
$$

$$
\begin{aligned}
\mathrm{P}_3^{\mathrm{dm}} = & -(-1+e)(1+e)(1+e)^{\gamma}\left\{11+e^2(-2+\gamma)^2\right. \\
& +3(-4+\gamma)\gamma+1-(-2+\gamma)\gamma(-5+2\gamma) \\
& -2e^3(-3+\gamma)(-2+\gamma)^2(-1+2\gamma) \\
& -2e(-3+\gamma)(-2+\gamma)(-1+\gamma)(-1+2\gamma) \\
& \left.-e^2(-2+\gamma)\left[-26+\gamma\left(63-42\gamma+8\gamma^2\right)\right]\right\},
\end{aligned} \tag{I.7}
$$

$$
\begin{aligned}
\mathrm{P}_4^{\mathrm{dm}} = & -(-1+e)(1+e)\{2(-1+e)(1-e)^{\gamma}(-3+\gamma)(-2+\gamma) \\
& \left[-1-2e(-1+\gamma)^2+\gamma+e^2(-2+\gamma)(-1+2\gamma)\right]\}.
\end{aligned} \tag{I.8}
$$

I.3 The Mean Anomaly at Epoch

The coefficients $\mathrm{H}_j^{\mathrm{dm}}, j = 1, 2, 3, 4$ of the precession of the mean anomaly at epoch of Equation (10.48) turn out to be

$$
\begin{aligned}
\mathrm{H}_1^{\mathrm{dm}} = & (-1+e)\left[(1+e)^{1+\gamma}\left[-1-2e(5+2e)+10\gamma+12e(2+e)\gamma\right.\right. \\
& \left.-9(1+e)^2\gamma^2+2(1+e)^2\gamma^3\right] \\
& +(1+e)^{\gamma}\left(-11+2e^4(-2+\gamma)^2(-1+\gamma)-3(-4+\gamma)\gamma\right. \\
& -e^3(-2+\gamma)^2(-1+2\gamma)+e\{11+\gamma[-34+3(9-2\gamma)\gamma]\} \\
& \left.\left.+e^2\{-86+\gamma[114+\gamma(-47+6\gamma)]\}\right)\right],
\end{aligned} \tag{I.9}
$$

$$
\begin{aligned}
\mathrm{H}_2^{\mathrm{dm}} = & -2(1+e)(1-e)^{\gamma}(-2+\gamma)\left[3+e^4(-2+\gamma)(-1+\gamma)\right. \\
& \left.-4\gamma+\gamma^2+e^2(-5+2\gamma)(-5+3\gamma)\right],
\end{aligned} \tag{I.10}
$$

$$
\begin{aligned}
\mathrm{H}_3^{\mathrm{dm}} = & (1+e)(1+e)^{\gamma}\left((-1+e)\left[11+e^2(-2+\gamma)^2+3(-4+\gamma)\gamma\right]\right. \\
& -1+e\{13+4e[12+e(13+2e)]\}+10\gamma \\
& -2e\{28+e[95+16e(5+e)]\}\gamma \\
& +(1+e)\{-9+e[74+3e(43+14e)]\}\gamma^2 \\
& \left.-2(1+e)^2[-1+e(16+11e)]\gamma^3+4e(1+e)^3\gamma^4\right),
\end{aligned} \tag{I.11}
$$

$$
\begin{aligned}
\mathrm{H}_4^{\mathrm{dm}} = & -2(-1+e)(1-e)^{\gamma}(-2+\gamma)\left(3-2e(-3+\gamma)(-1+\gamma)^2\right. \\
& -4\gamma+\gamma^2+e^4(-2+\gamma)(-1+\gamma)(-1+2\gamma) \\
& \left.-2e^3(-1+\gamma)^2(-7+3\gamma)+e^2\{-11+\gamma[41+6(-5+\gamma)\gamma]\}\right).
\end{aligned} \tag{I.12}
$$

Appendix J

Numerical Values of Relevant Astronomical Parameters

In Tables J.1–J.4, the numerical values of the physical parameters of interest are reported for some selected bodies of the solar system and for the double pulsar.

Table J.1 *Relevant physical parameters of the Sun (Pijpers, 1998; Seidelmann et al., 2007; Rozelot, 2009; Emilio et al., 2012; Park et al., 2017; Mecheri and Meftah, 2021; Park et al., 2021). RA and dec. of the north pole of rotation are ICRF equatorial coordinates at epoch J2000.0. The value for the ellipticity* $\varepsilon_\odot$ *is calculated from Equation (6.4) with the figures quoted for* $R_\mathrm{e}^\odot$ *and* $R_\mathrm{po}^\odot$.

Parameter	Units	Numerical Value
$\mu_\odot$	$\times 10^{20}\ \mathrm{m^3/s^2}$	1.32712440041279419 (Park et al., 2021)
$J_2^\odot$	$\times 10^{-7}$	2.2 (Park et al., 2017; Mecheri and Meftah, 2021)
$J_\odot$	$\times 10^{41}\ \mathrm{kg\ m^2/s}$	1.90 (Pijpers, 1998)
$\alpha_{J_\odot}$	$^\circ$	286.13 (Seidelmann et al., 2007)
$\delta_{J_\odot}$	$^\circ$	63.87 (Seidelmann et al., 2007)
$R_\mathrm{e}^\odot$	km	696342 (Emilio et al., 2012)
$R_\mathrm{po}^\odot$	km	$R_\mathrm{e}^\odot - 7.370$ (Rozelot, 2009)
$\varepsilon_\odot$	–	0.0046

Table J.2 *Relevant physical parameters of the Earth (Petit and Luzum, 2010; Seidelmann et al., 2007). RA and dec. of the north pole of rotation are ICRF equatorial coordinates at epoch J2000.0. The value of the angular momentum is inferred from that of the angular momentum per unit mass* $J_\oplus/M_\oplus \simeq 9\times 10^8\ \mathrm{m^2/s}$ *reported in Petit and Luzum (2010, p. 156). The value for the ellipticity* $\varepsilon_\oplus$ *is calculated from Equation (6.4) with the figures quoted for* $R_\mathrm{e}^\oplus$ *and* $R_\mathrm{po}^\oplus$.

Parameter	Units	Numerical Value
$\mu_\oplus$	$\times 10^{14}\ \mathrm{m^3/s^2}$	3.986004418 (Petit and Luzum, 2010)
$J_2^\oplus$	$\times 10^{-3}$	1.0826359 (Petit and Luzum, 2010)
$J_\oplus$	$\times 10^{33}\ \mathrm{kg\,m^2/s}$	$\simeq$ 5.85 (Petit and Luzum, 2010)
$\alpha_{J_\oplus}$	°	0.0 (Seidelmann et al., 2007)
$\delta_{J_\oplus}$	°	90.0 (Seidelmann et al., 2007)
$R_\mathrm{e}^\oplus$	km	6378.14 (Seidelmann et al., 2007)
$R_\mathrm{po}^\oplus$	km	6356.75 (Seidelmann et al., 2007)
$\varepsilon_\oplus$	–	0.082

Table J.3 *Relevant physical parameters of Jupiter (Soffel et al., 2003; Seidelmann et al., 2007; Petit and Luzum, 2010; Iess et al., 2018). RA and dec. of the north pole of rotation are ICRF equatorial coordinates at epoch J2000.0. The value for the ellipticity* $\varepsilon_{♃}$ *is calculated from Equation (6.4) with the figures quoted for* $R_\mathrm{e}^{♃}$ *and* $R_\mathrm{po}^{♃}$.

Parameter	Units	Numerical Value
$\mu_{♃}$	$\times 10^{17}\ \mathrm{m^3/s^2}$	1.26713 (Petit and Luzum, 2010)
$J_2^{♃}$	$\times 10^{-6}$	14696.572 (Iess et al., 2018)
$J_{♃}$	$\times 10^{38}\ \mathrm{kg\ m^2/s}$	6.9 (Soffel et al., 2003)
$\alpha_{J_{♃}}$	°	268.057132 (Iess et al., 2018)
$\delta_{J_{♃}}$	°	64.497159 (Iess et al., 2018)
$R_\mathrm{e}^{♃}$	km	71492 (Seidelmann et al., 2007)
$R_\mathrm{po}^{♃}$	km	66854 (Seidelmann et al., 2007)
$\varepsilon_{♃}$	–	0.354

Table J.4 *Relevant physical and orbital parameters of the double pulsar PSR J0737–3039 retrieved from Kramer et al. (2006), Kramer and Wex (2009), Iorio (2021c), and Silva et al. (2021). As far as the moment of inertia of B $\mathfrak{I}_{\mathrm{B}}$ is concerned, it is assumed equal to that of A (Iorio, 2021c). The quadrupole mass moment of B Q_2^{B}, calculated according to Iorio (2021c) and Silva et al. (2021), is several orders of magnitude smaller than Q_2^{A}. The semimajor axis a can be obtained from the third Kepler law.*

Parameter	Units	Numerical Value
M_{A}	$M_{\odot}$	1.3381 (Kramer et al., 2006; Kramer and Wex, 2009)
M_{B}	$M_{\odot}$	1.2489 (Kramer et al., 2006; Kramer and Wex, 2009)
P_{A}	s	0.022 (Kramer et al., 2006; Kramer and Wex, 2009)
P_{B}	s	2.773 (Kramer et al., 2006; Kramer and Wex, 2009)
$\mathfrak{I}_{\mathrm{A}}$	$\times 10^{38}\,\mathrm{kg\,m^2}$	1.6 (Silva et al., 2021)
Q_2^{A}	$\times 10^{34}\,\mathrm{kg\,m^2}$	−4.8 (Iorio, 2021c; Silva et al., 2021)
T_{K}	d	0.10225156248 (Kramer et al., 2006), (Kramer and Wex, 2009)
I	°	88.69 (Kramer et al., 2006)
e	–	0.0877775 (Kramer et al., 2006), (Kramer and Wex, 2009)

Appendix K
Post-Keplerian Orbital Effects: Numerical Values

Here, numerical values of the main pK orbital effects treated in the book are given for a variety of natural and artificial bodies in the solar system and elsewhere. In particular, the orbital precessions of the Sun's planets are listed in Section K.1. The net shifts per orbit of the spacecraft Juno, currently orbiting Jupiter, are collected in Section K.2, while the orbital precessions of the Earth's geodetic satellite LAGEOS are displayed in Section K.3. Section K.4 is devoted to the double pulsar PSR J0737–3039, while the possibility of using triple pulsars to measure the gravitomagnetic field due to the orbital angular momentum of the inner binary as source of a pN gravitomagnetic field is investigated in Section K.5. Various features of motion of the S star S4716 in the GC can be found in Section K.6.

In the following, $''$/cty stands for arcseconds per century, and mas is a shorthand for milliarcseconds, so that mas/cty and mas/yr refer to milliarcseconds per century and milliarcseconds per year, respectively. Finally, μas denotes microarcseconds, so that μas/cty are microarcseconds per century.

K.1 Solar System's Planets

The main pK orbital precessions (1pN gravitoelectric and Lense–Thirring, and Newtonian Sun's oblateness) of the planets of the solar system, referred to the ICRF, are numerically computed and listed in Tables K.1–K.4. The values of the relevant physical parameters of the Sun needed to compute them are retrieved from Table J.1.

K.1.1 The Inclination

While the 1pN gravitoelectric field of the Sun does not affect the inclinations of the orbital planes of the planets of the solar system, its 1pN gravitomagnetic and classical quadrupolar fields displace them by tiny amounts. For Mercury, the LT precession amounts to $\dot{I}^{\mathrm{LT}} = 0.06\,\mathrm{mas/cty} = 60\,\mu\mathrm{as/cty}$, while the Sun's

Table K.1 *Nominal values of the main pK precessions (1pN gravitoelectric, Lense–Thirring and Sun's oblateness) of the inclinations of the planets of the solar system referred to the mean Celestial Equator at J2000.0. The values of the relevant physical parameters of the Sun are retrieved from Table J.1.*

Planet	$\dot{I}^{\mathrm{1pN}}$ (″/cty)	$\dot{I}^{\mathrm{LT}}$ (mas/cty)	$\dot{I}^{J_2}$ (mas/cty)
Mercury	0	0.06	−1.73
Venus	0	0.009	−0.178
Earth	0	0.006	−0.112
Mars	0	0.001	−0.021
Jupiter	0	4×10^{-5}	-3.9×10^{-4}
Saturn	0	5×10^{-6}	-2.6×10^{-5}
Uranus	0	8×10^{-7}	-3.1×10^{-6}
Neptune	0	2×10^{-7}	-6×10^{-7}

Table K.2 *Nominal values of the main pK precessions (1pN gravitoelectric, Lense–Thirring and Sun's oblateness) of the nodes of the planets of the solar system referred to the mean Celestial Equator at J2000.0. The values of the relevant physical parameters of the Sun are retrieved from Table J.1.*

Planet	$\dot{\Omega}^{\mathrm{1pN}}$ (″/cty)	$\dot{\Omega}^{\mathrm{LT}}$ (mas/cty)	$\dot{\Omega}^{J_2}$ (mas/cty)
Mercury	0	0.09	−2.69
Venus	0	−0.008	0.177
Earth	0	−0.004	0.071
Mars	0	-5×10^{-4}	0.008
Jupiter	0	-4×10^{-5}	3×10^{-4}
Saturn	0	-9×10^{-6}	5×10^{-5}
Uranus	0	-6×10^{-7}	2×10^{-6}
Neptune	0	-3×10^{-7}	9×10^{-7}

oblateness affects it to the $\dot{I}^{J_2} = -1.73$ mas/cty level. For Mars, the LT precession is at the $\simeq$ μas/cty level.

K.1.2 The Longitude of the Ascending Node

While the 1pN gravitoelectric field of the Sun does not affect the nodes of the orbital planes of the planets of the solar system, its 1pN gravitomagnetic and

Table K.3 *Nominal values of the main pK precessions (1pN gravitoelectric, Lense–Thirring and Sun's oblateness) of the perihelia of the planets of the solar system referred to the mean Celestial Equator at J2000.0. The values of the relevant physical parameters of the Sun are retrieved from Table J.1.*

Planet	$\dot{\omega}^{1\text{pN}}$ ($''$/cty)	$\dot{\omega}^{\text{LT}}$ (mas/cty)	$\dot{\omega}^{J_2}$ (mas/cty)
Mercury	42.980	-2	30
Venus	8.624	-0.3	2.7
Earth	3.843	-0.1	0.8
Mars	1.350	-0.03	0.20
Jupiter	0.062	-7×10^{-4}	0.0025
Saturn	0.013	-1×10^{-4}	3×10^{-4}
Uranus	0.002	-1×10^{-5}	2×10^{-5}
Neptune	7×10^{-4}	-3×10^{-6}	5×10^{-6}

Table K.4 *Nominal values of the main pK precessions (1pN gravitoelectric, Lense–Thirring and Sun's oblateness) of the mean anomalies at epoch of the planets of the solar system referred to the mean Celestial Equator at J2000.0. The values of the relevant physical parameters of the Sun are retrieved from Table J.1.*

Planet	$\dot{\eta}^{1\text{pN}}$ ($''$/cty)	$\dot{\eta}^{\text{LT}}$ (mas/cty)	$\dot{\eta}^{J_2}$ (mas/cty)
Mercury	-127.984	0	27
Venus	-25.873	0	3
Earth	-11.528	0	0.9
Mars	-4.046	0	0.2
Jupiter	-0.187	0	0.003
Saturn	-0.040	0	3×10^{-4}
Uranus	-0.007	0	3×10^{-5}
Neptune	-0.002	0	6×10^{-6}

classical quadrupolar fields displace them by tiny amounts. For Mercury, the LT precession amounts to $\dot{\Omega}^{\text{LT}} = 0.09\,\text{mas/cty} = 90\,\mu\text{as/cty}$, while the Sun's oblateness affects it with a retrograde precession $\dot{\Omega}^{J_2} = -2.69\,\text{mas/cty}$. For the Earth, the LT precession is at the $\simeq \mu\text{as/cty}$ level.

K.1.3 The Argument of Perihelion

The 1pN gravitoelectric field of the Sun affects the perihelia of the planets of the solar system with relatively large precessions, while its 1pN gravitomagnetic and pK quadrupolar fields displace them by much smaller amounts. For Mercury, the 1pN gravitoelectric precession has the time-honoured value of $\dot{\omega}^{\mathrm{1pN}} = 42.98\,''/\mathrm{cty}$. On the other hand, the Hermean retrograde LT precession amounts to just $\dot{\omega}^{\mathrm{LT}} = -2\,\mathrm{mas/cty}$, while the Sun's oblateness affects it with a prograde rate as large as $\dot{\omega}^{J_2} = 30\,\mathrm{mas/cty}$. For Mars, the retrograde LT precession is at the $30\,\mu\mathrm{as/cty}$ level, while the prograde quadrupolar one is about ten times larger. For the ice giants of the solar system, the 1pN gravitoelectric precessions are at the $\simeq 1 - 10\,\mathrm{mas/cty}$ level, or even smaller, while the LT ones are quite negligible.

About the 2pN perihelion precession of Mercury, from Equation (4.41) it turns out that its value ranges from about -2 to $-15\,\mu$as/cty, depending on f_0.

K.1.4 The Mean Anomaly at Epoch

The 1pN gravitoelectric field of the Sun affects the mean anomalies at epoch of the planets of the solar system with relatively large precessions, while its Newtonian quadrupolar field displaces them by much smaller amounts. The 1pN gravitomagnetic field does not shift them. For Mercury, the retrograde 1pN gravitoelectric precession is as large as $\dot{\eta}^{\mathrm{1pN}} = -127.984\,''/\mathrm{cty}$. On the other hand, the Sun's oblateness affects it with a prograde rate as little as $\dot{\eta}^{J_2} = 27\,\mathrm{mas/cty}$. For Mars, the retrograde 1pN gravitoelectric precession is at the $4\,''/\mathrm{cty}$ level, while the prograde quadrupolar one is as little as $0.2\,\mathrm{mas/cty}$.

About the 2pN precession of the mean anomaly at epoch of Mercury, from Equation (4.42) it turns out that its value ranges from about 0.35 to 0.38 mas/cty, depending on f_0.

K.1.5 The Current Level of Accuracy in Determining the Orbital Precessions

Table 1 of Iorio (2019c) reports an evaluation of the present-day precision level in determining the planetary orbital precessions based on the EPM2017 ephemerides (Pitjeva and Pitjev, 2018). It should be stressed that the quoted figures are just representative of the mere *formal*, *statistical* errors; the *actual* accuracies may be up to one order of magnitude larger.

As far as Mercury is concerned, its inclination and node rates may be determined with a *formal* precision of $\sigma_{\dot{I}} = 0.003\,\mathrm{mas/cty}$ and $\sigma_{\dot{\Omega}} = 0.024\,\mathrm{mas/cty}$, respectively (Iorio, 2019c). Instead, the *formal* uncertainty in the precession of its longitude of perihelion ϖ is of the order of $\sigma_{\dot{\varpi}} = 0.008\,\mathrm{mas/cty}$ (Iorio, 2019c). Similar figures hold for Mars as well (Iorio, 2019c).

K.2 The Spacecraft Juno around Jupiter

Here, the net shifts per orbit due to some pK effects on the orbit of the Juno spacecraft currently orbiting Jupiter are calculated with respect to a planetocentric reference frame aligned with the ICRF. The timeframe considered ranges from

February 2024, when the orbital period of the probe was reduced to 32 days, to the scheduled end of the second extended mission in September 2025, covering the last 18 orbits.[1] The values of the orbital parameters of Juno are retrieved from the WEB interface Horizons System at https://ssd.jpl.nasa.gov/horizons/, maintained by the Jet Propulsion Laboratory (JPL) of the National Aeronautics and Space Administration (NASA), and accessed 7 March 2024. The relevant physical parameters of Jupiter needed to compute Juno's orbital shifts are taken from Table J.3. The pK features of motion considered are the classical ones due to the Jovian quadrupole mass moment, and the 1pN gravitoelectromagnetic effects induced by the mass monopole and quadrupole and the spin dipole and octupole moments of Jupiter. The Juno's node and perijove describe a full cycle in 132.3 yr and −37.05 yr, respectively, mainly due to the Jovian J_2. Thus, they are taken as fixed in calculating the shifts in Table K.5 since they can be considered as approximately constant during the year and a half or so taken into consideration.

From Table K.5, it turns out that, apart from the classical shifts due to J_2, the largest effects are due to the 1pN gravitoelectric monopole field of Jupiter, amounting to about 38 mas per revolution. The LT shifts are at the $\simeq$ mas level, while the other pN features of motion affect the Juno's orbital elements with signatures smaller by

Table K.5 *PK net shifts per orbit of the spacecraft Juno currently orbiting Jupiter calculated with respect to the ICRF. The adopted values of the probe's orbital parameters, retrieved from the WEB interface Horizons System at https://ssd.jpl.nasa.gov/horizons/, maintained by the Jet Propulsion Laboratory (JPL) of the National Aeronautics and Space Administration (NASA), and accessed 7 March 2024, refer to the last scheduled 18 orbits from February 2024 to September 2025, as per www.missionjuno.swri.edu/mission-perijoves. The relevant physical parameters of Jupiter are taken from Table J.3. The figures listed for the J_2/c^2 effects refer to the direct ones, as per Equations (8.7)–(8.12).*

Orbital element	J_2	1pN	LT	J_2/c^2	$J\varepsilon^2$
$\overline{\Delta a}$ (m)	0	0	0	−91.4	0
$\overline{\Delta e}$ (−)	0	0	0	-8×10^{-10}	-1.6×10^{-12}
$\overline{\Delta I}$ (mas)	3.85×10^5	0	0.89	−0.007	−0.021
$\overline{\Delta \Omega}$ (mas)	8.56×10^5	0	1.98	−0.027	−0.045
$\overline{\Delta \omega}$ (mas)	-3.06×10^6	37.8	1.07	0.018	−0.054
$\overline{\Delta \eta}$ (mas)	-7.3×10^5	−38.4	0	0.098	1×10^{-4}

[1] See www.missionjuno.swri.edu/mission-perijoves, accessed March 2024.

about two orders of magnitude. It is worth noticing that the semimajor axis of the probe is shifted by the 1pN oblateness field of Jupiter by almost 100 m. The total shifts nominally occurring until the scheduled end of the mission can be approximately[2] obtained by rescaling the figures in Table K.5 by 18, which is the number of full orbits to be completed with the current orbital configuration.

It should be noted that, since its inception in orbit in August 2016 to June 2021, Juno completed 33 orbits in a different orbital configuration corresponding to an orbital period of $T_{\rm K} = 53$ d; the relative net shifts per orbit are essentially equal to those quoted in Table K.5.

K.3 The Earth's Geodetic Satellite LAGEOS

The LAGEOS satellite (Cohen and Smith, 1985), launched in 1976, is one of the Earth's geodetic probes (Pearlman et al., 2019) currently tracked with the SLR technique used in tests of pN gravity (Cugusi and Proverbio, 1978; Ashby and Bertotti, 1984; Martin et al., 1985; Lucchesi et al., 2015, 2019a). It is a completely passive and dense object of spherical shape, entirely covered with retroreflectors (Lucchesi, 2004) which bounce back the laser pulses sent to it from SLR ground stations. Such features make relatively easy and accurate modelling the several non-gravitational perturbing accelerations (Milani et al., 1987) affecting its orbital motion (Farinella et al., 1996; Métris et al., 1997; Rubincam et al., 1997; Vokrouhlický and Farinella, 1997; Métris et al., 1999; Lucchesi, 2001, 2002, 2004; Andrés et al., 2006; Visco and Lucchesi, 2018).

Table K.6 displays the values of the rates of change of the same pK effects dealt with in Section K.2 calculated with respect to the ICRF. For LAGEOS, the orbital period is as short as $T_{\rm K} = 0.156\,\mathrm{d} = 3.7\,\mathrm{hr}$, while its node and perigee[3] describe a full cycle in just 2.8 yr and −4.7 yr, respectively. As far as the J_2/c^2 and $J\varepsilon^2$ effects are concerned, they include also harmonic signatures varying with twice the frequency of the perigee. Their amplitudes are not included in Table K.6 since their time series average out over just half the perigee period.

From Table K.6, it turns out that, apart from the classical precessions due to J_2, the largest effects are due to the Earth's 1pN gravitoelectric monopole field, amounting to several thousands of mas/yr. The LT precessions are at the $\simeq 30$ mas/yr level, while the other pN features of motion affect the orbital elements of LAGEOS with signatures smaller by about two to four orders of magnitude.

About the currently ongoing tests of the LT effect in the Earth's field with the geodetic satellites of the LAGEOS family, that is, LAGEOS itself, LAGEOS–2 (Ibba et al., 1989), LARES (Paolozzi et al., 2011), and LARES–2 (Paolozzi et al.,

[2] It is assumed that the perijove does not change too much due to J_2 over such a time span.
[3] From περί (+ accusative), meaning 'around, near, about, from', and γῆ, -ῆς, ἡ, 'earth'.

Table K.6 *PK orbital precessions of the Earth's geodetic satellite LAGEOS, currently tracked with the SLR technique, calculated with respect to the ICRF. The relevant physical parameters of the Earth are taken from Table J.2, while the orbital ones of LAGEOS are taken from Ciufolini et al. (2023). Only the secular parts of the rates of change of the J_2/c^2 (the direct ones as per Equations (8.7)–(8.12)) and $J\varepsilon^2$ effects are displayed because the harmonic ones, varying with a frequency double that of the perigee, average out over just a couple of years.*

Orbital element	J_2	1pN	LT	J_2/c^2	$J\varepsilon^2$
$\langle\dot{a}\rangle$ (m/yr)	0	0	0	–	0
$\langle\dot{e}\rangle$ (1/yr)	0	0	0	–	–
$\langle\dot{I}\rangle$ (mas/yr)	0	0	0	–	–
$\langle\dot{\Omega}\rangle$ (mas/yr)	4.50×10^{8}	0	30.66	−0.48	−0.007
$\langle\dot{\omega}\rangle$ (mas/yr)	-2.81×10^{8}	3278.76	31.22	0.14	-1×10^{-9}
$\langle\dot{\eta}\rangle$ (mas/yr)	-4.34×10^{8}	−9836.25	0	1.04	0.014

2019), mainly performed by Ciufolini and coworkers, the following remarks are in order (Iorio, 2023b).

For unknown reasons, the Earth's gravitomagnetic field has never been modeled so far, and no dedicated parameter(s) have ever been estimated, producing just time series of post-fit residuals of the satellites' nodes,[4] allegedly accounting in full for the unmodeled dynamics which includes the LT acceleration as well. Another puzzling issue is that there are several SLR stations scattered around the globe (Pearlman et al., 2002) where skilled teams of space geodesists routinely process laser ranging data from so many geodetic satellites with several dedicated softwares (Ebauer, 2017); yet, despite this, no one has ever tried to (correctly) perform LT tests independently of Ciufolini, or, if anyone has done so, they have not made their results public in the peer-reviewed literature. There are just some conference proceedings (Ries et al., 2003a,b; Ries et al., 2009) whose authors did *not* model and estimate the LT acceleration either. The same holds also for a few independent studies recently published in peer-reviewed journals by former coworkers of Ciufolini (Lucchesi et al., 2019b, 2020). In principle, there should be nothing easier for so many competent and expert people worldwide than adding one more acceleration in the data reduction softwares and estimating one more parameter.

[4] Some linear combinations of the time series of such node "residuals", suitably designed to cancel out the impact of the mismodelling of the first even zonal harmonics of the geopotential, have been used so far. For recent overviews, see, for example, Iorio et al. (2011), Ciufolini et al. (2013), Renzetti (2013b), and references therein.

Indeed, as explained in Chapter 1, the common practice in satellite geodesy, astrodynamics and astronomy is that, if one wants to put to the test a certain dynamical feature X they are interested in, they must do nothing more than explicitly model it along with the rest of the known dynamics and other pieces of the measurement process, and simultaneously estimate one or more parameters, characterizing it along with many other ones, taking into account other accelerations and so on, and inspecting the resulting covariance matrix to look at their mutual correlations. Looking at some sort of 'spurious' residuals constructed without including X in the models fit to a given set of observations is not a correct procedure since a possible signature with almost the same features of the expected one may be due just to some fortunate partial mutual cancellation of other effects having nothing to do with X itself. Furthermore, X may partly or totally be absorbed in the estimated values of other parameters solved for in the data reduction. In other words, the gravitomagnetic field of the Earth should be simultaneously estimated along with all the other coefficients characterizing the geopotential by using the same data sets which may be varied from time to time by their extension, starting date, and type of observations.

K.4 The Double Pulsar

The relevant pK orbital precessions of the double pulsar PSR J0737–3039, whose physical and orbital parameters can be found in Table J.4, are listed in Table K.7.

Extracting the LT contribution from the experimentally measured total periastron precession, a possibility first envisaged by Lyne et al. (2004), Lattimer and Schutz (2005), and Kramer et al. (2006), would allow one to get important insights on the equation of state (EOS) of the dense matter inside neutron stars. Indeed, by assuming the validity of GTR, the knowledge of the gravitomagnetic rate of change could be used to constrain the EOS through the determination of the moment of inertia $\mathfrak{I}_{\mathrm{A}}$ of PSR J0737–3039A. Conversely, if the former could be independently determined by other means, a test of the LT effect could be performed to some level of accuracy that Hu et al. (2020) quantified at $\simeq$ 7–11%. In principle, the quadrupole-induced orbital precessions may also serve the same purpose, but they are completely negligible (Hu et al., 2020; Iorio, 2021c).

From Table K.7, it turns out that the 1pN gravitoelectric two-body precessions are the largest ones, amounting to tens of degrees per year. The 2pN and the LT rates are about five orders of magnitude smaller; furthermore, the LT and the 2pN periastron precessions have almost the same magnitude. In order to extract $\langle\dot{\omega}\rangle^{\mathrm{LT}}$ from the total periastron precession, the 1pN and 2pN gravitoelectric precessions must be known with sufficiently high accuracy to be reliably subtracted; see Iorio (2009b), and the general discussion by Damour and Schäfer (1988) before the

Table K.7 *Relevant pK orbital precessions of the double pulsar PSR J0737–3039, whose physical and orbital parameters are retrieved from Table J.4. The column for the 2pN precessions, which depend on the true anomaly at epoch f_0 according to Equations (4.39)–(4.40), show their ranges of values for f_0 spanning from 0 to 360°. The LT precessions of the inclination and the node are completely negligible, amounting to a maximum of $\simeq 10^{-5}$ °/yr (Iorio, 2021c). The effects due to the quadrupole mass moment are even smaller, being of the order of $\simeq 10^{-8}$ °/yr (Iorio, 2021c).*

Orbital element	1pN	2pN	LT
$\langle\dot{\omega}\rangle$ (°/yr)	16.89	[−0.00079, −0.00045]	−0.0006
$\langle\dot{\eta}\rangle$ (°/yr)	−47.78	[0.076, 0.077]	0

discovery of the double pulsar. For a recent review on measuring the LT effect with binary pulsars, see Hu and Freire (2024), and references therein.

K.5 Triple Pulsars

PSR J0337+1715 (Ransom et al., 2014) is a hierarchical triple system made of one neutron star and two white dwarfs: an inner, tight binary $\mathcal{S}$ composed of a 2.7 ms pulsar A and a white dwarf B revolving one around each other in a circular relative orbit with orbital period $T_K = 1.6$ d, and another white dwarf C moving about $\mathcal{S}$ along a wider circular path with orbital period $T_K^{'} = 327$ d and coplanar with that of $\mathcal{S}$ itself, both inclined by $I = I^{'} = 39.2°$ to the plane of the sky, assumed as reference $\{x, y\}$ plane. Their masses are $M_A = 1.44M_\odot$, $M_B = 0.2M_\odot$ and $M_C = 0.4M_\odot$, respectively.

On the one hand, PSR J0337+1715 proved unsuitable, at least until now, to perform the usual tests of GTR done with some tight binary pulsars (Kramer et al., 2006; Wex and Kramer, 2020; Kramer et al., 2021) like, for example, the Hulse–Taylor pulsar PSR B1913+16 (Hulse and Taylor, 1975) and the double pulsar PSR J0737–3039 (Burgay et al., 2003; Lyne et al., 2004) because of its orbital configuration. Indeed, the argument of periastron $\omega^{'}$ is not well defined because of the almost vanishing eccentricity $e^{'}$ of the orbit of the outer white dwarf. Furthermore, M_B is far too small for the gravitational redshift of the pulsar signal to be measurable. Finally, the Shapiro delay is negligible; the radio waves travelling along the line of sight towards the Earth pass very distant from its companion because the orbital plane is not seen edge-on. On the other hand, PSR J0337+1715 was successfully used to put the tightest constraints so far on the (absence of the) Nordtvedt effect (Nordtvedt, 1968b,a). It is an orbital consequence of a possible violation of the strong EP which would occur should bodies with different amounts of gravitational self-energy, just like a neutron star and a white dwarf, fell with different accelerations in an external gravitational field. Then, if the falling objects orbit one around the other while moving altogether about a third body, their barycentric orbits should experience a differential elongation towards the source of the external field. Actually, no Nordtvedt effect was found in PSR J0337+1715 to a relative accuracy of two parts per million at 95% confidence level (Archibald et al., 2018; Voisin et al., 2020).

Triple pulsars offer, in principle, the possibility of testing the gravitomagnetic field due to the matter ring discussed in Section 5.1.3, provided that suitable systems, characterized by a not-too-large ratio $T_K^{'}/T_K$, eccentric and, hopefully, non-coplanar orbits as well of the outer companions, will be discovered.

In the case of PSR J0337+1715, the orbital angular momenta of the inner binary and of the outer companion are aligned (Ransom et al., 2014); thus, according to Equations (5.30)–(5.35), calculated by substituting[5] $\boldsymbol{J}$ with Equation (5.52), only the pericentre is shifted by the gravitomagnetic annular field. It turns out

$$H_b = 3.5 \times 10^{44}\,\mathrm{kg\,m^2/s}, \tag{K.1}$$

so that

$$\left\langle \dot{\omega}^{'} \right\rangle^{\mathrm{LT}} = -1.2\,\mathrm{mas/yr}. \tag{K.2}$$

while the uncertainty in measuring its periastron over 1.38 yr, during which

$$N_0 = 26280 \tag{K.3}$$

TOAs were collected, can be calculated from table 1 of Ransom et al. (2014) to be of the order of

$$\sigma_{\omega'} \simeq 63.9\,\mathrm{mas}. \tag{K.4}$$

By tentatively assuming that about the same number of TOAs as given by Equation (K.3) will be collected in 1.38 yr over, say, the next 10 years, the resulting accuracy would be improved by a factor of[6] $\simeq 640$ with respect to the figure quoted in Equation (K.4), while the total gravitomagnetic shift would amount to

$$\Delta\omega^{'}_{\mathrm{LT}} \simeq -12\,\mathrm{mas}. \tag{K.5}$$

Among the competing effects of classical and pN origin, the largest one is due to the quadrupolar term (Demetrian, 2006) of the expansion in multipoles of the

[5] In Equations (5.30)–(5.35), the orbital elements to be used are the primed ones referring to the motion of the outer companion about $\mathcal{S}$.

[6] Such an estimate is obtained by dividing Equation (K.4) by $\sqrt{N}$, where $N = (10/1.38) \times N_0 = 190435$.

Newtonian gravitational potential of a massive ring (Ciftja et al., 2009; Demetrian, 2006) which, in this case, represents the inner binary. It induces a periastron precession given by Equation (10.8), where $\mathtt{K} = (1/4)\mu_{\mathrm{b}} a_{\mathrm{b}}^2$ (Demetrian, 2006) and $a \rightarrow a^{'}, e \rightarrow e^{'}, n_{\mathrm{K}} \rightarrow n_{\mathrm{K}}^{'}$, whose mismodelling may reduce down to the $\simeq 0.5\,\mathrm{mas/yr}$ level in the next 10 years.

The situation would be more favourable for a hypothetical triple system, yet to be discovered, whose outer companion's orbit had a smaller size and was more eccentric, a scenario that should not be deemed as unrealistic. Furthermore, if $\hat{\boldsymbol{h}}$ and $\hat{\boldsymbol{h}}^{'}$ were misaligned, the inclination and the node precessions also would come into play, as per Equations (5.32)–(5.33).

K.6 The Star S4716 around Sgr A*

Here, the S star S4716, orbiting the SMBH in Sgr A* at the GC is taken into account. The upper bounds of the absolute values of the pK shifts of some orbital features of motion and the corresponding values of the hole's spin axis angles and the argument of latitude at epoch are found.

About the role of the LT orbital effects, it should be recalled that a measurement of the BH's spin parameter χ_g by means of the former ones is important for two reasons (Will and Yunes, 2020). First, it would allow one to check if naked singularities, corresponding to $\chi_g > 1$, exist or not. Second, important insights about the formation of the BH would be gained. Indeed, if it were formed from the merger of, say, two pre-existing smaller rotating BHs, a relatively large value of χ_g may be expected, just as when two ice skaters pulling together at the end of a dance finally rotate faster. On the other hand, if the hole were the result of the steady accretion of stars and gases from arbitrary directions, a rather small value of χ_g should occur because of a mutual cancellation, on average, of the spin contributions of the randomly infalling matter.

Furthermore, measuring the quadrupole-driven precessions would allow to put to the test the 'no–hair' theorems, treated in Chapter 5, which are a key prediction of GTR (Will and Yunes, 2020). Indeed, in studying the Kerr metric, it was realized that it is the only possible solution of the Einstein equations for an isolated, stationary BH; all the details of its external gravitational field, encoded by its multipole moments, depend only on its mass $M_\bullet$ and angular momentum $\boldsymbol{J}_\bullet$, as per Equation (5.17). Various types of exotic compact objects (ECOs) have been proposed so far as a possible alternative to general relativistic BHs (Cardoso and Pani, 2019; Psaltis, 2023): 'boson stars', made of concentrations of heavy, dark matter bosons or fermions (Torres et al., 2000), 'fermion balls' (Viollier et al., 1993; Tsiklauri and Viollier, 1998; Becerra–Vergara et al., 2020), 'grava–stars', which would be alleged stars supported by negative vacuum pressure (Mazur and Mottola, 2004;

Cardoso and Pani, 2019), or 'wormholes' (Morris and Thorne, 1988; Cardoso and Pani, 2019).

K.6.1 The Net Shifts per Orbit of the Orbital Elements

Tables K.8–K.11 refer to the net shifts per orbit of the inclination, the node, the periastron, and the mean anomaly at epoch.

Table K.8 *Largest absolute values of the LT and Q_2 net shifts per orbit of the inclination of S4716 and corresponding values of the BH's spin axis angles. For the hole's mass, the value $M_\bullet = 4.1 \times 10^6\, M_\odot$ is used (Peißker et al., 2022), while the spin parameter χ_g is left as a free parameter.*

	$\left\lvert\overline{\Delta I}^{\mathrm{LT}}\right\rvert_{\max}$ ($''$)	$\left\lvert\overline{\Delta I}^{Q_2}\right\rvert_{\max}$ ($''$)
Spin axis angles	χ_g 10.14	χ_g^2 0.06
$i_\bullet^{\max}$ (°)	90	47.9
$\zeta_\bullet^{\max}$ (°)	151.5	169.37

Table K.9 *Largest absolute values of the LT and Q_2 net shifts per orbit of the node of S4716 and corresponding values of the BH's spin axis angles. For the hole's mass, the value $M_\bullet = 4.1 \times 10^6\, M_\odot$ is used (Peißker et al., 2022), while the spin parameter χ_g is left as a free parameter.*

	$\left\lvert\overline{\Delta \Omega}^{\mathrm{LT}}\right\rvert_{\max}$ ($''$)	$\left\lvert\overline{\Delta \Omega}^{Q_2}\right\rvert_{\max}$ ($''$)
Spin axis angles	χ_g 31.5	χ_g^2 0.18
$i_\bullet^{\max}$ (°)	108.76	26.24
$\zeta_\bullet^{\max}$ (°)	241.54	61.54

Table K.10 *Largest absolute values of the 1pN gravitoelectric, LT and Q_2 net shifts per orbit of the perinigricon of S4716 and corresponding values of the BH's spin axis angles. For the hole's mass, the value $M_\bullet = 4.1 \times 10^6\, M_\odot$ is used (Peißker et al., 2022), while the spin parameter χ_g is left as a free parameter.*

	$\left\lvert\overline{\Delta \omega}^{\mathrm{1pN}}\right\rvert_{\max}$ ($''$)	$\left\lvert\overline{\Delta \omega}^{\mathrm{LT}}\right\rvert_{\max}$ ($''$)	$\left\lvert\overline{\Delta \omega}^{Q_2}\right\rvert_{\max}$ ($''$)
Spin axis angles	965.65	χ_g 36.1	χ_g^2 0.22
$i_\bullet^{\max}$ (°)	–	142.9	12.7
$\zeta_\bullet^{\max}$ (°)	–	241.54	61.54

Table K.11 *Largest absolute values of the 1pN gravitoelectric and Q_2 net shifts per orbit of the mean anomaly at epoch of S4716 and corresponding values of the BH's spin axis angles. For the hole's mass, the value $M_\bullet = 4.1 \times 10^6\, M_\odot$ is used (Peißker et al., 2022), while the spin parameter χ_g is left as a free parameter.*

	$\left\|\overline{\Delta\eta}^{\text{1pN}}\right\|_{\max}$ ($''$)	$\left\|\overline{\Delta\eta}^{\text{LT}}\right\|_{\max}$ ($''$)	$\left\|\overline{\Delta\eta}^{Q_2}\right\|_{\max}$ ($''$)
Spin axis angles	−2332.24	0	$\chi_g^2\, 0.069$
$i_\bullet^{\max}$ (°)	–	–	170.8
$\zeta_\bullet^{\max}$ (°)	–	–	360

Table K.12 *Largest absolute values of the LT and Q_2 net shifts per orbit of the radial velocity semiamplitude K of S4716 and corresponding values of the BH's spin axis angles. For the hole's mass, the value $M_\bullet = 4.1\times10^6\, M_\odot$ is used (Peißker et al., 2022), while the spin parameter χ_g is left as a free parameter.*

	$\left\|\overline{\Delta K}^{\text{1pN}}\right\|_{\max}$ (km/s)	$\left\|\overline{\Delta K}^{\text{LT}}\right\|_{\max}$ (km/s)	$\left\|\overline{\Delta K}^{Q_2}\right\|_{\max}$ (km/s)
Spin axis angles	0	$\chi_g\, 0.22$	$\chi_g^2\, 0.001$
$i_\bullet^{\max}$ (°)	–	90	47.96
$\zeta_\bullet^{\max}$ (°)	–	151.54	169.36

It turns out that the largest effects occur for the 1pN gravitoelectric shifts of the periastron and the mean anomaly at epoch, being of the order of $\simeq 1{,}000 - 2{,}000\,''$ = 0.3 – 0.5° per revolution. The maximum values of the magnitude of the LT net shifts per orbit, to be scaled by χ_g, amount to a few tens of arcseconds. The upper bounds on the sizes of the quadrupole-driven effects, to be scaled by χ_g^2, are as little as $\simeq 0.1 - 0.01\,''$. It should be noted that they do not hold simultaneously, occurring for different pairs of values of the angles of the hole's spin axis.

K.6.2 The Net Shift per Orbit of the Radial Velocity Semiamplitude

Table K.12 displays the upper bounds on the magnitude of the non-vanishing LT and quadrupole-driven net shifts per orbit of the radial velocity semiamplitude. While for the LT effect the maximum value, to be scaled by χ_g, is 0.22 km/s, the upper limit of the magnitude of the $Q_2^\bullet$ shift, to be scaled by χ_g^2 is as little as 1 m/s.

Appendix L

A Cursory Overview on Some Proposed Spacecraft-Based Missions

Here, some recently proposed spacecraft-based missions aimed to measure some pN orbital effects in various astronomical scenarios in the solar system are cursorily reviewed: Highly Elliptical Relativity Orbiter (HERO) in Section L.1, In-Orbit Relativity Iuppiter Observatory, or IOvis Relativity In-Orbit Observatory (IORIO) in Section L.2, Elliptical Uranian Relativity Orbiter (EURO) in Section L.3, LEnse–Thirring Sun–Geo Orbiter (LETSGO) in Section L.4, and ELXIS in Section L.5. Further past and recent space-based proposals[1] by other authors to measure, among other things, pN orbital effects are briefly reviewed in Section L.6.

L.1 HERO

The space-based mission concept Highly Elliptical Relativity Orbiter (HERO) (Iorio, 2019a) is aimed to perform several tests of pN gravity around the Earth with a preferably drag-free spacecraft moving along a highly elliptical orbit fixed within its plane which undergoes a relatively fast secular precession. Two possible scenarios are considered: (a) a fast, 4-h orbit with high perigee height of 1047 km, and (b) a slow, 21-h path with a low perigee height of 642 km. In both cases, the orbital plane is inclined to the Earth's equator by an amount known as critical inclination since it allows one to cancel out the main competing classical perturbation on the perigee due to J_2. Thus, the J_2/c^2 signatures, which generally depend on the perigee's frequency, resemble linear trends. HERO may detect, for the first time, the pN orbital effects induced by the mass quadrupole moment J_2 of the Earth which, among other things, affects the semimajor axis a via a secular trend of $\simeq$ 4–12 cm/yr. Recently, the secular decay of the semimajor axis of the passive satellite LARES was measured with an error as little as 0.7 cm/yr (Lucchesi et al., 2019a).

[1] Only missions exploiting orbital motions are mentioned.

L.2 IORIO

In view of its characteristics, Jupiter, whose relevant physical parameters are listed in Table J.3, seems to be the ideal candidate, at least in principle, to try to measure several pN orbital effects. Suffice it to say that it has often been considered for testing various aspects of gravitomagnetism over the years; see, for example, Haas and Ross (1975), Mashhoon (2000), Tartaglia (2000c,a), Iorio (2010b), Schärer et al. (2017), and Iorio (2019g, 2024a). The mission concept IORIO, acronym of In-Orbit Relativity Iuppiter[2] Observatory, or, equally well, of IOvis[3] Relativity In-orbit Observatory (Iorio, 2019g,d, 2024a), relies upon a dedicated Juno-like spacecraft circling Jupiter along a highly elliptical, polar orbit to measure, among other things, the 1pN mass quadrupole and spin octupole effects. Although more difficult to be practically implemented, the case of a less elliptical orbit[4] also is being considered since it yields much larger figures for the relativistic effects of interest.

L.3 EURO

Elliptical Uranian Relativity Orbiter (EURO) (Iorio et al., 2023) is a preliminary mission concept investigating the possibility of dynamically measuring the angular momentum of Uranus by means of the LT effect affecting a putative planet's orbiter. It is possible, at least in principle, to separate the relativistic precessions of the orbital inclination to the Celestial Equator and of the longitude of the ascending node of the spacecraft from its classical rates of the pericentre induced by the multipoles of the planet's gravity field by adopting an appropriate orbital configuration. For a wide and elliptical 2,000 $\times$ 100,000 km orbit, the gravitomagnetic signatures amount to tens of mas/yr, while, for a suitable choice of the initial conditions, the peak-to-peak amplitude of the range rate shift can reach the level of $\simeq 1.5 \times 10^{-3}$ mm/s in a single pericentre passage of a few hours. By lowering the apocentre height to 10,000 km, the LT precessions are enhanced to the level of hundreds of mas/yr. The uncertainties in the orientation of the planetary spin axis and in the inclination are major sources of systematic bias; it turns out that they should be determined with accuracies as good as $\simeq$ 0.1–1 and $\simeq$ 1–10 mas, respectively.

L.4 LETSGO

LEnse–Thirring Sun-Geo Orbiter (LETSGO) (Iorio, 2013c) is a proposed space-based mission involving the use of a spacecraft moving along a highly eccentric heliocentric orbit perpendicular to the ecliptic. It aims to accurately measure some

[2] *Iuppĭtĕr* is one of the forms of the Latin noun of the god Jupiter.

[3] In Latin, *Iŏvis* means 'of Jupiter'.

[4] Inserting a spacecraft into a moderately elliptical orbit around Jupiter is a very daunting task because of the exceedingly large amount of fuel required.

important physical properties of the Sun and to test some pN features of its gravitational field by continuously monitoring the Earth–probe range. Preliminary sensitivity analyses show that, by assuming a cm-level accuracy in ranging to the spacecraft, it would be possible to test, in principle, the LT effect at a $\simeq 10^{-2}$ level over a timescale of 2 years, while the larger 1pN gravitoelectric component of the solar gravitational field may be sensed with a relative accuracy of about 10^{-8}–10^{-9} during the same temporal interval. The competing range perturbation due to the Sun's oblateness would be a source of systematic error, but it turns out that all the three dynamical features of motion examined affect the Earth–probe range in different ways, allowing for separating them in real data analyses. The high eccentricity would help in reducing the impact of the non-gravitational perturbations whose disturbance effect would certainly be severe when LETSGO approaches the Sun at just a few solar radii. It can be preliminarily argued that a drag-free apparatus should perform at a 10^{-8}–10^{-9} $\mathrm{m/s^2/\sqrt{Hz}}$ level for frequencies of about 10^{-7} Hz. Further studies should be devoted to investigate both the consequences of the non-conservative forces and the actual measurability of the effects of interest by means of extensive numerical data simulations, parameter estimations, and covariance analyses. Also an alternative, fly-by configuration is worthy of consideration.

L.5 ELXIS

ELXIS[5] (Iorio, 2019f,e) is a hypothetical new terrestrial artificial satellite, to be placed in a circular path in an orbital plane displaced by $\Omega = 90^\circ$ with respect to the reference direction of the Vernal Equinox Υ perpendicularly to the Earth's equator, aimed at measuring the de Sitter precession of the orbital inclination to the terrestrial equator with a possible relative accuracy level of 10^{-5}. A rather strict polar orbital configuration, with departures as little as $\simeq 10^{-3}$–$10^{-5\,\circ}$, would be required to reduce the impact of the aliasing perturbations due to the solid and ocean components of the K_1 tide, which would be one of the major sources of systematic errors, especially if not-too-high altitudes were to be adopted. The long-term rates of change of I due to the even and odd zonal harmonics of the geopotential vanish for the orbital geometry proposed. It is assumed that the data analysis would be performed in a geocentric kinematically rotating and dynamically non–rotating reference frame.

L.6 Other Proposed Space-Based Missions

As far as further space-based proposals by other authors are concerned, see Anderson et al. (1997) and Clark et al. (2003) for mission concepts to Mercury aimed

[5] From ἕλξις, -εως, ἡ, meaning 'dragging', 'trailing'.

at testing also some orbital effects due to the 1pN gravitoelectric field of the Sun. OPTIS was a proposed spaceborne mission to accurately measure, among other things, the Lense–Thirring effect in the field of the Earth with a drag-free, laser-ranged satellite (Iorio et al., 2004). The Laser Astrometric Test Of Relativity (LATOR) was a space-based mission concept to probe several aspects of relativistic gravity in the solar system by means of laser interferometry between two microspacecraft whose lines of sight pass close by the Sun (Turyshev et al., 2004). A close concept was the Astrodynamical Space Test of Relativity using Optical Devices (ASTROD) mission series aiming at high-precision measurements in interplanetary space by means of optical devices and drag-free control of some spacecraft in the areas of fundamental physics, GTR as well as Sun and solar system research (Selig et al., 2013). Buscaino et al. (2015) suggested to test long–distance modifications of gravity to 100 au with a drag-free cruising spacecraft. A recent proposal by Turyshev et al. (2024) to test new physics in the solar system envisages the use of tetrahedral[6] formations of drag-free spacecraft in heliocentric, eccentric orbits. In January 2024, the European Space Agency (ESA) formally approved[7] the Laser Interferometer Space Antenna (LISA) mission (Amaro–Seoane et al., 2012) aimed at detecting low-frequency gravitational waves in space with a constellation of three drag-free spacecraft in heliocentric orbits mutually exchanging laser beams on a continuous basis. A similar mission concept is TianQuin (Luo et al., 2016).

For a review of testing fundamental physics with space missions, see, for example, Lämmerzahl and Dittus (2008), and references therein.

6 From τετράεδρον ('triangle-based pyramid'); it comes from τετράς ('four'), and ἕδρα, -ας, ἡ, meaning, among other things, 'face of a regular solid'.

7 www.esa.int/Science_Exploration/Space_Science/Capturing_the_ripples_of_spacetime_LISA_gets_go-ahead. Accessed 4 April 2024.

References

Adams, F. C., and Laughlin, G. 2006a. Effects of Secular Interactions in Extrasolar Planetary Systems. *Astrophys. J.*, **649**, 992–1003.

Adams, F. C., and Laughlin, G. 2006b. Long-Term Evolution of Close Planets Including the Effects of Secular Interactions. *Astrophys. J.*, **649**, 1004–1009.

Adams, F. C., and Laughlin, G. 2006c. Relativistic Effects in Extrasolar Planetary Systems. *Int. J. Mod. Phys. D*, **15**, 2133–2140.

Adelberger, E. G., Heckel, B. R., and Nelson, A. E. 2003. Tests of the Gravitational Inverse-Square Law. *Annu. Rev. Nucl. Part. Sci.*, **53**, 77–121.

Adelberger, E. G., Heckel, B. R., Hoedl, S., et al. 2007. Particle-Physics Implications of a Recent Test of the Gravitational Inverse-Square Law. *Phys. Rev. Lett.*, **98**, 131104.

Adkins, G. S., and McDonnell, J. 2007. Orbital Precession due to Central-Force Perturbations. *Phys. Rev. D*, **75**, 082001.

Adler, C. G., and Coulter, B. L. 1978. Galileo and the Tower of Pisa experiment. *Am. J. Phys.*, **46**, 199–201.

Adler, R., Bazin, M., and Schiffer, M. 1975. *Introduction to General Relativity. Second Edition*. McGraw-Hill.

Adler, S. L. 2009. Planet-Bound Dark Matter and the Internal Heat of Uranus, Neptune, and Hot-Jupiter Exoplanets. *Phys. Lett. B*, **671**, 203–206.

Albrecht, S. H., Dawson, R. I., and Winn, J. N. 2022. Stellar Obliquities in Exoplanetary Systems. *Publ. Astron. Soc. Pac.*, **134**, 082001.

Ali, A. F., and Khalil, M. M. 2016. Black Hole with Quantum Potential. *Nucl. Phys. B*, **909**, 173–185.

Ali, B., Paul, D., Eckart, A., et al. 2020. Kinematic Structure of the Galactic Center S Cluster. *Astrophys. J.*, **896**, 100.

Amaro-Seoane, P., Aoudia, S., Babak, S., et al. 2012. Low-Frequency Gravitational-Wave Science with eLISA/NGO. *Class. Quantum Gravit.*, **29**, 124016.

Amelin, V. M. 1966. Determination of the Quasi-Nodal Period of the Satellite 1960 ε 3 from Simultaneous Visual Tracking Data. *Pages 15–18 of* Kovalevsky, J. (ed.), *Trajectories of Artificial Celestial Bodies as Determined from Observations/Trajectoires des Corps Celestes Artificiels Déterminées D'après les Observations*. Springer.

Anderson, J. D., Campbell, J. K., Jurgens, R. F., and Lau, E. L. 1993. Recent Developments in Solar-System Tests of General Relativity. *Pages 353–355 of* Sato, F., and Nakamura, T. (eds.), *Proceedings of the Sixth Marcel Grossmann Meeting on General Relativity*. World Scientific.

Anderson, J. D., Keesey, M. S. W., Lau, E. L., Standish, Jr., E. M., and Newhall, X. X. 1978. Tests of General Relativity Using Astrometric and Radio-metric Observations of the Planets. *Acta Astronaut.*, **5**, 43–61.

Anderson, J. D., Lau, E. L., Krisher, T. P., et al. 1995. Improved Bounds on Nonluminous Matter in Solar Orbit. *Astrophys. J.*, **448**, 885–892.

Anderson, J. D., Lau, E. L., Taylor, A. H., et al. 1989. Bounds on Dark Matter in Solar Orbit. *Astrophys. J.*, **342**, 539–544.

Anderson, J. D., Turyshev, S. G., Asmar, S. W., et al. 1997. Radio-Science Investigation on a Mercury Orbiter Mission. *Planet. Space Sci.*, **45**, 21–29.

Andrés, J. I., Noomen, R., and Vecellio None, S. 2006. Numerical Simulation of the LAGEOS Thermal Behavior and Thermal Accelerations. *J. Geophys. Res.*, **111**, B09406.

Angélil, R., Saha, P., and Merritt, D. 2010. Toward Relativistic Orbit Fitting of Galactic Center Stars and Pulsars. *Astrophys. J.*, **720**, 1303–1310.

Antoniadis, J., Bassa, C. G., Wex, N., Kramer, M., and Napiwotzki, R. 2011. A White Dwarf Companion to the Relativistic Pulsar PSR J1141-6545. *Mon. Not. Roy. Astron. Soc.*, **412**, 580–584.

Antoniadis, J., Freire, P. C. C., Wex, N., et al. 2013. A Massive Pulsar in a Compact Relativistic Binary. *Science*, **340**, 6131.

Antoniciello, G., Borsato, L., Lacedelli, G., et al. 2021. Detecting General Relativistic Orbital Precession in Transiting Hot Jupiters. *Mon. Not. Roy. Astron. Soc.*, **505**, 1567–1574.

Archibald, A. M., Gusinskaia, N. V., Hessels, J. W. T., et al. 2018. Universality of free fall from the orbital motion of a pulsar in a stellar triple system. *Nature*, **559**, 73–76.

Armitage, P. J., and Natarajan, P. 1999. Lense–Thirring Precession of Accretion Disks around Compact Objects. *Astrophys. J.*, **525**, 909–914.

Asada, H., and Futamase, T. 1997. Chapter 2. Post-Newtonian Approximation: Its Foundation and Applications. *Prog. Theor. Phys. Supp.*, **128**, 123–181.

Asensio-Torres, R., Janson, M., Bonavita, M., et al. 2018. SPOTS: The Search for Planets Orbiting Two Stars. III. Complete sample and statistical analysis. *Astron. Astrophys.*, **619**, A43.

Ashby, N., and Allison, T. 1993. Canonical Planetary Perturbation Equations for Velocity-Dependent Forces, and the Lense–Thirring Precession. *Celest. Mech. Dyn. Astr.*, **57**, 537–585.

Ashby, N., and Bertotti, B. 1984. Relativistic Perturbations of an Earth Satellite. *Phys. Rev. Lett.*, **52**, 485–488.

Ashtekar, A. 2017. Implications of a Positive Cosmological Constant for General Relativity. *Rep. Prog. Phys.*, **80**, 102901.

Astier, P., and Pain, R. 2012. Observational Evidence of the Accelerated Expansion of the Universe. *C. R. Phys.*, **13**, 521–538.

Bailey, Q. G. 2010. Lorentz-Violating Gravitoelectromagnetism. *Phys. Rev. D*, **82**, 065012.

Bailey, Q. G., and Kostelecký, V. A. 2006. Signals for Lorentz Violation in Post-Newtonian Gravity. *Phys. Rev. D*, **74**, 045001.

Balogh, A., Grard, R., Solomon, S. C., et al. 2007. Missions to Mercury. *Space Sci. Rev.*, **132**, 611–645.

Banik, I., and Zhao, H. 2022. From Galactic Bars to the Hubble Tension: Weighing Up the Astrophysical Evidence for Milgromian Gravity. *Symmetry*, **14**, 1331.

Bardeen, J. M., and Petterson, J. A. 1975. The Lense–Thirring Effect and Accretion Disks around Kerr Black Holes. *Astrophys. J. Lett.*, **195**, L65–L67.

Barker, B. M., and O'Connell, R. F. 1975. Gravitational Two-Body Problem with Arbitrary Masses, Spins, and Quadrupole Moments. *Phys. Rev. D*, **12**, 329–335.

Barker, B. M., and O'Connell, R. F. 1979. The Gravitational Interaction: Spin, Rotation, and Quantum Effects – a Review. *Gen. Relativ. Gravit.*, **11**, 149–175.

Barnes, E. W. 1908. A New Development in the Theory of the Hypergeometric Functions. *Proc. London Math. Soc.*, **6**, 141–177.

Basu, S. 2016. Global Seismology of the Sun. *Living Rev. Solar Phys.*, **13**, 2. arXiv: https://arxiv.org/abs/1606.07071.

Becerra-Vergara, E. A., Argüelles, C. R., Krut, A., Rueda, J. A., and Ruffini, R. 2020. Geodesic Motion of S2 and G2 as a Test of the Fermionic Dark Matter Nature of our Galactic Core. *Astron. Astrophys.*, **641**, A34.

Bejger, M., and Haensel, P. 2002. Moments of Inertia for Neutron and Strange Stars: Limits Derived for the Crab Pulsar. *Astron. Astrophys.*, **396**, 917–921.

Bekenstein, J. D. 2009. Relativistic MOND as an Alternative to the Dark Matter Paradigm. *Nucl. Phys. A*, **827**, 555–560.

Benkhoff, J., van Casteren, J., Hayakawa, H., et al. 2010. BepiColombo: Comprehensive Exploration of Mercury: Mission overview and Overview and Science Goals. *Planet. Space Sci.*, **58**, 2–20.

Benkhoff, J., Murakami, G., Baumjohann, W., et al. 2021. BepiColombo: Mission Overview and Science Goals. *Space Sci. Rev.*, **217**, 90.

Benn, I. M., and Tucker, R. W. 1987. *An Introduction to Spinors and Geometry with Applications in Physics*. Adam Hilger.

Bennett, D. P., Rhie, S. H., Udalski, A., et al. 2016. The First Circumbinary Planet Found by Microlensing: OGLE-2007-BLG-349L(AB)c. *Astron J.*, **152**, 125.

Bertolami, O., and Páramos, J. 2005. Astrophysical Constraints on Scalar Field Models. *Phys. Rev. D*, **71**, 023521.

Bertolami, O., Páramos, J., and Turyshev, S. G. 2007. General Theory of Relativity: Will It Survive the Next Decade? *Pages 27–74 of* Dittus, H., Lämmerzahl, C., and Turyshev, S. G. (eds.), *Lasers, Clocks, and Drag-Free Control: Exploration of Relativistic Gravity in Space*. Astrophysics and Space Science Library, vol. 349. Springer.

Bertolami, O., Francisco, F., Gil, P. J. S., and Páramos, J. 2008. Thermal Analysis of the Pioneer Anomaly: A Method to Estimate Radiative Momentum Transfer. *Phys. Rev. D*, **78**, 103001.

Bertotti, B. 1973. Is the Solar System Gravitationally Closed? *Astrophys. Lett.*, **14**, 51–53.

Bertotti, B., Ciufolini, I., and Bender, P. L. 1987. New Test of General Relativity: Measurement of De Sitter Geodetic Precession Rate for Lunar Perigee. *Phys. Rev. Lett.*, **58**, 1062–1065.

Bertotti, B., Farinella, P., and Vokrouhlický, D. 2003. *Physics of the Solar System*. Kluwer.

Beskin, V. S., Gurevich, A. V., and Istomin, Ia. N. 1988. Theory of the Radio Emission of Pulsars. *Astrophys. Space Sci.*, **146**, 205–281.

Beuermann, K., Hessman, F. V., Dreizler, S., et al. 2010. Two Planets Orbiting the Recently Formed Post-Common Envelope Binary NN Serpentis. *Astron. Astrophys.*, **521**, L60.

Bhattacharya, S., and Chakraborty, S. 2017. Constraining Some Horndeski Gravity Theories. *Phys. Rev. D*, **95**, 044037.

Bini, D., and Jantzen, R. T. 2003. Gravitomagnetic Clock Effects in Black Hole Spacetimes. *Pages 17–28 of:* Marmo, G., Rubano, C., and Scudellaro, P. (eds.), *General Relativity, Cosmology and Gravitational Lensing*. Napoli series on Physics and Astrophysics, vol. 6. Bibliopolis.

Bini, D., de Felice, F., and Geralico, A. 2004. Spinning Test Particles and Clock Effect in Kerr Spacetime. *Class. Quantum Gravit.*, **21**, 5441–5456.

Bini, D., Geralico, A., and Jantzen, R. T. 2005. Kerr Metric, Static Observers and Fermi Coordinates. *Class. Quantum Gravit.*, **22**, 4729–4742.

Bini, D., Mashhoon, B., and Obukhov, Y. N. 2022. Gravitomagnetic Helicity. *Phys. Rev. D*, **105**, 064028.

Bishop, R. L., and Goldberg, S. I. 1980. *Tensor Analysis on Manifolds*. Dover.

Biskupek, L., Müller, J., and Torre, J.-M. 2021. Benefit of New High-Precision LLR Data for the Determination of Relativistic Parameters. *Universe*, **7**, 34.

Blanchet, L. 2003. On the Accuracy of the Post-Newtonian Approximation. *Pages 411–430 of:* Ciufolini, I., Dominici, D., and Lusanna, L. (eds.), *2001: A Relativistic Spacetime Odyssey*. World Scientific.

Blanchet, L. 2006. Gravitational Radiation from Post-Newtonian Sources and Inspiralling Compact Binaries. *Living Rev. Relativ.*, **9**, 4.

Blanchet, L., Hébrard, G., and Larrouturou, F. 2019. Detecting the General Relativistic Orbital Precession of the Exoplanet HD 80606b. *Astron. Astrophys.*, **628**, A80.

Blandford, R. D. 2015. A Century of General Relativity: Astrophysics and Cosmology. *Science*, **347**, 1103–1108.

Blandford, R. D., and Znajek, R. L. 1977. Electromagnetic Extraction of Energy from Kerr Black Holes. *Mon. Not. Roy. Astron. Soc.*, **179**, 433–456.

Boehle, A., Ghez, A. M., Schödel, R., et al. 2016. An Improved Distance and Mass Estimate for Sgr A* from a Multistar Orbit Analysis. *Astrophys. J.*, **830**, 17.

Bogorodskii, A. F. 1959. Relativistic Effects in the Motion of an Artificial Earth Satellite. *Soviet Astron.*, **3**, 857–862.

Bolton, S. (ed). 2018. *The Juno Mission*. Springer.

Bolton, S. J., Lunine, J., Stevenson, D., et al. 2017. The Juno Mission. *Space Sci. Rev.*, **213**, 5–37.

Bonanno, A., and Reuter, M. 2000. Renormalization Group Improved Black Hole Spacetimes. *Phys. Rev. D*, **62**, 043008.

Bond, V. R., and Janin, G. 1981. Canonical Orbital Elements in Terms of an Arbitrary Independent Variable. *Celest. Mech. Dyn. Astr.*, **23**, 159–172.

Bonnor, W. B., and Steadman, B. R. 1999. The Gravitomagnetic Clock Effect. *Class. Quantum Gravit.*, **16**, 1853–1861.

Boshkayev, K., Quevedo, H., and Zhami, B. 2017. *I*–Love–*Q* Relations for White Dwarf Stars. *Mon. Not. Roy. Astron. Soc.*, **464**, 4349–4359.

Braginsky, V. B., Caves, C. M., and Thorne, K. S. 1977. Laboratory Experiments to Test Relativistic Gravity. *Phys. Rev. D*, **15**, 2047–2068.

Brahm, R., Espinoza, N., Jordán, A., et al. 2018. K2–232 b: A Transiting Warm Saturn on an Eccentric $P = 11.2$ d Orbit around a $V = 9.9$ star. *Mon. Not. Roy. Astron. Soc.*, **477**, 2572–2581.

Brax, P., van de Bruck, C., and Davis, A.-C. 2004. Brane World Cosmology. *Rep. Prog. Phys.*, **67**, 2183–2231.

Breton, R. P., Kaspi, V. M., Kramer, M., et al. 2008. Relativistic Spin Precession in the Double Pulsar. *Science*, **321**, 104.

Breu, C., and Rezzolla, L. 2016. Maximum Mass, Moment of Inertia and Compactness of Relativistic Stars. *Mon. Not. Roy. Astron. Soc.*, **459**, 646–656.

Broderick, A. E., Fish, V. L., Doeleman, S. S., and Loeb, A. 2009. Estimating the Parameters of Sagittarius A*'s Accretion Flow via Millimeter VLBI. *Astrophys. J.*, **697**, 45–54.

Broderick, A. E., Fish, V. L., Doeleman, S. S., and Loeb, A. 2011. Evidence for Low Black Hole Spin and Physically Motivated Accretion Models from Millimeter-VLBI Observations of Sagittarius A*. *Astrophys. J.*, **735**, 110.

Broucke, R. A., and Cefola, P. J. 1972. On the Equinoctial Orbit Elements. *Celest. Mech. Dyn. Astr.*, **5**, 303–310.

Brouwer, D., and Clemence, G. M. 1961. *Methods of Celestial Mechanics*. Academic Press.

Brownstein, J. R., and Moffat, J. W. 2006a. Galaxy Cluster Masses without Nonbaryonic Dark Matter. *Mon. Not. Roy. Astron. Soc.*, **367**, 527–540.

Brownstein, J. R., and Moffat, J. W. 2006b. Galaxy Rotation Curves without Nonbaryonic Dark Matter. *Astrophys. J.*, **636**, 721–741.

Brumberg, V. A. 1991. *Essential Relativistic Celestial Mechanics*. Adam Hilger.

Brumberg, V. A. 2010. Relativistic Celestial Mechanics. *Scholarpedia*, **5**, 10669.

Brumberg, V. A., and Kopeikin, S. M. 1989a. Relativistic Reference Systems and Motion of Test Bodies in the Vicinity of the Earth. *Nuovo Cim. B*, **103**, 63–98.

Brumberg, V. A., and Kopeikin, S. M. 1989b. Relativistic Theory of Celestial Reference Frames. *Pages 115–142 of:* Kovalevsky, J., Mueller, I. I., and Kolaczek, B. (eds.), *Reference Frames in Astronomy and Geophysics*. Kluwer.

Brush, S. G. 1989. Prediction and Theory Evaluation: The Case of Light Bending. *Science*, **246**, 1124–1129.

Bugg, D. V. 2015. MOND–A Review. *Can. J. Phys.*, **93**, 119–125.

Buldgen, G., Salmon, S., and Noels, A. 2019. Progress in Global Helioseismology: A New Light on the Solar Modelling Problem and Its Implications for Solar–Like Stars. *Front. Astron. Space Sci.*, **6**, 42.

Bullock, J. S., and Boylan-Kolchin, M. 2017. Small-Scale Challenges to the ΛCDM Paradigm. *Annu. Rev. Astron. Astr.*, **55**, 343–387.

Burgay, M., D'Amico, N., Possenti, A., et al. 2003. An Increased Estimate of the Merger Rate of Double Neutron Stars from Observations of a Hghly Relativistic System. *Nature*, **426**, 531–533.

Burgess, C. P., and Cloutier, J. 1988. Astrophysical Evidence for a Weak New Force? *Phys. Rev. D*, **38**, 2944–2950.

Buscaino, B., DeBra, D., Graham, P. W., Gratta, G., and Wiser, T. D. 2015. Testing Long-Distance Modifications of Gravity to 100 Astronomical Units. *Phys. Rev. D*, **92**, 104048.

Cai, Y.-F., Capozziello, S., De Laurentis, M., and Saridakis, E. N. 2016. $f(T)$ Teleparallel Gravity and Cosmology. *Rep. Prog. Phys.*, **79**, 106901.

Capderou, M. 2005. *Satellites: Orbits and Missions*. Springer.

Capitaine, N., and Soffel, M. 2015. On the Definition and Use of the Ecliptic in Modern Astronomy. *Pages 61–64 of:* Malkin, N., and Capitaine, N. (eds.), *Journées 2014 'Systèmes de référence spatio–temporels'*. Pulkovo observatory.

Capozziello, S., and de Laurentis, M. 2011. Extended Theories of Gravity. *Phys. Rep.*, **509**, 167–321.

Capozziello, S., Harko, T., Koivisto, T. S., Lobo, F. S. N., and Olmo, G. J. 2015. Hybrid Metric-Palatini Gravity. *Universe*, **1**, 199–238.

Capozziello, S., Sarracino, G., and De Somma, G. 2024. A Critical Discussion on the H_0 Tension. *Universe*, **10**, 140.

Capuzzo-Dolcetta, R., and Sadun-Bordoni, M. 2023. Orbital Precession of Stars in the Galactic Centre. *Mon. Not. Roy. Astron. Soc.*, **522**, 5828–5839.

Cardoso, V., and Pani, P. 2019. Testing the Nature of Dark Compact Objects: A Status Report. *Living Rev. Relativ.*, **22**, 4.

Carroll, S. M. 2019. *Spacetime and Geometry*. Cambridge University Press.

Carter, B. 1971. Axisymmetric Black Hole Has Only Two Degrees of Freedom. *Phys. Rev. Lett.*, **26**, 331–333.

Carter, J. A., Yee, J. C., Eastman, J., Gaudi, B. S., and Winn, J. N. 2008. Analytic Approximations for Transit Light–Curve Observables, Uncertainties, and Covariances. *Astrophys. J.*, **689**, 499–512.

Casotto, S. 1993. Position and Velocity Perturbations in the Orbital Frame in Terms of Classical Element Perturbations. *Celest. Mech. Dyn. Astr.*, **55**, 209–221.

Cattaneo, C. 1958. General Relativity: Relative Standard Mass, Momentum, Energy and Gravitational Field in a General System of Reference. *Nuovo Cim.*, **10**, 318–337.

Cervantes-Cota, J. L., Galindo-Uribarri, S., and Smoot, G. F. 2016. A Brief History of Gravitational Waves. *Universe*, **2**, 22.

Cervantes-Cota, J. L., Galindo-Uribarri, S., and Smoot, G. F. 2023. The Unsettled Number: Hubble's Tension. *Universe*, **9**, 501.

Chakraborty, S., and SenGupta, S. 2017. Strong Gravitational Lensing – A Probe for Extra Dimensions and Kalb-Ramond Field. *J. Cosmol. Astropart. Phys.*, **2017**, 045.

Champion, D. J., Lorimer, D. R., McLaughlin, M. A., et al. 2004. PSR J1829+2456: a Relativistic Binary Pulsar. *Mon. Not. Roy. Astron. Soc.*, **350**, L61–L65.

Chandrasekhar, S. 1983. *The Mathematical Theory of Black Holes*. Oxford University Press.

Chandrasekhar, S. 1995. *Newton's Principia for the Common Reader*. Clarendon Press.

Charbonneau, D., Brown, T. M., Latham, D. W., and Mayor, M. 2000. Detection of Planetary Transits Across a Sun–like Star. *Astrophys. J. Lett.*, **529**, L45–L48.

Charlot, P., Jacobs, C. S., Gordon, D., et al. 2020. The Third Realization of the International Celestial Reference Frame by Very Long Baseline Interferometry. *Astron. Astrophys.*, **644**, A159.

Cheng, M. K. 2002. Gravitational Perturbation Theory for Intersatellite Tracking. *J. Geod.*, **76**, 169–185.

Cheng, T.-P. 2009. *Relativity, Gravitation and Cosmology. A Basic Introduction. Second Edition*. Oxford University Press.

Chicone, C., and Mashhoon, B. 2002. The Generalized Jacobi Equation. *Class. Quantum Gravit.*, **19**, 4231–4248.

Chicone, C., Mashhoon, B., and Retzloff, D. G. 1996a. Gravitational Ionization: Periodic Orbits of Binary Systems Perturbed by Gravitational Radiation. *Ann. Inst. Henri Poincaré Phys. Théor.*, **64**, 87–125.

Chicone, C., Mashhoon, B., and Retzloff, D. G. 1996b. On the Ionization of a Keplerian Binary System by Periodic Gravitational Radiation. *J. Math. Phys.*, **37**, 3997–4016.

Ciftja, O., Babineaux, A., and Hafeez, N. 2009. The Electrostatic Potential of a Uniformly Charged Ring. *Eur. J. Phys.*, **30**, 623–627.

Ciufolini, I., Lucchesi, D. M., Vespe, F., and Mandiello, A. 1996. Measurement of Dragging of Inertial Frames and Gravitomagnetic Field Using Laser-Ranged Satellites. *Nuovo Cim. A*, **109A**, 575–590.

Ciufolini, I., Paolozzi, A., Koenig, R., et al. 2013. Fundamental Physics and General Relativity with the LARES and LAGEOS satellites. *Nucl. Phys. B Proc. Suppl.*, **243**, 180–193.

Ciufolini, I., Paolozzi, A., Pavlis, E. C., et al. 2023. The LARES 2 Satellite, General Relativity and Fundamental Physics. *Eur. Phys. J. C*, **83**, 87.

Claret, A. 1995. Stellar Models for a Wide Range of Initial Chemical Compositions until Helium Burning. I. From $X = 0.60$ to $X = 0.80$ for $Z = 0.02$. *Astron. Astrophys. Suppl. Ser.*, **109**, 441–446.

Claret, A., Giménez, A., Baroch, D., et al. 2021. Analysis of Apsidal Motion in Eclipsing Binaries Using TESS Data. II. A Test of Internal Stellar Structure. *Astron. Astrophys.*, **654**, A17.

Clark, P. E., Curtis, S. A., Marr, G., Reuter, D., and McKenna-Lawlor, S. 2003. LUGH/Mercury Express: A Proposed Multi-platform Flyby Mission to Mercury. *Acta Astronaut.*, **52**, 181–188.

Clark, S. J., and Tucker, R. W. 2000. Gauge Symmetry and Gravitoelectromagnetism. *Class. Quantum Gravit.*, **17**, 4125–4157.

Clemence, G. M. 1947. The Relativity Effect in Planetary Motions. *Rev. Mod. Phys.*, **19**, 361–364.

Clifton, T., Ferreira, P. G., Padilla, A., and Skordis, C. 2012. Modified gravity and cosmology. *Phys. Rep.*, **513**, 1–189.

Cohen, J. M., and Mashhoon, B. 1993. Standard Clocks, Interferometry, and Gravitomagnetism. *Phys. Lett. A*, **181**, 353–358.

Cohen, S. C., and Smith, D. E. 1985. LAGEOS Scientific Results: Introduction. *J. Geophys. Res.*, **90**, 9217–9220.

Colladay, D., and Kostelecký, V. A. 1997. CPT Violation and the Standard Model. *Phys. Rev. D*, **55**, 6760–6774.

Colladay, D., and Kostelecký, V. A. 1998. Lorentz-Violating Extension of the Standard Model. *Phys. Rev. D*, **58**, 116002.

Constantin, D. R. 2010. On the First Determination of Mercury's Perihelion Advance. *Ann. Univ. Bucarest (math. ser.)*, **1 (LIX)**, 229–238.

Correia, A. C. M., and Rodríguez, A. 2013. On the Equilibrium Figure of Close-in Planets and Satellites. *Astrophys. J.*, **767**, 128.

Correia, A. C. M., Udry, S., Mayor, M., et al. 2005. The CORALIE Survey for Southern Extra-solar Planets. XIII. A Pair of Planets around HD 202206 or a Circumbinary Planet? *Astron. Astrophys.*, **440**, 751–758.

Costa, L. F. O., Natário, J., and Santos, N. O. 2021. Gravitomagnetism in the Lewis Cylindrical Metrics. *Class. Quantum Gravit.*, **38**, 055003.

Costa, L. Filipe O., and Herdeiro, C. A. R. 2008. Gravitoelectromagnetic Analogy Based on Tidal Tensors. *Phys. Rev. D*, **78**, 024021.

Costa, L. Filipe O., and Natário, J. 2014. Gravito-electromagnetic Analogies. *Gen. Relativ. Gravit.*, **46**, 1792.

Costa, L. Filipe. O., and Natário, J. 2021. Frame-Dragging: Meaning, Myths, and Misconceptions. *Universe*, **7**, 388.

Coulot, D., Deleflie, F., Bonnefond, P., et al. 2011. Satellite Laser Ranging. *Pages 1049–1055 of:* Gupta, H. K. (ed.), *Encyclopedia of Solid Earth Geophysics*. Encyclopedia of Earth Sciences Series. Springer.

Crass, J., Gaudi, B. S., Leifer, S., et al. 2021. Extreme Precision Radial Velocity Working Group Final Report. *arXiv e-prints*, arXiv:2107.14291.

Crease, R. P. 2006. The Legend of the Leaning Tower. *Pages 10–14 of:* Hall, L. E. (ed), *The Laws of Motion. An Anthology of Current Thought*. The Rosen Publishing Group.

Cugusi, L., and Proverbio, E. 1978. Relativistic Effects on the Motion of Earth's Artificial Satellites. *Astron. Astrophys.*, **69**, 321–325.

Cui, W., Zhang, S. N., and Chen, W. 1998. Evidence for Frame Dragging around Spinning Black Holes in X–Ray Binaries. *Astrophys. J. Lett.*, **492**, L53–L57.

Cui, Y., Hada, K., Kawashima, T., et al. 2023. Precessing Jet Nozzle Connecting to a Spinning Black Hole in M87. *Nature*, **621**, 711–715.

Daly, R. A., Donahue, M., O'Dea, C. P., et al. 2024. New Black Hole Spin Values for Sagittarius A* Obtained with the Outflow Method. *Mon. Not. Roy. Astron. Soc.*, **527**, 428–436.

Damiani, C., and Lanza, A. F. 2011. Prospecting Transit Duration Variations in Extrasolar Planetary Systems. *Astron. Astrophys.*, **535**, A116.

Damour, T. 1987. The Problem of Motion in Newtonian and Einsteinian Gravity. *Pages 128–198 of:* Hawking, S., and Israel, W., (eds.), *Three Hundred Years of Gravitation*. Cambridge University Press.

Damour, T. 2007. General Relativity Today. *Pages 1–49 of:* Damour, T., Duplantier, B., and Rivasseau, V. (eds.), *Gravitation and Experiment: Poincaré Seminar 2006*. Progress in Mathematical Physics, vol. 52. Birkhäuser.

Damour, T., and Deruelle, N. 1985. General Relativistic Celestial Mechanics of Binary Systems. I. The post-Newtonian motion. *Ann. Inst. Henri Poincaré Phys. Théor.*, **43**, 107–132.

Damour, T., and Deruelle, N. 1986. General Relativistic Celestial Mechanics of Binary Systems. II. The Post-Newtonian Timing Formula. *Ann. Inst. Henri Poincaré Phys. Théor.*, **44**, 263–292.

Damour, T., and Ruffini, R. 1974. Sur certaines vérifications nouvelles de la Relativité générale rendues possibles par la découverte d'un pulsar membre d'un système binaire. *C. R. Acad. Sc. Paris, Série A*, **279**, 971–973.

Damour, T., and Schäfer, G. 1988. Higher-Order Relativistic Periastron Advances and Binary Pulsars. *Nuovo Cim. B*, **101**, 127–176.

Damour, T., and Schäfer, G. 1991. New Tests of the Strong Equivalence Principle Using Binary-Pulsar Data. *Phys. Rev. Lett.*, **66**, 2549–2552.

Damour, T., and Taylor, J. H. 1992. Strong-Field Tests of Relativistic Gravity and Binary Pulsars. *Phys. Rev. D*, **45**, 1840–1868.

Damour, T., Soffel, M., and Xu, C. 1991. General-Relativistic Celestial Mechanics. I. Method and Definition of Reference Systems. *Phys. Rev. D*, **43**, 3273–3307.

Damour, T., Soffel, M., and Xu, C. 1994. General-Relativistic Celestial Mechanics. IV. Theory of Satellite Motion. *Phys. Rev. D*, **49**, 618–635.

Das, S., and Sur, S. 2022. Dark Matter or Strong Gravity? *Int. J. Mod. Phys. D*, **31**, 2242020–290.

De Felice, A., and Tsujikawa, S. 2010. $f(R)$ Theories. *Living Rev. Relativ.*, **13**, 3.

de Felice, F. 1995. Circular Orbits: A New Relativistic Effect in the Weak Gravitational Field of a Rotating Source. *Class. Quantum Gravit.*, **12**, 1119–1126.

de Martino, I., Chakrabarty, S. S., Cesare, V., et al. 2020. Dark Matters on the Scale of Galaxies. *Universe*, **6**, 107.

de Rham, C. 2014. Massive Gravity. *Living Rev. Relativ.*, **17**, 7.

De Risi, G., Harko, T., and Lobo, F. S. N. 2012. Solar System Constraints on Local Dark Matter Density. *J. Cosmol. Astropart. Phys.*, **2012**, 047.

de Sitter, W. 1916a. Einstein's Theory of Gravitation and Its Astronomical Consequences. *Mon. Not. Roy. Astron. Soc.*, **76**, 699–728.

de Sitter, W. 1916b. On Einstein's Theory of Gravitation and Its Astronomical Consequences. Second Paper. *Mon. Not. Roy. Astron. Soc.*, **77**, 155–184.

Debono, I., and Smoot, G. F. 2016. General Relativity and Cosmology: Unsolved Questions and Future Directions. *Universe*, **2**, 23.

Deeg, H. J., and Belmonte, J. A. 2018. *Handbook of Exoplanets*. Springer.

Demetrian, M. 2006. Motion of a Testing Particle in Gravitational Field of a Ring. *arXiv e-prints*, physics/0602034.

Dey, L., Valtonen, M. J., Gopakumar, A., et al. 2018. Authenticating the Presence of a Relativistic Massive Black Hole Binary in OJ 287 Using Its General Relativity Centenary Flare: Improved Orbital Parameters. *Astrophys. J.*, **866**, 11.

Dey, L., Gopakumar, A., Valtonen, M., et al. 2019. The Unique Blazar OJ 287 and Its Massive Binary Black Hole Central Engine. *Universe*, **5**, 108.

Dicke, R. H. 1970. The Rotation of the Sun. *Pages 289–317 of:* Slettebak, A. (ed.), *Stellar Rotation*. Kluwer.

d'Inverno, R., and Vickers, J. 2022. The Energy-Momentum Tensor. *Pages 203–216 of: Introducing Einstein's Relativity: A Deeper Understanding. Second Edition*. Oxford University Press.

Dirkx, D., Prochazka, I., Bauer, S., et al. 2019. Laser and Radio Tracking for Planetary Science Missions: A Comparison. *J. Geod.*, **93**, 2405–2420.

Doeleman, S., Agol, E., Backer, D., et al. 2009. Imaging an Event Horizon: submm-VLBI of a Super Massive Black Hole. *Page 68 of: astro2010: The Astronomy and Astrophysics Decadal Survey*, vol. 2010.

Doyle, L. R., Carter, J. A., Fabrycky, D. C., et al. 2011. Kepler-16: A Transiting Circumbinary Planet. *Science*, **333**, 1602–1606.

Drake, S. 1978. *Galileo at Work: His Scientific Biography*. University of Chicago Press.

Duerr, P. M., and Wolf, J. W. 2023. Methodological Reflections on the MOND/Dark Matter Debate. *Stud. Hist. Philos. Sci.*, **101**, 1–23.

Dumusque, X., Borsa, F., Damasso, M., et al. 2017. Radial-Velocity Fitting Challenge. II. First Results of the Analysis of the Data Set. *Astron. Astrophys.*, **598**, A133.

Dymnikova, I. G. 1986. REVIEWS OF TOPICAL PROBLEMS: Motion of particles and Photons in the Gravitational Field of a Rotating Body (In Memory of Vladimir Afanas'evich Ruban). *Sov. Phys. Usp.*, **29**, 215–237.

Eastman, J. D., Rodriguez, J. E., Agol, E., et al. 2019. EXOFASTv2: A Public, Generalized, Publication-Quality Exoplanet Modeling Code. *arXiv e-prints*, arXiv:1907.09480.

Ebauer, K. 2017. Development of a Software Package for Determination of Geodynamic Parameters from Combined Processing of SLR Data from LAGEOS and LEO. *Geod. Geodyn.*, **8**, 213–220.

Eckart, A., and Genzel, R. 1996. Observations of Stellar Proper Motions Near the Galactic Centre. *Nature*, **383**, 415–417.

Eckart, A., Genzel, R., Ott, T., and Schödel, R. 2002. Stellar Orbits near Sagittarius A*. *Mon. Not. Roy. Astron. Soc.*, **331**, 917–934.

Efroimsky, M. 2005. Gauge Freedom in Orbital Mechanics. *Ann. N. Y. Acad. Sci.*, **1065**, 346–374.

Egorov, V. A. 1958. Definition of the True Anomaly in Perturbed Motion. *Sov. Astron.*, **2**, 147–149.

Eibe, M. T., Cuesta, L., Ullán, A., Pérez-Verde, A., and Navas, J. 2012. Analysis of Variations in Transit Time and Transit Duration in WASP–3. Evidence of Secular Perturbations Reconsidered. *Mon. Not. Roy. Astron. Soc.*, **423**, 1381–1389.

Einasto, J. 1965. On the Construction of a Composite Model for the Galaxy and on the Determination of the System of Galactic Parameters. *Tr. Astrofiz. Inst. Alma Ata*, **5**, 87–100.

Einstein, A. 1905. Zur Elektrodynamik bewegter Körper. *Ann. Phys.-Berlin*, **322**, 891–921.

Einstein, A. 1915a. Die Feldgleichungen der Gravitation. *Sitzber. Preuss. Akad.*, 844–847.

Einstein, A. 1915b. Erklärung der Perihelbewegung des Merkur aus der allgemeinen Relativitätstheorie. *Sitzber. Preuss. Akad.*, **47**, 831–839.

Einstein, A. 1915c. Zur allgemeinen Relativitätstheorie. *Sitzber. Preuss. Akad.*, 778–786.

Einstein, A. 1915d. Zur allgemeinen Relativitätstheorie (Nachtrag). *Sitzber. Preuss. Akad.*, 799–801.

Einstein, A. 1916. Die Grundlage der allgemeinen Relativitätstheorie. *Ann. Phys.-Berlin*, **354**, 769–822.

Einstein, A. 1955. *The Meaning of Relativity*. Princeton University Press.

Eisenhauer, F., Abuter, R., Bickert, K., et al. 2003. SINFONI - Integral Field Spectroscopy at 50 Milli-arcsecond Resolution with the ESO VLT. *Pages 1548–1561 of:* Iye, M., and Moorwood, A. F. M. (eds.), *Instrument Design and Performance for Optical/Infrared Ground-based Telescopes*. Society of Photo-Optical Instrumentation Engineers (SPIE) Conference Series, vol. 4841.

Eling, C., Jacobson, T., and D., Mattingly. 2006. Einstein-Aether Theory. *Pages 163–179 of:* Liu, J. T., Duff, M. J., and Stelle, K. S. (eds.), *Deserfest: A Celebration of the Life and Works of Stanley Deser*. World Scientific.

Emilio, M., Kuhn, J. R., Bush, R. I., and Scholl, I. F. 2012. Measuring the Solar Radius from Space during the 2003 and 2006 Mercury Transits. *Astrophys. J.*, **750**, 135.

Event Horizon Telescope Collaboration. 2019. First M87 Event Horizon Telescope Results. I. The Shadow of the Supermassive Black Hole. *Astrophys. J. Lett.*, **875**, L1.

Event Horizon Telescope Collaboration. 2022. First Sagittarius A* Event Horizon Telescope Results. I. The Shadow of the Supermassive Black Hole in the Center of the Milky Way. *Astrophys. J. Lett.*, **930**, L12.

Everitt, C. W. F. 1974. The Gyroscope Experiment - I: General Description and Analysis of Gyroscope Performance. *Pages 331–360 of:* Bertotti, B. (ed.), *Proceedings of the International School of Physics 'Enrico Fermi'. Course LVI. Experimental Gravitation*. Academic Press.

Everitt, C. W. F., Buchman, S., Debra, D. B., et al. 2001. Gravity Probe B: Countdown to Launch. *Pages 52–82 of:* Lämmerzahl, C., Everitt, C. W. F., and Hehl, F. W. (eds.), *Gyros, Clocks, Interferometers ...: Testing Relativistic Gravity in Space*. Lecture Notes in Physics, vol. 562. Springer.

Everitt, C. W. F., Debra, D. B., Parkinson, B. W., et al. 2011. Gravity Probe B: Final Results of a Space Experiment to Test General Relativity. *Phys. Rev. Lett.*, **106**, 221101.

Everitt, C. W. F., Muhlfelder, B., Debra, D. B., et al. 2015. The Gravity Probe B Test of General Relativity. *Class. Quantum Gravit.*, **32**, 224001.

Fabrycky, D. C., Ford, E. B., Steffen, J. H., et al. 2012. Transit Timing Observations from *Kepler*. IV. Confirmation of Four Multiple-planet Systems by Simple Physical Models. *Astrophys. J.*, **750**, 114.

Falanga, M., Melia, F., Tagger, M., Goldwurm, A., and Bélanger, G. 2007. General Relativistic Flux Modulations from Disk Instabilities in Sagittarius A*. *Astrophys. J. Lett.*, **662**, L15–L18.

Famaey, B., and McGaugh, S. S. 2012. Modified Newtonian Dynamics (MOND): Observational Phenomenology and Relativistic Extensions. *Living Rev. Relativ.*, **15**, 10.

Farinella, P., Vokrouhlicky, D., and Barlier, F. 1996. The Rotation of LAGEOS and Its Long-Term Semimajor Axis Decay: A Self-Consistent Solution. *J. Geophys. Res.*, **101**, 17861–17872.

Faruque, S. B. 2004. Gravitomagnetic Clock Effect in the Orbit of a Spinning Particle Orbiting the Kerr Black Hole. *Phys. Lett. A*, **327**, 95–97.

Fermi, E. 1922. Sopra i fenomeni che avvengono in vicinanza di una linea oraria. *Rend. Lincei Sci. Fis. Nat.*, **31**, 21–23; 51–52.

Ferraro, R. 2012. $f(R)$ and $f(T)$ Theories of Modified Gravity. *Pages 103–110 of:* Alcaniz, J., Carneiro, S., Chimento, L. P., et al. (eds.), *I Cosmosul: Cosmology and Gravitation in the Southern Cone*. American Institute of Physics Conference Series, vol. 1471. AIP Publishing.

Fienga, A., and Minazzoli, O. 2024. Testing Theories of Gravity with Planetary Ephemerides. *Living Rev. Relativ.*, **27**, 1.

Fischbach, E., Krause, D. E., Mostepanenko, V. M., and Novello, M. 2001. New Constraints on Ultrashort-Ranged Yukawa Interactions from Atomic Force Microscopy. *Phys. Rev. D*, **64**, 075010.

Fischell, R. E., and Pisacane, V. L. 1978. A Drag-Free Lo-Lo Satellite System for Improved Gravity Field Measurements. *Pages 213–219 of:* Müller, I. I. (ed.), *Applications of Geodesy to Geodynamics. Proceedings of the 9th GEOP Conference. Rep. 280 Dep. of Geod. Sci.* Ohio State University.

Fischer, D. A., Anglada-Escude, G., Arriagada, P., et al. 2016. State of the Field: Extreme Precision Radial Velocities. *Publ. Astron. Soc. Pac.*, **128**, 066001.

Fok, V. A. 1959. *The Theory of Space, Time, and Gravitation*. Pergamon Press.

Fokker, A. D. 1921. The Geodesic Precession: A Consequence of Einstein's Theory of Gravitation. *Proc. K. Ned. Akad. Wet.*, **23**, 729–738.

Ford, E. B., Quinn, S. N., and Veras, D. 2008. Characterizing the Orbital Eccentricities of Transiting Extrasolar Planets with Photometric Observations. *Astrophys. J.*, **678**, 1407–1418.

Forward, R. L. 1961. General Relativity for the Experimentalist. *Proc. IRE*, **49**, 892–904.

Fragione, G., and Loeb, A. 2020. An Upper Limit on the Spin of SgrA* Based on Stellar Orbits in Its Vicinity. *Astrophys. J. Lett.*, **901**, L32.

Fragione, G., and Loeb, A. 2022. Implication of Spin Constraints by the Event Horizon Telescope on Stellar Orbits in the Galactic Center. *Astrophys. J. Lett.*, **932**, L17.

Franchini, A., Lodato, G., and Facchini, S. 2016. Lense–Thirring Precession around Supermassive Black Holes during Tidal Disruption Events. *Mon. Not. Roy. Astron. Soc.*, **455**, 1946–1956.

Francisco, F., Bertolami, O., Gil, P. J. S., and Páramos, J. 2012. Modelling the Reflective Thermal Contribution to the Acceleration of the Pioneer Spacecraft. *Phys. Lett. B*, **711**, 337–346.

Frère, J. M., Ling, F. S., and Vertongen, G. 2008. Bound on the Dark Matter Density in the Solar System from Planetary Motions. *Phys. Rev. D*, **77**, 083005.

Frutos–Alfaro, F., and Soffel, M. H. 2018. On Relativistic Multipole Moments of Stationary Space-times. *R. Soc. Open Sci.*, **5**, 180640.

Fukui, A., Narita, N., Tristram, P. J., et al. 2011. Measurements of Transit Timing Variations for WASP–5b. *Publ. Astron. Soc. Pac.*, **63**, 287–300.

Futamase, T., and Itoh, Y. 2007. The Post-Newtonian Approximation for Relativistic Compact Binaries. *Living Rev. Relativ.*, **10**, 2.

Futamase, T., and Matsuda, T. 1979. Resonance between Primordial Gravitational Waves and Gravitationally Bound Systems. *Prog. Theor. Phys.*, **61**, 86–93.

Garrett, K., and Duda, G. 2011. Dark Matter: A Primer. *Adv. Astron.*, **2011**, 968283.

Genzel, R., Eisenhauer, F., and Gillessen, S. 2010. The Galactic Center Massive Black Hole and Nuclear Star Cluster. *Rev. Mod. Phys.*, **82**, 3121–3195.

Genzel, R., Eisenhauer, F., and Gillessen, S. 2024. Experimental Studies of Black Holes: Status and Future Prospects. *Astron. Astrophys. Rev.*, arXiv:2404.03522.

Gergely, L. A. 2010. Spinning Compact Binary Inspiral. II: Conservative Angular Dynamics. *Phys. Rev. D*, **82**, 104031.

Geroch, R. 1970. Multipole Moments. II: Curved Space. *J. Math. Phys.*, **11**, 2580–2588.

Getley, A. K., Carter, B., King, R., and O'Toole, S. 2017. Evidence for a Planetary Mass Third Body Orbiting the Binary Star KIC 5095269. *Mon. Not. Roy. Astron. Soc.*, **468**, 2932–2937.

Gezari, S., Ghez, A. M., Becklin, E. E., et al. 2002. Adaptive Optics Near-Infrared Spectroscopy of the Sagittarius A* Cluster. *Astrophys. J.*, **576**, 790–797.

Ghez, A. M., Duchêne, G., Matthews, K., et al. 2003. The First Measurement of Spectral Lines in a Short-Period Star Bound to the Galaxy's Central Black Hole: A Paradox of Youth. *Astrophys. J. Lett.*, **586**, L127–L131.

Ghez, A. M., Salim, S., Weinberg, N. N., et al. 2008. Measuring Distance and Properties of the Milky Way's Central Supermassive Black Hole with Stellar Orbits. *Astrophys. J.*, **689**, 1044–1062.

Gilbertson, C., Ford, E. B., Jones, D. E., and Stenning, D. C. 2020. Toward Extremely Precise Radial Velocities. II: A Tool for Using Multivariate Gaussian Processes to Model Stellar Activity. *Astrophys. J.*, **905**, 155.

Gill, E., Soffel, M., Ruder, H., and Schneider, M. 1992. *Relativistic Motion of Gyroscopes and Space Gradiometry*. German Geodetic Commission, Series A Theoretical Geodesy, vol. 107. Bavarian Academy of Sciences.

Gillessen, S., Eisenhauer, F., Trippe, S., et al. 2009. Monitoring Stellar Orbits around the Massive Black Hole in the Galactic Center. *Astrophys. J.*, **692**, 1075–1109.

Gillessen, S., Plewa, P. M., Eisenhauer, F., et al. 2017. An Update on Monitoring Stellar Orbits in the Galactic Center. *Astrophys. J.*, **837**, 30.

Ginzburg, V. L. 1957. The Use of Artificial Earth Satellites for Verifying the General Theory of Relativity. *Usp. Fiz. Nauk*, **63**, 119–122.

Ginzburg, V. L. 1959. Artificial Satellites and the Theory of Relativity. *Sci. Am.*, **200**, 149–160.

Giraud, Q., and Pétri, J. 2021. Frequency-Resolved Radio and High-Energy Emission of Pulsars. *Astron. Astrophys.*, **654**, A86.

Goldreich, P. 1966. Final Spin States of Planets and Satellites. *Astron J.*, **71**, 1.

Goldstein, H. 1980. *Classical Mechanics. Second Edition*. Addison Wesley.

Gou, X., Pan, X., and Wang, L. 2021. General Relativity Testing in Exoplanetary Systems. *IOP Conf. Ser.: Earth Environ. Sci.*, **658**, 012051.

Gouttenoire, Y. 2023. Standard Model of Elementary Particles. *Pages 9–71 of: Beyond the Standard Model Cocktail: A Modern and Comprehensive Review of the Major Open Puzzles in Theoretical Particle Physics and Cosmology with a Focus on Heavy Dark Matter*. Springer Theses. Springer.

GRAVITY Collaboration. 2019. A Geometric Distance Measurement to the Galactic Center Black Hole With 0.3% Uncertainty. *Astron. Astrophys.*, **625**, L10.

GRAVITY Collaboration, Abuter, R., Amorim, A., et al. 2020. Detection of the Schwarzschild Precession in the Orbit of the Star S2 Near the Galactic Centre Massive Black Hole. *Astron. Astrophys.*, **636**, L5.

Gray, D. F. 1982. The Rotation of Cool Main-Sequence Stars. *Astrophys. J.*, **261**, 259–264.

Greif, S. K., Hebeler, K., Lattimer, J. M., Pethick, C. J., and Schwenk, A. 2020. Equation of State Constraints from Nuclear Physics, Neutron Star Masses, and Future Moment of Inertia Measurements. *Astrophys. J.*, **901**, 155.

Grenier, S., Baylac, M.-O., Rolland, L., et al. 1999. Radial Velocities: Measurements of 2800 B2–F5 Stars for HIPPARCOS. *Astron. Astrophys. Suppl. Ser.*, **137**, 451–456.

Grishchuk, L. P., and Polnarev, A. G. 1980. Gravitational Waves and Their Interaction with Matter and Fields. *Pages 393–434 of:* Held, A. (ed.), *General Relativity and Gravitation: One Hundred Years after the Birth of Albert Einstein. Volume II*, vol. 2. Plenum Press.

Grøn, O., and Soleng, H. H. 1996. Experimental Limits on the Density of Dark Matter in the Solar System. *Astrophys. J.*, **456**, 445.

Gronwald, F., Gruber, E., Lichtenegger, H. I. M., and Puntigam, R. A. 1997. Gravity Probe C(lock)-Probing the Gravitomagnetic Field of the Earth by Means of a Clock Experiment. *Pages 29–35 of:* Wilson, A. (ed), *Proceedings of the Alpbach Summer*

School 1997 on Fundamental Physics in Space, Alpbach, Austria, 22-31 July 1997. ESA-SP420. ESA Publications Division.

Gurfil, P. 2004. Analysis of J_2-Perturbed Motion Using Mean Non-osculating Orbital Elements. *Celest. Mech. Dyn. Astr.*, **90**, 289–306.

Gurfil, P., and Efroimsky, M. 2022. Analysis of the PPN Two-Body Problem Using Non-osculating Orbital Elements. *Adv. Space Res.*, **69**, 538–553.

Gurfil, P., and Seidelmann, P. K. 2016. *Celestial Mechanics and Astrodynamics: Theory and Practice*. Springer.

Haas, M. R., and Ross, D. K. 1975. Measurement of the Angular Momentum of Jupiter and the Sun by Use of the Lense–Thirring Effect. *Astrophys. Space Sci.*, **32**, 3–11.

Habibi, M., Gillessen, S., Martins, F., et al. 2017. Twelve Years of Spectroscopic Monitoring in the Galactic Center: The Closest Look at S-Stars Near the Black Hole. *Astrophys. J.*, **847**, 120.

Hackmann, E., and Lämmerzahl, C. 2014. Generalized Gravitomagnetic Clock Effect. *Phys. Rev. D*, **90**, 044059.

Haghighipour, N. (ed.). 2010. *Planets in Binary Star Systems*. Astrophysics and Space Science Library, vol. 366. Springer.

Han, W.-B. 2014. Possibility of Measuring Spin Precession of the Nearest Supermassive Black Hole by Using S Stars. *Res. Astron. Astrophys.*, **14**, 1415–1422.

Hansen, R. O. 1974. Multipole Moments of Stationary Space-Times. *J. Math. Phys.*, **15**, 46–52.

Hara, N. C., and Ford, E. B. 2023. Statistical Methods for Exoplanet Detection with Radial Velocities. *Annu. Rev. Stat. Appl.*, **10**, 623–649.

Harikumar, S. 2022. Moffat MOdified Gravity (MOG). *Universe*, **8**, 259.

Harris, E. G. 1991. Analogy between General Relativity and Electromagnetism for Slowly Moving Particles in Weak Gravitational Fields. *Am. J. Phys.*, **59**, 421–425.

Harvey, A. 2009. Dark Energy and the Cosmological Constant: A Brief Introduction. *Eur. Phys. J. Plus*, **30**, 877–889.

Haswell, C. A. 2010. *Transiting Exoplanets*. Cambridge University Press.

Hawking, S. W., and Ellis, G. F. R. 1973. *The Large Scale Structure of Space-Time.* Cambridge University Press.

Hébrard, G., Bouchy, F., Pont, F., et al. 2008. Misaligned Spin-Orbit in the XO–3 Planetary System? *Astron. Astrophys.*, **488**, 763–770.

Heimberger, J., Soffel, M., and Ruder, H. 1989. Relativistic Effects in the Motion of Artificial Satellites: The Oblateness of the Central Body II. *Celest. Mech. Dyn. Astr.*, **47**, 205–217.

Hellard, H., Csizmadia, S., Padovan, S., Rauer, H., Cabrera, J., Sohl, F., Spohn, T., and Breuer, D. 2019. Retrieval of the Fluid Love Number k_2 in Exoplanetary Transit Curves. *Astrophys. J.*, **878**, 119.

Heller, R., Leconte, J., and Barnes, R. 2011. Tidal Obliquity Evolution of Potentially Habitable Planets. *Astron. Astrophys.*, **528**, A27.

Henry, G. W., Marcy, G. W., Butler, R. P., and Vogt, S. S. 2000. A Transiting '51 Peg-like' Planet. *Astrophys. J. Lett.*, **529**, L41–L44.

Hirano, T., Sanchis-Ojeda, R., Takeda, Y., et al. 2014. Measurements of Stellar Inclinations for Kepler Planet Candidates. II: Candidate Spin-Orbit Misalignments in Single- and Multiple-Transiting Systems. *Astrophys. J.*, **783**, 9.

Hoffmeister, C. 1930. 93 neue Veränderliche. *Astron. Nachr.*, **240**, 193–198.

Hofmann, F., and Müller, J. 2018. Relativistic Tests with Lunar Laser Ranging. *Class. Quantum Gravit.*, **35**, 035015.

Hu, H., and Freire, Paulo C. C. 2024. Measuring the Lense–Thirring Precession and the Neutron Star Moment of Inertia with Pulsars. *Universe*, **10**, 160.

Hu, H., Kramer, M., Wex, N., Champion, D. J., and Kehl, M. S. 2020. Constraining the Dense Matter Equation-of-State with Radio Pulsars. *Mon. Not. Roy. Astron. Soc.*, **497**, 3118–3130.

Hu, J.-P., and Wang, F.-Y. 2023. Hubble Tension: The Evidence of New Physics. *Universe*, **9**, 94.

Huang, C., and Liu, L. 1992. Analytical Solutions to the Four Post-Newtonian Effects in a Near-Earth Satellite Orbit. *Celest. Mech. Dyn. Astr.*, **53**, 293–307.

Huang, C., Ries, J. C., Tapley, B. D., and Watkins, M. M. 1990. Relativistic Effects for Near-Earth Satellite Orbit Determination. *Celest. Mech. Dyn. Astr.*, **48**, 167–185.

Hulse, R. A., and Taylor, J. H. 1975. Discovery of a Pulsar in a Binary System. *Astrophys. J. Lett.*, **195**, L51–L53.

Ibba, R., Rum, G., Varesio, F., and Bussolino, L. 1989. IRIS–LAGEOS 2 Mission. *Acta Astronaut.*, **19**, 521–527.

Iess, L., and Boscagli, G. 2001. Advanced Radio Science Instrumentation for the Mission BepiColombo to Mercury. *Planet. Space Sci.*, **49**, 1597–1608.

Iess, L., Folkner, W. M., Durante, D., et al. 2018. Measurement of Jupiter's Asymmetric Gravity Field. *Nature*, **555**, 220–222.

Ingram, A., Done, C., and Fragile, P. C. 2009. Low-Frequency Quasi-Periodic Oscillations Spectra and Lense–Thirring Precession. *Mon. Not. Roy. Astron. Soc.*, **397**, L101–L105.

Ingram, A., van der Klis, M., Middleton, M., et al. 2016. A Quasi-Periodic Modulation of the Iron Line Centroid Energy in the Black Hole Binary H1743–322. *Mon. Not. Roy. Astron. Soc.*, **461**, 1967–1980.

Iorio, L. 2001a. Satellite Gravitational Orbital Perturbations and the Gravitomagnetic Clock Effect. *Int. J. Mod. Phys. D*, **10**, 465–476.

Iorio, L. 2001b. Satellite Non-gravitational Orbital Perturbations and the Detection of the Gravitomagnetic Clock Effect. *Class. Quantum Gravit.*, **18**, 4303–4310.

Iorio, L. 2005. Is It Possible to Measure the Lense–Thirring Effect on the Orbits of the Planets in the Gravitational Field of the Sun? *Astron. Astrophys.*, **431**, 385–389.

Iorio, L. 2006a. Are We Far from Testing General Relativity with the Transmitting Extrasolar Planet HD 209458b 'Osiris'? *New Astron.*, **11**, 490–494.

Iorio, L. 2006b. COMMENTS, REPLIES AND NOTES: A Note on the Evidence of the Gravitomagnetic Field of Mars. *Class. Quantum Gravit.*, **23**, 5451–5454.

Iorio, L. 2006c. Solar System Planetary Orbital Motions and Dark Matter. *J. Cosmol. Astropart. Phys.*, **2006**, 002.

Iorio, L. 2009a. Mars and Frame-Dragging: Study for a Dedicated Mission. *Gen. Relativ. Gravit.*, **41**, 1273–1284.

Iorio, L. 2009b. Prospects for Measuring the Moment of Inertia of Pulsar J0737-3039A. *New Astron.*, **14**, 40–43.

Iorio, L. 2010a. Effect of Sun and Planet-Bound Dark Matter on Planet and Satellite Dynamics in the Solar System. *J. Cosmol. Astropart. Phys.*, **2010**, 018.

Iorio, L. 2010b. Juno, the Angular Momentum of Jupiter and the Lense–Thirring Effect. *New Astron.*, **15**, 554–560.

Iorio, L. 2011a. Classical and Relativistic Long-Term Time Variations of Some Observables for Transiting Exoplanets. *Mon. Not. Roy. Astron. Soc.*, **411**, 167–183.

Iorio, L. 2011b. Classical and Relativistic Node Precessional Effects in WASP–33b and Perspectives for Detecting Them. *Astrophys. Space Sci.*, **331**, 485–496.

Iorio, L. 2011c. Long-Term Classical and General Relativistic Effects on the Radial Velocities of the Stars Orbiting Sgr A*. *Mon. Not. Roy. Astron. Soc.*, **411**, 453–463.

Iorio, L. 2012a. Constraining the Angular Momentum of the Sun with Planetary Orbital Motions and General Relativity. *Sol. Phys.*, **281**, 815–826.

Iorio, L. 2012b. Constraining the Electric Charges of Some Astronomical Bodies in Reissner–Nordström Spacetimes and Generic r^{-2}-type power-law potentials from orbital motions. *Gen. Relativ. Gravit.*, **44**, 1753–1767.

Iorio, L. 2012c. Constraints from Orbital Motions around the Earth of the Environmental Fifth-Force Hypothesis for the OPERA Superluminal Neutrino Phenomenology. *J. High Energy Phys.*, **2012**, 73.

Iorio, L. 2012d. Model-Independent Constraints on r^{-3} Extra-interactions from Orbital Motions. *Ann. Phys.-Berlin*, **524**, 371–377.

Iorio, L. 2012e. Orbital Effects of Lorentz-Violating Standard Model Extension Gravitomagnetism around a Static Body: A Sensitivity Analysis. *Class. Quantum Gravit.*, **29**, 175007.

Iorio, L. 2013a. A Possible New Test of General Relativity with Juno. *Class. Quantum Gravit.*, **30**, 195011.

Iorio, L. 2013b. Exact Expressions for the Pericenter Precession Caused by Some Dark Matter Distributions and Constraints on Them from Orbital Motions in the Solar System, in the Double Pulsar and in the Galactic Center. *Galaxies*, **1**, 6–30.

Iorio, L. 2013c. LETSGO: A Spacecraft-Based Mission to Accurately Measure the Solar Angular Momentum with Frame-Dragging. *Acta Astronaut.*, **86**, 149–157.

Iorio, L. 2014a. Orbital Effects of a Monochromatic Plane Gravitational Wave with Ultra-Low Frequency Incident on a Gravitationally Bound Two-Body System. *ScienceOpen Res.*, **2014**, 1.

Iorio, L. 2014b. Orbital Motions as Gradiometers for Post-Newtonian Tidal Effects. *Front. Astron. Space Sci.*, **1**, 3.

Iorio, L. 2015a. Editorial for the Special Issue 100 Years of Chronogeometrodynamics: The Status of the Einstein's Theory of Gravitation in Its Centennial Year. *Universe*, **1**, 38–81.

Iorio, L. 2015b. Post-Newtonian Direct and Mixed Orbital Effects due to the Oblateness of the Central Body. *Int. J. Mod. Phys. D*, **24**, 1550067–59.

Iorio, L. 2016a. Accurate Characterization of the Stellar and Orbital parameters of the Exoplanetary System WASP–33 b from Orbital Dynamics. *Mon. Not. Roy. Astron. Soc.*, **455**, 207–213.

Iorio, L. 2016b. Post-Keplerian Corrections to the Orbital Periods of a Two-Body System and Their Measurability. *Mon. Not. Roy. Astron. Soc.*, **460**, 2445–2452.

Iorio, L. 2017. Post-Keplerian Perturbations of the Orbital Time Shift in Binary Pulsars: An Analytical Formulation with Applications to the Galactic Center. *Eur. Phys. J. C*, **77**, 439.

Iorio, L. 2018. Analytically Calculated Post-Keplerian Range and Range-Rate Perturbations: The Solar Lense–Thirring Effect and BepiColombo. *Mon. Not. Roy. Astron. Soc.*, **476**, 1811–1825.

Iorio, L. 2019a. A HERO for General Relativity. *Universe*, **5**, 165.

Iorio, L. 2019b. A Post-Newtonian Gravitomagnetic Effect on the Orbital Motion of a Test Particle around Its Primary Induced by the Spin of a Distant Third Body. *Universe*, **5**, 87.

Iorio, L. 2019c. Calculation of the Uncertainties in the Planetary Precessions with the Recent EPM2017 Ephemerides and Their Use in Fundamental Physics and Beyond. *Astron J.*, **157**, 220.

Iorio, L. 2019d. Erratum: The Post-Newtonian Gravitomagnetic Spin-Octupole Moment of an Oblate Rotating Body and Its Effects on an Orbiting Test Particle; Are They Measurable in the Solar System? *Mon. Not. Roy. Astron. Soc.*, **485**, 4090–4090.

Iorio, L. 2019e. Measuring General Relativistic Dragging Effects in the Earth's Gravitational Field with ELXIS: A Proposal. *Class. Quantum Gravit.*, **36**, 035002.

Iorio, L. 2019f. Measuring the De Sitter Precession with a New Earth's Satellite to the $\simeq 10^{-5}$ Level: A Proposal. *Eur. Phys. J. C*, **79**, 64.

Iorio, L. 2019g. The Post-Newtonian Gravitomagnetic Spin-Octupole Moment of an Oblate Rotating Body and Its Effects on an Orbiting Test Particle; Are They Measurable in the Solar System? *Mon. Not. Roy. Astron. Soc.*, **484**, 4811–4832.

Iorio, L. 2020a. A Comment on 'Lense–Thirring Frame Dragging Induced by a Fast-Rotating White Dwarf in a Binary Pulsar System' by V. Venkatraman Krishnan et al. *Mon. Not. Roy. Astron. Soc.*, **495**, 2777–2785.

Iorio, L. 2020b. Revisiting the 2PN Pericenter Precession in View of Possible Future Measurements. *Universe*, **6**, 53.

Iorio, L. 2020c. The Short-Period S-Stars S4711, S62, S4714 and the Lense–Thirring Effect due to the Spin of Sgr A*. *Astrophys. J.*, **904**, 186.

Iorio, L. 2021a. On the 2PN Periastron Precession of the Double Pulsar PSR J0737-3039A/B. *Universe*, **7**, 443.

Iorio, L. 2021b. On the 2PN Pericentre Precession in the General Theory of Relativity and the Recently Discovered Fast-Orbiting S-Stars in Sgr A*. *Universe*, **7**, 37.

Iorio, L. 2021c. The Impact of the Spin-Orbit Misalignment and of the Spin of B on the Lense–Thirring Orbital Precessions of the Double Pulsar PSR J0737-3039A/B. *Mon. Not. Roy. Astron. Soc.*, **507**, 421–430.

Iorio, L. 2022. Frame-Dragging in Extrasolar Circumbinary Planetary Systems. *Universe*, **8**, 546.

Iorio, L. 2023a. Is It Possible to Measure the Lense–Thirring Orbital Shifts of the Short-Period S Star S4716 Orbiting Sgr A*? *Astrophys. J.*, **954**, 219.

Iorio, L. 2023b. Limitations in Testing the Lense–Thirring Effect with LAGEOS and the Newly Launched Geodetic Satellite LARES 2. *Universe*, **9**, 211.

Iorio, L. 2023c. Post-Newtonian Effects on Some Characteristic Time-Scales of Transiting Exoplanets. *Mon. Not. Roy. Astron. Soc.*, **518**, 2599–2613.

Iorio, L. 2023d. The Lense–Thirring Effect on the Galilean Moons of Jupiter. *Universe*, **9**, 304.

Iorio, L. 2023e. The Post-Newtonian Motion around an Oblate Spheroid: The Mixed Orbital Effects due to the Newtonian Oblateness and the Post-Newtonian Mass Monopole Accelerations. *Gen. Relativ. Gravit.*, **55**, 136.

Iorio, L. 2024a. Post–Newtonian Orbital Effects Induced by the Mass Quadrupole and Spin Octupole Moments of an Axisymmetric Body. *Astron J.*, **167**, 78.

Iorio, L. 2024b. Revisiting the Gravitomagnetic Clock Effect. *Eur. Phys. J. C*, **84**, 280.

Iorio, L., and Lichtenegger, H. I. M. 2005. On the Possibility of Measuring the Gravitomagnetic Clock Effect in an Earth Space-Based Experiment. *Class. Quantum Gravit.*, **22**, 119–132.

Iorio, L, and Mashhoon, B. 2024. Review of the Gravitomagnetic Clock Effect. *Ann. Phys.-Berlin*, **536**, 2300466.

Iorio, L., and Ruggiero, M. L. 2010. Constraining the Kehagias–Sfetsos Solution of the Hořava–Lifshitz Modified Gravity with Extrasolar Planets. *The Open Astronomy Journal*, **3**, 167–171.

Iorio, L., and Ruggiero, M. L. 2018. Constraining some r^{-n} Extra-potentials in Modified Gravity Models with LAGEOS-type Laser-Ranged Geodetic Satellites. *J. Cosmol. Astropart. Phys.*, **2018**, 021.

Iorio, L., and Saridakis, E. N. 2012. Solar System Constraints on $f(T)$ Gravity. *Mon. Not. Roy. Astron. Soc.*, **427**, 1555–1561.

Iorio, L., Ciufolini, I., Pavlis, E.C., et al. 2004. On the Possibility of Measuring the Lense Thirring Effect with a LAGEOS LAGEOS II OPTIS Mission. *Class. Quantum Gravit.*, **21**, 2139–2151.

Iorio, L., Lichtenegger, H. I. M., Ruggiero, M. L., and Corda, C. 2011. Phenomenology of the Lense–Thirring Effect in the Solar System. *Astrophys. Space Sci.*, **331**, 351–395.

Iorio, L., Ruggiero, M. L., and Corda, C. 2013. Novel Considerations about the Error Budget of the LAGEOS–Based Tests of Frame-Dragging with GRACE Geopotential Models. *Acta Astronaut.*, **91**, 141–148.

Iorio, L., Rieutord, M., Rozelot, J.-P., and Domiciano de Souza, A. 2019. Classical and General Relativistic Post-Keplerian Effects in Binary Pulsars Hosting Fast Rotating Main Sequence Stars. *Eur. Phys. J. C*, **79**, 690.

Iorio, L., Girija, A. P., and Durante, D. 2023. One EURO for Uranus: The Elliptical Uranian Relativity Orbiter Mission. *Mon. Not. Roy. Astron. Soc.*, **523**, 3595–3614.

Israel, W. 1967. Event Horizons in Static Vacuum Space-Times. *Phys. Rev.*, **164**, 1776–1779.

Ivashchenko, A. V. 1987. Variation of the Keplerian Elements of a Planetary Orbit under the Action of a Gravitational Wave. *Sov. Astron.*, **31**, 76.

Jacobson, T. 2008. Einstein–Æther Gravity: Theory and Observational Constraints. *Pages 92–99 of:* Kostelecký, V.A. (ed), *CPT and Lorentz Symmetry. Proceedings of the Fourth Meeting, Bloomington, USA, 8–11 August 2007*. World Scientific.

Jacobson, T., and Mattingly, D. 2004. Einstein–Aether Waves. *Phys. Rev. D*, **70**, 024003.

Jain, C., Paul, B., Sharma, R., Jaleel, A., and Dutta, A. 2017. Indication of a Massive Circumbinary Planet Orbiting the Low-Mass X-Ray Binary MXB 1658–298. *Mon. Not. Roy. Astron. Soc.*, **468**, L118–L122.

Jantzen, R. T., Carini, P., and Bini, D. 1992a. Gravitoelectromagnetism: Relativity of Splitting Formalisms. *J. Korean Phys. Soc.*, **25**, S230–S240.

Jantzen, R. T., Carini, P., and Bini, D. 1992b. The Many Faces of Gravitoelectromagnetism. *Ann. Phys. (N Y)*, **215**, 1–50.

Jaroszynski, M. 1998. Relativistic Effects in Proper Motions of Stars Surrounding the Galactic Center. *Acta Astronaut.*, **48**, 653–665.

Jiale Li, K., Wu, K., Younsi, Z., Teixeira, J., and Singh, D. 2024. Generic Gravito-magnetic Clock Effects. *Mon. Not. Roy. Astron. Soc.*, **530**, 1118–1128.

Jordán, A., and Bakos, G. Á. 2008. Observability of the General Relativistic Precession of Periastra in Exoplanets. *Astrophys. J.*, **685**, 543–552.

Jordán, A., and Bakos, G. Á. 2009. Observability of the General Relativistic Precession of Periastra in Exoplanets. *Pages 492–495 of:* Pont, F., Sasselov, D., and Holman, M. (eds.), *Transiting Planets. Proceedings IAU Symposium No. 253*. Proceedings of the International Astronomical Union Symposia and Colloquia, vol. 253. Cambridge University Press.

Jørgensen, J. H., Bjælde, O. E., and Hannestad, S. 2016. Probing the Spin of the Central Black Hole in the Galactic Centre with Secondary Images. *Mon. Not. Roy. Astron. Soc.*, **458**, 3614–3618.

Joyce, A., Lombriser, L., and Schmidt, F. 2016. Dark Energy versus Modified Gravity. *Annu. Rev. Nucl. Part. Sci.*, **66**, 95–122.

Jurgenson, C., Fischer, D., McCracken, T., et al. 2016. EXPRES: A Next Generation RV Spectrograph in the Search for Earth-like Worlds. *Page 99086T of:* Evans, C. J., Simard, L., and Takami, H. (eds.), *Ground-Based and Airborne Instrumentation for Astronomy VI*. Society of Photo–Optical Instrumentation Engineers (SPIE) Conference Series, vol. 9908.

Kalitzin, N. St. 1959. Über die Bewegung der rotierenden Satelliten und Doppelsterne nach der Einsteinschen Gravitationstheorie. *Nuovo Cim.*, **11**, 178–185.

Kaluzienski, L. J., and Holt, S. S. 1977. Variable X-Ray Sources. *IAU Circ.*, **3099**, 3.

Kane, S. R., Horner, J., and von Braun, K. 2012. Cyclic Transit Probabilities of Long-Period Eccentric Planets due to Periastron Precession. *Astrophys. J.*, **757**, 105.

Kaplan, G. H. 2015. Coordinate System Issues in Binary Star Computations. Division A poster 1.05. *Page 2255042 of: IAU General Assembly*, vol. 29.

Kaspi, V. M., Lyne, A. G., Manchester, R. N., et al. 2000. Discovery of a Young Radio Pulsar in a Relativistic Binary Orbit. *Astrophys. J.*, **543**, 321–327.

Kassimenko, T. V. 1966. Evaluation of the Satellite Period on the Base of Simultaneous Visual Tracking from Two Given Stations. *Pages 19–22 of:* Kovalevsky, J. (ed.), *Trajectories of Artificial Celestial Bodies as Determined from Observations/Trajectoires des Corps Celestes Artificiels Déterminées D'après les Observations*. Springer.

Kasting, J. F., Whitmire, D. P., and Reynolds, R. T. 1993. Habitable Zones around Main Sequence Stars. *Icarus*, **101**, 108–128.

Kaula, W. M. 2000. *Theory of Satellite Geodesy: Applications of Satellites to Geodesy*. Dover.

Kehl, M. S., Wex, N., Kramer, M., and Liu, K. 2017. Future Measurements of the Lense–Thirring Effect in the Double Pulsar. *Pages 1860–1865 of:* Bianchi, M., Jantzen, R., and Ruffini, R. (eds.), *The Fourteenth Marcel Grossmann Meeting. Proceedings of the MG14 Meeting on General Relativity*. World Scientific.

Kellermann, C., Becker, A., and Redmer, R. 2018. Interior Structure Models and Fluid Love Numbers of Exoplanets in the Super-Earth Regime. *Astron. Astrophys.*, **615**, A39.

Kenyon, I. R. 2023. *Introduction to General Relativity and Cosmology (Second Edition)*. IOP Publishing.

Kerr, A. W., Hauck, J. C., and Mashhoon, B. 2003. Standard Clocks, Orbital Precession and the Cosmological Constant. *Class. Quantum Gravit.*, **20**, 2727–2736.

Kerr, R. P. 1963. Gravitational Field of a Spinning Mass as an Example of Algebraically Special Metrics. *Phys. Rev. Lett.*, **11**, 237–238.

Khriplovich, I. B. 2007. Density of Dark Matter in the Solar System and Perihelion Precession of Planets. *Int. J. Mod. Phys. D*, **16**, 1475–1478.

Khriplovich, I. B., and Pitjeva, E. V. 2006. Upper Limits on Density of Dark Matter in Solar System. *Int. J. Mod. Phys. D*, **15**, 615–618.

Kidder, L. E. 1995. Coalescing Binary Systems of Compact Objects to (post)$^{5/2}$–Newtonian Order. V: Spin effects. *Phys. Rev. D*, **52**, 821–847.

Kipping, D. M., Hartman, J., Bakos, G. Á., et al. 2011. HAT-P-31b,c: A Transiting, Eccentric, Hot Jupiter and a Long-Period, Massive Third Body. *Astron J.*, **142**, 95.

Kitchin, C. 2012. *Exoplanets: Finding, Exploring, and Understanding Alien Worlds*. Springer.

Klioner, S. A., and Kopeikin, S. M. 1994. The Post-Keplerian Orbital Representations of the Relativistic Two-Body Problem. *Astrophys. J.*, **427**, 951.

Kobayashi, T. 2019. Horndeski Theory and Beyond: A Review. *Rep. Prog. Phys.*, **82**, 086901.

Kochkin, R. A., and Sbytov, Y. G. 1987. Influence of a Periodic Gravitational Wave on the Parameters of a Binary System. *Sov. Astron.*, **31**, 537.

Konacki, M., Maciejewski, A. J., and Wolszczan, A. 2000. Improved Timing Formula for the PSR B1257+12 Planetary System. *Astrophys. J.*, **544**, 921–926.

Königsdörffer, C., and Gopakumar, A. 2006. Parametric Derivation of the Observable Relativistic Periastron Advance for Binary Pulsars. *Phys. Rev. D*, **73**, 044011.

Kopal, Z. 1959. *Close binary systems*. Chapman & Hall.

Kopeikin, S. M., and Potapov, V. A. 1994. Relativistic Shift of the Periastron of a Double Pulsar in the Post-Post-Newtonian Approximation of General Relativity. *Astron. Rep.*, **38**, 104–114.

Kopeikin, S. M., Efroimsky, M., and Kaplan, G. 2011. *Relativistic Celestial Mechanics of the Solar System*. Wiley.

Kornfeld, R. P., Arnold, B. W., Gross, M. A., et al. 2019. GRACE-FO: The Gravity Recovery and Climate Experiment Follow-On Mission. *J. Spacecr. Rockets*, **56**, 931–951.

Kostelecký, V. A. 2004. Gravity, Lorentz Violation, and the Standard Model. *Phys. Rev. D*, **69**, 105009.

Kostelecký, V. A., and Potting, R. 1995. CPT, Strings, and Meson Factories. *Phys. Rev. D*, **51**, 3923–3935.

Kostelecký, V. A., and Potting, R. 2005. Gravity from Local Lorentz Violation. *Int. J. Mod. Phys. D*, **14**, 2341–2346.

Kostelecký, V. A., and Potting, R. 2009. Gravity from Spontaneous Lorentz Violation. *Phys. Rev. D*, **79**, 065018.

Kostov, V. B., McCullough, P. R., Carter, J. A., et al. 2014. Kepler-413b: A Slightly Misaligned, Neptune-Size Transiting Circumbinary Planet. *Astrophys. J.*, **784**, 14.

Kostov, V. B., Orosz, J. A., Welsh, W. F., et al. 2016. Kepler–1647b: The Largest and Longest-Period *Kepler* Transiting Circumbinary Planet. *Astrophys. J.*, **827**, 86.

Kostov, V. B., Powell, B. P., Orosz, J. A., et al. 2021. TIC 172900988: A Transiting Circumbinary Planet Detected in One Sector of TESS Data. *Astron J.*, **162**, 234.

Kottler, F. 1918. Über die physikalischen Grundlagen der Einsteinschen Gravitationstheorie. *Ann. Phys.-Berlin*, **361**, 401–462.

Kozak, A., and Wojnar, A. 2021. Metric-Affine Gravity Effects on Terrestrial Exoplanet Profiles. *Phys. Rev. D*, **104**, 084097.

Kraft, R. P. 1969. Stellar rotation. *Pages 317–367 of:* Chiu, H. Y., Warasila, R. L., and Remo, J. L. (eds.), *Stellar Astronomy. Vol. 2*. Gordon & Breach.

Kraft, R. P. 1970. Stellar Rotation. *Pages 385–422 of:* Herbig, G. H., and Struve, O. (eds.), *Spectroscopic Astrophysics: An Assessment of the Contributions of Otto Struve*. University of California Press.

Kramer, M, and Wex, N. 2009. TOPICAL REVIEW: The Double Pulsar System: A Unique Laboratory for Gravity. *Class. Quantum Gravit.*, **26**, 073001.

Kramer, M., Stairs, I. H., Manchester, R. N., et al. 2006. Tests of General Relativity from Timing the Double Pulsar. *Science*, **314**, 97–102.

Kramer, M., Stairs, I. H., Manchester, R. N., et al. 2021. Strong-Field Gravity Tests with the Double Pulsar. *Phys. Rev. X*, **11**, 041050.

Kraniotis, G. V. 2007. Periapsis and Gravitomagnetic Precessions of Stellar Orbits in Kerr and Kerr–de Sitter Black Hole Spacetimes. *Class. Quantum Gravit.*, **24**, 1775–1808.

Kraus, A. L., Ireland, M. J., Cieza, L. A., et al. 2014. Three Wide Planetary-Mass Companions to FW Tau, ROXs 12, and ROXs 42b. *Astrophys. J.*, **781**, 20.

Krogh, K. 2007. COMMENTS, REPLIES AND NOTES: Comment on 'Evidence of the Gravitomagnetic Field of Mars'. *Class. Quantum Gravit.*, **24**, 5709–5715.

Kubo, Y. 2022. Attempt to Obtain the General Relativistic Planet's Motion by Special Relativity Techniques. *Astron. Nachr.*, **343**, e20038.

Kuntz, A. 2022. Precession Resonances in Hierarchical Triple Systems. *Phys. Rev. D*, **105**, 024017.

Kuntz, A., Serra, F., and Trincherini, E. 2021. Effective Two-Body Approach to the Hierarchical Three-Body Problem. *Phys. Rev. D*, **104**, 024016.

Laarakkers, W. G., and Poisson, E. 1999. Quadrupole Moments of Rotating Neutron Stars. *Astrophys. J.*, **512**, 282–287.

Lahav, O. 2020. Dark Energy: Is It 'Just' Einstein's Cosmological Constant Λ? *Contemp. Phys.*, **61**, 132–145.

Lämmerzahl, C., and Dittus, H. 2008. Fundamental Physics, Space, Missions and Technologies. *Pages 3–25 of:* Dittus, H., Lämmerzahl, C., and Turyshev, S.G. (eds), *Lasers, Clocks and Drag-Free Control.* Astrophysics and Space Science Library, vol. 249. Springer.

Lattimer, J. M., and Prakash, M. 2001. Neutron Star Structure and the Equation of State. *Astrophys. J.*, **550**, 426–442.

Lattimer, J. M., and Schutz, B. F. 2005. Constraining the Equation of State with Moment of Inertia Measurements. *Astrophys. J.*, **629**, 979–984.

Le Verrier, U. J. 1859a. Lettre de M. Le Verrier à M. Faye sur la théorie de Mercure et sur le mouvement du périhélie de cette planète. *Cr. Hebd. Acad. Sci.*, **49**, 379–383.

Le Verrier, U. J. 1859b. Théorie du mouvement de Mercure. *Ann. Obs. Paris*, **5**, 51–103.

Leconte, J., Lai, D., and Chabrier, G. 2011. Distorted, Nonspherical Transiting Planets: Impact on the Transit Depth and on the Radius Determination. *Astron. Astrophys.*, **528**, A41.

Lee, J. W., Kim, S.-L., Kim, C.-H., et al. 2009. The sdB+M Eclipsing System HW Virginis and Its Circumbinary Planets. *Astron J.*, **137**, 3181–3190.

Lense, J., and Thirring, H. 1918. Über den Einfluß der Eigenrotation der Zentralkörper auf die Bewegung der Planeten und Monde nach der Einsteinschen Gravitationstheorie. *Phys. Z*, **19**, 156–163.

Levi-Civita, T. 1926. Sur l' écart géodésique. *Math. Ann.*, **97**, 291–320.

Li, L.-S. 2012. Parameterized Post-Newtonian Orbital Effects in Extrasolar Planets. *Astrophys. Space Sci.*, **341**, 323–330.

Liang, Y., Winn, J. N., and Albrecht, S. H. 2022. DI Herculis Revisited: Starspots, Gravity Darkening, and 3D Obliquities. *Astrophys. J.*, **927**, 114.

Lichtenegger, H. I. M., Gronwald, F., and Mashhoon, B. 2000. On detecting the gravitomagnetic field of the Earth by means of orbiting clocks. *Adv. Space Res.*, **25**, 1255–1258.

Lichtenegger, H. I. M., Iorio, L., and Mashhoon, B. 2006. The Gravitomagnetic Clock Effect and Its Possible Observation. *Ann. Phys.-Berlin*, **518**, 868–876.

LIGO Scientific Collaboration, and Virgo Collaboration. 2016. Observation of Gravitational Waves from a Binary Black Hole Merger. *Phys. Rev. Lett.*, **116**, 061102.

LIGO Scientific Collaboration, and Virgo Collaboration. 2017. GW170817: Observation of Gravitational Waves from a Binary Neutron Star Inspiral. *Phys. Rev. Lett.*, **119**, 161101.

Lim, H., and Rodriguez, C. L. 2020. Relativistic Three-Body Effects in Hierarchical Triples. *Phys. Rev. D*, **102**, 064033.

Lindegren, L., and Dravins, D. 2003. The Fundamental Definition of 'Radial Velocity'. *Astron. Astrophys.*, **401**, 1185–1201.

Linet, B. 1982a. Absorption of Gravitational Waves by Nearly Newtonian Systems. *Gen. Relativ. Gravit.*, **14**, 479–492.

Linet, B. 1982b. Interaction of a Gravitational Wave with a Newtonian System in the Brans–Dicke Theory. *Phys. Lett. A*, **92**, 63–64.

Lorimer, D., and Kramer, M. 2005. *Handbook of Pulsar Astronomy*. Cambridge Observing Handbooks for Research Astronomers. Cambridge University Press.

Lorimer, D. R. 2008. Binary and Millisecond Pulsars. *Living Rev. Relativ.*, **11**, 8.

Love, A. E. H. 1911. *Some Problems of Geodynamics*. Cambridge University Press.

Lovis, C., and Fischer, D. 2010. Radial Velocity Techniques for Exoplanets. *Pages 27–53 of:* Seager, S. (ed.), *Exoplanets*. University of Arizona Press.

Lucchesi, D. M. 2001. Reassessment of the Error Modelling of Non-gravitational Perturbations on LAGEOS II and Their Impact in the Lense–Thirring Determination. Part I. *Planet. Space Sci.*, **49**, 447–463.

Lucchesi, D. M. 2002. Reassessment of the Error Modelling of Non-gravitational Perturbations on LAGEOS II and Their Impact in the Lense–Thirring Derivation. Part II. *Planet. Space Sci.*, **50**, 1067–1100.

Lucchesi, D. M. 2004. LAGEOS Satellites Germanium Cube–Corner-Retroreflectors and the Asymmetric Reflectivity Effect. *Celest. Mech. Dyn. Astr.*, **88**, 269–291.

Lucchesi, D. M. 2007. The LAGEOS Satellites Orbital Residuals Determination and the Way to Extract Gravitational and Non-gravitational Unmodeled Perturbing Effects. *Adv. Space Res.*, **39**, 1559–1575.

Lucchesi, D. M., and Balmino, G. 2006. The LAGEOS Satellites Orbital Residuals Determination and the Lense Thirring Effect Measurement. *Planet. Space Sci.*, **54**, 581–593.

Lucchesi, D. M., and Peron, R. 2010. Accurate Measurement in the Field of the Earth of the General-Relativistic Precession of the LAGEOS II Pericenter and New Constraints on Non-Newtonian Gravity. *Phys. Rev. Lett.*, **105**, 231103.

Lucchesi, D. M., and Peron, R. 2014. LAGEOS II Pericenter General Relativistic Precession (1993–2005): Error Budget and Constraints in Gravitational Physics. *Phys. Rev. D*, **89**, 082002.

Lucchesi, D. M., Anselmo, L., Bassan, M., et al. 2015. Testing the Gravitational Interaction in the Field of the Earth via Satellite Laser Ranging and the Laser Ranged Satellites Experiment (LARASE). *Class. Quantum Gravit.*, **32**, 155012.

Lucchesi, D. M., Anselmo, L., Bassan, M., et al. 2019a. General Relativity Measurements in the Field of Earth with Laser-Ranged Satellites: State of the Art and Perspectives. *Universe*, **5**, 141.

Lucchesi, D. M., Visco, M., Peron, R., et al. 2020. A 1% Measurement of the Gravitomagnetic Field of the Earth with Laser-Tracked Satellites. *Universe*, **6**, 139.

Luo, J., Chen, L.-S., Duan, H.-Z., et al. 2016. TianQin: A Space-Borne Gravitational Wave Detector. *Class. Quantum Gravit.*, **33**, 035010.

Lyne, A. G., Burgay, M., Kramer, M., et al. 2004. A Double–Pulsar System: A Rare Laboratory for Relativistic Gravity and Plasma Physics. *Science*, **303**, 1153–1157.

Lyutikov, M., Blandford, R. D., and Machabeli, G. 1999. On the nature of pulsar radio emission. *Mon. Not. Roy. Astron. Soc.*, **305**, 338–352.

Ma, C., Arias, E. F., Eubanks, T. M., et al. 1998. The International Celestial Reference Frame as Realized by Very Long Baseline Interferometry. *Astron. J.*, **116**, 516–546.

Maartens, R., and Bassett, B. A. 1998. Gravito-electromagnetism. *Class. Quantum Gravit.*, **15**, 705–717.

Maartens, R., and Koyama, K. 2010. Brane-World Gravity. *Living Rev. Relativ.*, **13**, 5.

MacDonald, D., and Thorne, K. S. 1982. Black-Hole Electrodynamics: An Absolute-Space/Universal-Time Formulation. *Mon. Not. Roy. Astron. Soc.*, **198**, 345–382.

Maeda, H., and Dadhich, N. 2007. Matter without Matter: Kaluza–Klein Spacetime in Einstein–Gauss–Bonnet Gravity. *Phys. Rev. D*, **75**, 044007.

Maggiore, M. 2017. Nonlocal Infrared Modifications of Gravity: A Review. *Pages 221–281 of:* Bagla, J., and Engineer, S. (eds.), *Gravity and the Quantum*. Fundamental Theories of Physics, vol. 187. Springer.

Mariani, V., Minazzoli, O., Fienga, A., Laskar, J., and Gastineau, M. 2023. Bayesian Test of Brans–Dicke Theories with Planetary Ephemerides: Investigating the Strong Equivalence Principle. *arXiv e-prints*, arXiv:2310.00719.

Marković , D., and Lamb, F. K. 1998. Lense–Thirring Precession and Quasi-Periodic Oscillations in X–Ray Binaries. *Astrophys. J.*, **507**, 316–326.

Martin, C. F., Torrence, M. H., and Misner, C. W. 1985. Relativistic Effects on an Earth-orbiting Satellite in the Barycenter Coordinate System. *J. Geophys. Res.*, **90**, 9403–9410.

Martins, F., Gillessen, S., Eisenhauer, F., et al. 2008. On the Nature of the Fast–Moving Star S2 in the Galactic Center. *Astrophys. J. Lett.*, **672**, L119.

Marzari, F., and Nagasawa, M. 2019. Influence of General-Relativity Effects, Dynamical Tides, and Collisions on Planet–Planet Scattering Close to the Star. *Astron. Astrophys.*, **625**, A121.

Mashhoon, B. 1978. On tidal resonance. *Astrophys. J.*, **223**, 285–298.

Mashhoon, B. 1979. On the detection of gravitational radiation by the Doppler tracking of spacecraft. *Astrophys. J.*, **227**, 1019–1036.

Mashhoon, B. 1997. Clocks and General Relativity. *Pages 41–48 of:* Maleki, L. (ed.), *Proc. Workshop on the Scientific Applications of Clocks in Space. (Pasadena, November 7–8, 1996)*. JPL-Publ-97-15. Jet Propulsion Laboratory.

Mashhoon, B. 2000. Gravitational Couplings of Intrinsic Spin. *Class. Quantum Gravit.*, **17**, 2399–2409.

Mashhoon, B. 2001. Gravitoelectromagnetism. *In:* Pascual-Sánchez, J. F., Floría, L., San Miguel, A., and Vicente, F. (eds.), *Reference Frames and Gravitomagnetism*. World Scientific.

Mashhoon, B. 2007. Gravitoelectromagnetism: A Brief Review. *Pages 29–39 of:* Iorio, L. (ed.), *The Measurement of Gravitomagnetism: A Challenging Enterprise*. Nova Science.

Mashhoon, B. 2017. *Nonlocal Gravity*. Oxford University Press.

Mashhoon, B., and Santos, N. O. 2000. Rotating Cylindrical Systems and Gravitomagnetism. *Ann. Phys.-Berlin*, **512**, 49–63.

Mashhoon, B., and Singh, D. 2006. Dynamics of Extended Spinning Masses in a Gravitational Field. *Phys. Rev. D*, **74**, 124006.

Mashhoon, B., Carr, B. J., and Hu, B. L. 1981. The Influence of Cosmological Gravitational Waves on a Newtonian Binary System. *Astrophys. J.*, **246**, 569–591.

Mashhoon, B., Hehl, F. W., and Theiss, D. S. 1984. On the Gravitational Effects of Rotating Masses: The Thirring–Lense Papers. *Gen. Relativ. Gravit.*, **16**, 711–750.

Mashhoon, B., Paik, H. J., and Will, C. M. 1989. Detection of the Gravitomagnetic Field Using an Orbiting Superconducting Gravity Gradiometer. Theoretical Principles. *Phys. Rev. D*, **39**, 2825–2838.

Mashhoon, B., Gronwald, F., and Theiss, D. S. 1999. Measuring Gravitomagnetism via Spaceborne Clocks: A Gravitomagnetic Clock Effect. *Ann. Phys.-Berlin*, **511**, 135–152.

Mashhoon, B., Gronwald, F., and Lichtenegger, H. I. M. 2001. Gravitomagnetism and the Clock Effect. *Pages 83–108 of:* Lämmerzahl, C., Everitt, C. W. F., and Hehl,

F. W. (eds.), *Gyros, Clocks, Interferometers ...: Testing Relativistic Gravity in Space*. Lecture Notes in Physics, vol. 562. Springer.

Mason, J. W. 2008. *Exoplanets: Detection, Formation, Properties, Habitability*. Springer.

Massi, M., and Zimmermann, L. 2010. Feasibility study of Lense–Thirring precession in LS I +61°303. *Astron. Astrophys.*, **515**, A82.

Mathew, A., and Nandy, M. K. 2017. General Relativistic Calculations for White Dwarfs. *Res. Astron. Astrophys.*, **17**, 061.

Matsuo, T., Greene, T. P., Qezlou, M., et al. 2022. Densified Pupil Spectrograph as High-Precision Radial Velocimetry: From Direct Measurement of the Universe's Expansion History to Characterization of Nearby Habitable Planet Candidates. *Astron J.*, **163**, 63.

Mattingly, D. 2005. Modern Tests of Lorentz Invariance. *Living Rev. Relativ.*, **8**, 5.

Mazur, P. O., and Mottola, E. 2004. Gravitational Vacuum Condensate Stars. *Proc. Natl. Acad. Sci.*, **101**, 9545–9550.

McGaugh, S. S. 2015. A Tale of Two Paradigms: The Mutual Incommensurability of ΛCDM and MOND. *Can. J. Phys.*, **93**, 250–259.

McLaughlin, D. B. 1924. Some Results of a Spectrographic Study of the Algol System. *Astrophys. J.*, **60**, 22–31.

Mecheri, R., and Meftah, M. 2021. Updated Values of Solar Gravitational Moments J_{2n} Using HMI Helioseismic Inference of Internal Rotation. *Mon. Not. Roy. Astron. Soc.*, **506**, 2671–2676.

Meichsner, J., and Soffel, M. H. 2015. Effects on Satellite Orbits in the Gravitational Field of an Axisymmetric Central Body with a Mass Monopole and Arbitrary Spin Multipole Moments. *Celest. Mech. Dyn. Astr.*, **123**, 1–12.

Merkowitz, S. M. 2010. Tests of Gravity Using Lunar Laser Ranging. *Living Rev. Relativ.*, **13**, 7.

Merrifield, M. R. 2005. Dark Matter on Galactic Scales (Or the Lack Thereof). *Pages 49–58 of:* Spooner, N. J. C., and Kudryavtsev, V. (eds.), *The Identification of Dark Matter*. World Scientific.

Merritt, D., Graham, A. W., Moore, B., Diemand, J., and Terzić, B. 2006. Empirical Models for Dark Matter Halos. I. Nonparametric Construction of Density Profiles and Comparison with Parametric Models. *Astron J.*, **132**, 2685–2700.

Merritt, D., Alexander, T., Mikkola, S., and Will, C. M. 2010. Testing Properties of the Galactic Center Black Hole Using Stellar Orbits. *Phys. Rev. D*, **81**, 062002.

Métris, G., Vokrouhlický, D., Ries, J. C., and Eanes, R. J. 1997. Nongravitational effects and the LAGEOS eccentricity excitations. *J. Geophys. Res.*, **102**, 2711–2730.

Métris, G., Vokrouhlický, D., Ries, J. C., and Eanes, R. J. 1999. LAGEOS Spin Axis and Non-gravitational Excitations of Its Orbit. *Adv. Space Res.*, **23**, 721–725.

Meyer, L., Schödel, R., Eckart, A., et al. 2007. On the orientation of the Sagittarius A* system. *Astron. Astrophys.*, **473**, 707–710.

Michalska, R. 1960. The Equations of Motion of Rotating Oblate Bodies in the General Theory of Relativity. *Bull. Acad. Polon. Sci., Ser. Math. Astr. Phys.*, **8**, 247–253.

Milani, A., Nobili, A.M., and Farinella, P. 1987. *Non-gravitational Perturbations and Satellite Geodesy*. Adam Hilger.

Milgrom, M. 1983a. A Modification of the Newtonian Dynamics as a Possible Alternative to the Hidden Mass Hypothesis. *Astrophys. J.*, **270**, 365–370.

Milgrom, M. 1983b. A Modification of the Newtonian Dynamics: Implications for Galaxies. *Astrophys. J.*, **270**, 371–383.

Milgrom, M. 1983c. A Modification of the Newtonian Dynamics: Implications for Galaxy Systems. *Astrophys. J.*, **270**, 384–389.

Milgrom, M. 2014. The MOND Paradigm of Modified Dynamics. *Scholarpedia*, **9**, 31410.

Mioc, V., and Radu, E. 1977. The Influence of Direct Solar Radiation Pressure on the Nodal Period of Artificial Earth Satellites. *Astron. Nachr.*, **298**, 107–110.

Mioc, V., and Radu, E. 1979. Perturbations in the Anomalistic Period of Artificial Satellites Caused by the Direct Solar Radiation Pressure. *Astron. Nachr.*, **300**, 313–315.

Misner, C. W., Thorne, K. S., and Wheeler, J. A. 2017. *Gravitation*. Princeton University Press.

Mitskevich, N. V., and Pulido Garcia, I. 1970. Motion of Test Masses in the Gravitational Field of a Rotating Body. *Dokl. Akad. Nauk SSSR*, **192**, 1263–1265.

Modenini, D., and Tortora, P. 2014. Pioneer 10 and 11 Orbit Determination Analysis Shows No Discrepancy with Newton-Einstein Laws of Gravity. *Phys. Rev. D*, **90**, 022004.

Moffat, J. W. 2006. Scalar Tensor Vector Gravity Theory. *J. Cosmol. Astropart. Phys.*, **2006**, 004.

Moffat, J. W., and Toth, V. T. 2009. Fundamental Parameter-free Solutions in Modified Gravity. *Class. Quantum Gravit.*, **26**, 085002.

Montenbruck, O., and Gill, E. 2000. *Satellite Orbits*. Spinger, Berlin.

Montenbruck, O., Kirschner, M., D'Amico, S., and Bettadpur, S. 2006. E/I-Vector Separation for Safe Switching of the GRACE Formation. E/I–Vektor Trennung für den sicheren Wechsel der GRACE Formation. *Aerosp. Sci. Technol.*, **10**, 628–635.

Morris, M. S., and Thorne, K. S. 1988. Wormholes in Spacetime and Their Use for Interstellar Travel: A Tool for Teaching General Relativity. *Am. J. Phys.*, **56**, 395–412.

Moyer, T. D. 2003. *Formulation for Observed and Computed Values of Deep Space Network Data Types for Navigation*. JPL Deep Space Communications and Navigation Series. Wiley.

Müller, J., Murphy, T. W., Schreiber, U., et al. 2019. Lunar Laser Ranging: A Tool for General Relativity, Lunar Geophysics and Earth Science. *J. Geod.*, **93**, 2195–2210.

Murray, C. D., and Dermott, S. F. 1999. *Solar System Dynamics*. Cambridge University Press.

Nash, G. 2019. Modified General Relativity. *Gen. Relativ. Gravit.*, **51**, 53.

Nash, G. 2023. Modified General Relativity and Dark Matter. *Int. J. Mod. Phys. D*, **32**, 2350031–168.

Nelson, L. A., and Chau, W. Y. 1982. Orbital Perturbations of a Gravitationally Bound Two-Body System with the Passage of Gravitational Waves. *Astrophys. J.*, **254**, 735–747.

Nesvorný, D., Kipping, D. M., Buchhave, L. A., et al. 2012. The Detection and Characterization of a Nontransiting Planet by Transit Timing Variations. *Science*, **336**, 1133.

Nesvorný, D., Kipping, D., Terrell, D., et al. 2013. KOI-142, The King of Transit Variations, Is a Pair of Planets Near the 2:1 Resonance. *Astrophys. J.*, **777**, 3.

Newcomb, S. 1895. *Tables of the Four Inner Planets*. Government Printing Office, Washington.

Newton, I. 1687. *Philosophiæ Naturalis Principia Mathematica*. For the Royal Society by Joseph Streater, London.

Nieto, M. M., and Goldman, T. 1991. The Arguments against 'Antigravity' and the Gravitational Acceleration of Antimatter. *Phys. Rep.*, **205**, 221–281.

Nishiyama, S., Saida, H., Takamori, Y., et al. 2018. Radial Velocity Measurements of an Orbiting Star around Sgr A*. *Publ. Astron. Soc. Jpn.*, **70**, 74.

Nobili, A. M., and Anselmi, A. 2018. Relevance of the Weak Equivalence Principle and Experiments to Test It: Lessons from the Past and Improvements Expected in Space. *Phys. Lett. A*, **382**, 2205–2218.

Nobili, A. M., and Will, C. M. 1986. The Real Value of Mercury's Perihelion Advance. *Nature*, **320**, 39–41.

Nojiri, S., and Odintsov, S. D. 2007. Introduction to Modified Gravity and Gravitational Alternative for Dark Energy. *Int. J. Geom. Meth. Mod. Phys.*, **04**, 115–145.

Nordström, G. 1918. On the Energy of the Gravitation field in Einstein's Theory. *Proc. Sect. Sci. K. Ned. Akad. Wet. Amst.*, **20**, 1238–1245.

Nordtvedt, K. Jr. 2000. Improving Gravity Theory Tests with Solar System 'grand fits'. *Phys. Rev. D*, **61**, 122001.

Nordtvedt, K. L. 1968a. Equivalence Principle for Massive Bodies. I: Phenomenology. *Phys. Rev.*, **169**, 1014–1016.

Nordtvedt, K. L. 1968b. Testing Relativity with Laser Ranging to the Moon. *Phys. Rev.*, **170**, 1186–1187.

Nordtvedt, K. L. 1999. 30 Years of Lunar Laser Ranging and the Gravitational Interaction. *Class. Quantum Gravit.*, **16**, A101–A112.

Noutsos, A., Desvignes, G., Kramer, M., et al. 2020. Understanding and Improving the Timing of PSR J0737–3039B. *Astron. Astrophys.*, **643**, A143.

Ochocimskij, D. E., Eneev, T. M., and Taratynova, G. P. 1959. Bestimmung der Lebensdauer eines künstlichen Erdsatelliten und Untersuchung der säkularen Störungen seiner Bahn. *Fortschritte der Phys.*, **7**, 34–54.

Ohanian, H. C., and Ruffini, R. J. 2013. *Gravitation and Spacetime. Third Edition.* Cambridge University Press.

O'Leary, J. 2021. *General Relativistic and Post–Newtonian Dynamics for Near-Earth Objects and Solar System Bodies*. Springer Theses. Springer.

Oloff, R. 2023. *The Geometry of Spacetime. A Mathematical Introduction to Relativity Theory*. Springer.

Olver, F. W. J., Lozier, D. W., Boisvert, R. F., and Clark, C. W. (eds). 2010. *NIST Handbook of Mathematical Functions*. Cambridge University Press.

O'Raifeartaigh, C., O'Keeffe, M., Nahm, W., and Mitton, S. 2018. One Hundred Years of the Cosmological Constant: from 'Superfluous Stunt' to Dark Energy. *Eur. Phys. J. H*, **43**.

Orosz, J. A., Welsh, W. F., Carter, J. A., et al. 2012a. Kepler–47: A Transiting Circumbinary Multiplanet System. *Science*, **337**, 1511.

Orosz, J. A., Welsh, W. F., Carter, J. A., et al. 2012b. The Neptune–Sized Circumbinary Planet Kepler–38b. *Astrophys. J.*, **758**, 87.

Padnanabhan, T. 2010. *Gravitation: Foundations and Frontiers*. Cambridge University Press.

Pál, A., and Kocsis, B. 2008. Periastron precession measurements in transiting extrasolar planetary systems at the level of general relativity. *Mon. Not. Roy. Astron. Soc.*, **389**, 191–198.

Panhans, M., and Soffel, M. H. 2014. Gravito-magnetism of an Extended Celestial Body. *Class. Quantum Gravit.*, **31**, 245012.

Paolozzi, A., Ciufolini, I., and Vendittozzi, C. 2011. Engineering and Scientific Aspects of LARES Satellite. *Acta Astronaut.*, **69**, 127–134.

Paolozzi, A., Sindoni, G., Felli, F., et al. 2019. Studies on the Materials of LARES 2 Satellite. *J. Geod.*, **93**, 2437–2446.

Park, R. S., Folkner, W. M., Konopliv, A. S., et al. 2017. Precession of Mercury's Perihelion from Ranging to the MESSENGER Spacecraft. *Astron J.*, **153**, 121.

Park, R. S., Folkner, W. M., Williams, J. G., and Boggs, D. H. 2021. The JPL Planetary and Lunar Ephemerides DE440 and DE441. *Astron J.*, **161**, 105.

Parker, L., and Christensen, S.M. 1994a. The Ricci, Einstein, and Weyl Tensors. *Pages 30–32 of: MathTensor: A System for Doing Tensor Analysis by Computer*. Addison-Wesley.

Parker, L., and Christensen, S.M. 1994b. The Riemann Curvature Tensor. *Pages 28–32 of: MathTensor: A System for Doing Tensor Analysis by Computer*. Addison-Wesley.

Pascoli, G. 2024. A Comparative Study of MOND and MOG Theories versus the κ-Model: An Application to Galaxy Clusters. *Can. J. Phys.*, **102**, 69–84.

Pearlman, M., Arnold, D., Davis, M., et al. 2019. Laser Geodetic Satellites: A High-Accuracy Scientific Tool. *J. Geod.*, **93**, 2181–2194.

Pearlman, M. R., Degnan, J. J., and Bosworth, J. M. 2002. The International Laser Ranging Service. *Adv. Space Res.*, **30**, 135–143.

Peebles, P. J., and Ratra, B. 2003. The Cosmological Constant and Dark Energy. *Rev. Mod. Phys.*, **75**, 559–606.

Peißker, F., Eckart, A., Zajaček, M., Ali, B., and Parsa, M. 2020. S62 and S4711: Indications of a Population of Faint Fast-Moving Stars inside the S2 Orbit–S4711 on a 7.6 yr Orbit around Sgr A*. *Astrophys. J.*, **899**, 50.

Peißker, F., Eckart, A., Zajaček, M., and Britzen, S. 2022. Observation of S4716-a Star with a 4 yr Orbit around Sgr A*. *Astrophys. J.*, **933**, 49.

Penrose, R. 2002. 'Golden Oldie': Gravitational Collapse: The Role of General Relativity. *Gen. Relativ. Gravit.*, **7**, 1141–1165.

Penrose, R., and Floyd, R. M. 1971. Extraction of Rotational Energy from a Black Hole. *Nature Physical Science*, **229**, 177–179.

Pepe, F. A., Cristiani, S., Rebolo Lopez, R., et al. 2010. ESPRESSO: The Echelle Spectrograph for Rocky Exoplanets and Stable Spectroscopic Observations. *Page 77350F of:* McLean, Ian S., Ramsay, Suzanne K., and Takami, Hideki (eds), *Ground–based and Airborne Instrumentation for Astronomy III*. Society of Photo-Optical Instrumentation Engineers (SPIE) Conference Series, vol. 7735.

Perlmutter, S., Aldering, G., Goldhaber, G., et al. 1999. Measurements of Ω and Λ from 42 High-Redshift Supernovae. *Astrophys. J.*, **517**, 565–586.

Perryman, M. 2018. *The Exoplanet Handbook. Second edition*. Cambridge University Press.

Petit, G., and Luzum, B. (eds) 2010. *IERS Conventions (2010)*. IERS Technical Note, vol. 36. Verlag des Bundesamts für Kartographie und Geodäsie, Frankfurt am Main.

Pfister, H. 2007. On the History of the So-Called Lense–Thirring Effect. *Gen. Relativ. Gravit.*, **39**, 1735–1748.

Pfister, H. 2008. The History of the So–Called Lense–Thirring Effect. *Pages 2456–2458 of:* Kleinert, H., Jantzen, R. T., and Ruffini, R. (eds.), *The Eleventh Marcel Grossmann Meeting On Recent Developments in Theoretical and Experimental General Relativity, Gravitation and Relativistic Field Theories*. World Scientific.

Pfister, H. 2014. Gravitomagnetism: From Einstein's 1912 Paper to the Satellites LAGEOS and Gravity Probe B. *Pages 191–197 of:* Bičák, J., and Ledvinka, T. (eds.), *Relativity and Gravitation*. Springer Proceedings in Physics,, vol. 157. Springer.

Pijpers, F. P. 1998. Helioseismic Determination of the Solar Gravitational Quadrupole Moment. *Mon. Not. Roy. Astron. Soc.*, **297**, L76–L80.

Pireaux, S., and Rozelot, J.-P. 2003. Solar Quadrupole Moment and Purely Relativistic Gravitation Contributions to Mercury's Perihelion Advance. *Astrophys. Space Sci.*, **284**, 1159–1194.

Pitjev, N. P., and Pitjeva, E. V. 2013. Constraints on Dark Matter in the Solar System. *Astron. Lett.*, **39**, 141–149.

Pitjeva, E. V., and Pitjev, N. P. 2018. Masses of the Main Asteroid Belt and the Kuiper Belt from the Motions of Planets and Spacecraft. *Astron. Lett.*, **44**, 554–566.

Pogossian, S. P. 2022. Comparative Study of Mercury's Perihelion Advance. *Celest. Mech. Dyn. Astr.*, **134**, 33.

Poisson, E., and Will, C. M. 2014. *Gravity: Newtonian, Post-Newtonian, Relativistic.* Cambridge University Press.

Preto, M., and Saha, P. 2009. On Post-Newtonian Orbits and the Galactic-center Stars. *Astrophys. J.*, **703**, 1743–1751.

Provost, J.-P. 2017. A Brief History of the Energy-Momentum Tensor 1912–1915: Einstein Physicist's Logic in Compromising Gravitation with Relativity. *Pages 3348–3353 of:* Bianchi, M., Jantzen, R. T., and Ruffini, R. J. (eds.), *The Fourteenth Marcel Grossmann Meeting. On Recent Developments in Theoretical and Experimental General Relativity, Astrophysics, and Relativistic Field Theories*. World Scientific.

Psaltis, D. 2023. Black Holes in Classical General Relativity and Beyond. *Pages 1–25 of:* Haiman, Z. (ed), *The Encyclopedia of Cosmology. Set 2: Frontiers in Cosmology. Volume 3: Black Holes*. World Scientific.

Psaltis, D., Wex, N., and Kramer, M. 2016. A Quantitative Test of the No-Hair Theorem with Sgr A* Using Stars, Pulsars, and the Event Horizon Telescope. *Astrophys. J.*, **818**, 121.

Pugh, G. E. 1959. *Proposal for a Satellite Test of the Coriolis Prediction of General Relativity*. Research Memorandum. Weapons Systems Evaluation Group, The Pentagon, Washington D.C.

Qian, S.-B., Liao, W.-P., Zhu, L.-Y., and Dai, Z.-B. 2010. Detection of a Giant Extrasolar Planet Orbiting the Eclipsing Polar DP Leo. *Astrophys. J. Lett.*, **708**, L66–L68.

Qian, S.-B., Liu, L., Zhu, L.-Y., et al. 2012a. A circumbinary planet in orbit around the short–period white dwarf eclipsing binary RR Cae. *Mon. Not. Roy. Astron. Soc.*, **422**, L24–L27.

Qian, S.-B., Zhu, L.-Y., Dai, Z.-B., et al. 2012b. Circumbinary Planets Orbiting the Rapidly Pulsating Subdwarf B–Type Binary NY Vir. *Astrophys. J. Lett.*, **745**, L23.

Queloz, D., Eggenberger, A., Mayor, M., et al. 2000. Detection of a Spectroscopic Transit by the Planet Orbiting the Star HD209458. *Astron. Astrophys.*, **359**, L13–L17.

Ragozzine, D., and Wolf, A. S. 2009. Probing the Interiors of Very Hot Jupiters Using Transit Light Curves. *Astrophys. J.*, **698**, 1778–1794.

Rainer, M., Desidera, S., Borsa, F., et al. 2023. The GAPS Programme at TNG. XLIV. Projected Rotational Velocities of 273 Exoplanet-Host Stars Observed with HARPS–N. *Astron. Astrophys.*, **676**, A90.

Ransom, S. M., Stairs, I. H., Archibald, A. M., et al. 2014. A Millisecond Pulsar in a Stellar Triple System. *Nature*, **505**, 520–524.

Ravenhall, D. G., and Pethick, C. J. 1994. Neutron Star Moments of Inertia. *Astrophys. J.*, **424**, 846–851.

Rees, M. J. 1978. Relativistic Jets and Beams in Radio Galaxies. *Nature*, **275**, 516–517.

Rees, M. J. 1984. Black Hole Models for Active Galactic Nuclei. *Annu. Rev. Astron. Astr.*, **22**, 471–506.

Reigber, Ch. 1978. Improvements of the Gravity Field from Satellite Techniques as Proposed to the European Space Agency. *Pages 221–232 of:* Müller, I. I. (ed), *Applications of Geodesy to Geodynamincs. Proceedings of the 9th GEOP Conference. Rep. 280 Dep. of Geod. Sci.* Ohio State University, Columbus.

Reissner, H. 1916. Über die Eigengravitation des elektrischen Feldes nach der Einsteinschen Theorie. *Ann. Phys.-Berlin*, **355**, 106–120.

Renzetti, G. 2012a. Are Higher Degree even Zonals Really Harmful for the LARES/LAGEOS Frame-Dragging Experiment? *Can. J. Phys.*, **90**, 883–888.

Renzetti, G. 2012b. Exact Geodetic Precession of the Orbit of a Two-Body Gyroscope in Geodesic Motion about a Third Mass. *Earth Moon Planets*, **109**, 55–59.

Renzetti, G. 2013a. First Results from LARES: An Analysis. *New Astron.*, **23**, 63–66.

Renzetti, G. 2013b. History of the Attempts to Measure Orbital Frame-Dragging with Artificial Satellites. *Centr. Eur. J. Phys.*, **11**, 531–544.

Renzetti, G. 2014. Some Reflections on the Lageos Frame-Dragging Experiment in View of Recent Data Analyses. *New Astron.*, **29**, 25–27.

Renzetti, G. 2015. On Monte Carlo Simulations of the LAser RElativity Satellite Experiment. *Acta Astronaut.*, **113**, 164–168.

Reynaud, S., and Jaekel, M.-T. 2005. Testing the Newton Law at Long Distances. *Int. J. Mod. Phys. A*, **20**, 2294–2303.

Ries, J. C., Eanes, R. J., and Tapley, B. D. 2003a. Lense–Thirring Precession Determination from Laser Ranging to Artificial Satellites. *Pages 201–211 of:* Ruffini, R. J., and Sigismondi, C. (eds.), *Nonlinear Gravitodynamics. The Lense–Thirring Effect*. World Scientific.

Ries, J. C., Eanes, R. J., Tapley, B. D., and Peterson, G. E. 2003b. Prospects for an Improved Lense–Thirring Test with SLR and the GRACE Gravity Mission. *In:* Noomen, R., Klosko, S., Noll, C., and Pearlman, M. (eds.), *Proceedings of the 13th International Laser Ranging Workshop, Washington DC, October 7-11, 2002*, vol. NASA CP (2003-212248). NASA Goddard, Greenbelt.

Ries, J. C., Eanes, R. J., and Watkins, M. 2009. Confirming the Frame-Dragging Effect with Satellite Laser Ranging. *Page 128 of:* Schillak, S. (ed.), *Proceedings of the 16th International Workshop on Laser Ranging. October 2008, Poznań, Poland. Volume 1*. Space Research Centre, Polish Academy of Sciences.

Riess, A. G. 2000. The Case for an Accelerating Universe from Supernovae. *Publ. Astron. Soc. Pac.*, **112**, 1284–1299.

Riess, A. G., Filippenko, A. V., Challis, P., et al. 1998. Observational Evidence from Supernovae for an Accelerating Universe and a Cosmological Constant. *Astron J.*, **116**, 1009–1038.

Rievers, B., and Lämmerzahl, C. 2011. High Precision Thermal Modeling of Complex Systems with Application to the Flyby and Pioneer Anomaly. *Ann. Phys.-Berlin*, **523**, 439–449.

Rievers, B., Lämmerzahl, C., List, M., Bremer, S., and Dittus, H. 2009. New Powerful Thermal Modelling for High-Precision Gravity Missions with Application to Pioneer 10/11. *New J. Phys.*, **11**, 113032.

Rievers, B., Lämmerzahl, C., and Dittus, H. 2010. Modeling of Thermal Perturbations Using Raytracing Method with Preliminary Results for a Test Case Model of the Pioneer 10/11 Radioisotopic Thermal Generators. *Space Sci. Rev.*, **151**, 123–133.

Rindler, W. 2001. *Relativity: Special, General, and Cosmological*. Oxford University Press.

Robinson, D. C. 1975. Uniqueness of the Kerr Black Hole. *Phys. Rev. Lett.*, **34**, 905–906.

Roseveare, N. T. 1982. *Mercury's Perihelion, from Le Verrier to Einstein*. Clarendon Press.

Roshan, M., and Mashhoon, B. 2022. Nonlocal Gravity: Modification of Newtonian Gravitational Force in the Solar System. *Universe*, **8**, 470.

Rossiter, R. A. 1924. On the Detection of an Effect of Rotation during Eclipse in the Velocity of the Brighter Component of Beta Lyrae, and on the Constancy of Velocity of This System. *Astrophys. J.*, **60**, 15–21.

Roy, A. E. 2005. *Orbital Motion. Fourth Edition*. IOP Publishing.

Rozelot, J.-P. 2009. What Is Coming: Issues Raised from Observation of the Shape of the Sun. *Pages 15–43 of:* Rozelot, J.-P., and Neiner, C. (eds.), *The Rotation of Sun and Stars*. Lecture Notes in Physics, vol. 765. Springer.

Rubincam, D. P., Currie, D. G., and Robbins, J. W. 1997. LAGEOS I Once-per-Revolution Force due to Solar Heating. *J. Geophys. Res.*, **102**, 585–590.

Rudenko, V. N. 1975. Test Bodies under the Effect of Gravitational Radiation. *Sov. Astron.*, **19**, 270.

Ruggiero, M. L. 2016. Gravitomagnetic Field of Rotating Rings. *Astrophys. Space Sci.*, **361**, 140.

Ruggiero, M. L. 2021. A Note on the Gravitoelectromagnetic Analogy. *Universe*, **7**, 451.

Ruggiero, M. L., and Iorio, L. 2020. Probing a r^{-n} Modification of the Newtonian Potential with Exoplanets. *J. Cosmol. Astropart. Phys.*, **2020**, 042.

Ruggiero, M. L., and Radicella, N. 2015. Weak-Field Spherically Symmetric Solutions in $f(T)$ Gravity. *Phys. Rev. D*, **91**, 104014.

Ruggiero, M. L., and Tartaglia, A. 2002. Gravitomagnetic Effects. *Nuovo Cim. B*, **117**, 743.

Saadat, H., Mousavi, S. N., Saadat, M., Saadat, N., and Saadat, A. M. 2010. The Effect of Dark Matter on Solar System and Perihelion Precession of Earth Planet. *Int. J. Theor. Phys.*, **49**, 2506–2511.

Saida, H. 2017. How to Measure a Black Hole's Mass, Spin, and Direction of Spin Axis in the Kerr Lens Effect 1: Test Case with Simple Source Emission Near a Black Hole. *Prog. Theor. Exp. Phys.*, **2017**, 053E02.

Sanders, R. H., and McGaugh, S. S. 2002. Modified Newtonian Dynamics as an Alternative to Dark Matter. *Annu. Rev. Astron. Astr.*, **40**, 263–317.

Schäfer, G. 2004. Gravitomagnetic Effects. *Gen. Relativ. Gravit.*, **36**, 2223–2235.

Schäfer, G. 2009. Gravitomagnetism in Physics and Astrophysics. *Space Sci. Rev.*, **148**, 37–52.

Schäfer, G., and Wex, N. 1993. Second Post-Newtonian Motion of Compact Binaries. *Phys. Lett. A*, **174**, 196–205.

Schanner, M., and Soffel, M. H. 2018. Relativistic Satellite Orbits: Central Body with Higher Zonal Harmonics. *Celest. Mech. Dyn. Astr.*, **130**, 40.

Schärer, A., Bondarescu, R., Saha, P., et al. 2017. Prospects for Measuring Planetary Spin and Frame-Dragging in Spacecraft Timing Signals. *Front. Astron. Space Sci.*, **4**, 11.

Scheffer, L. K. 2003. Conventional Forces Can Explain the Anomalous Acceleration of Pioneer 10. *Phys. Rev. D*, **67**, 084021.

Scheumann, J., Philipp, D., Herrmann, S., et al. 2023. Gravitomagnetic Clock Effect: Using GALILEO to Explore General Relativity. *arXiv e-prints*, arXiv:2311.12018.

Schiff, L. 1960. Possible New Experimental Test of General Relativity Theory. *Phys. Rev. Lett.*, **4**, 215–217.

Schlaufman, K. C. 2010. Evidence of Possible Spin-Orbit Misalignment along the Line of Sight in Transiting Exoplanet Systems. *Astrophys. J.*, **719**, 602–611.

Schmid, C. 2023. The Gravito-Maxwell Equations of General Relativity in the Local Reference Frame of a GR-Noninertial Observer. *Class. Quantum Gravit.*, **40**, 055005.

Schmidt, B. P. 2012. Nobel Lecture: Accelerating Expansion of the Universe through Observations of Distant Supernovae. *Rev. Mod. Phys.*, **84**, 1151–1163.

Schouten, J. A. 1918. Over het ontstaan eener praecessiebeweging tengevolge van het niet euklidisch zijn der ruimte in de nabijheiud van zon. *Versl. Kon. Ak. Wet.*, **27**, 214–220.

Schutz, B. 2022. *A First Course in General Relativity. Third Edition*. Cambridge University Press.

Schwab, C., Rakich, A., Gong, Q., et al. 2016. Design of NEID, an Extreme Precision Doppler Spectrograph for WIYN. *Page 99087H of:* Evans, C. J., Simard, L., and Takami, H. (eds), *Ground-Based and Airborne Instrumentation for Astronomy VI*. Society of Photo-Optical Instrumentation Engineers (SPIE) Conference Series, vol. 9908.

Schwamb, M. E., Orosz, J. A., Carter, J. A., et al. 2013. Planet Hunters: A Transiting Circumbinary Planet in a Quadruple Star System. *Astrophys. J.*, **768**, 127.

Seager, S. 2011. *Exoplanets*. University of Arizona Press.

Segre, M. 1989. Galileo, Viviani and the Tower of Pisa. *Stud. Hist. Philos. Sci.*, **20**, 435–451.

Seidelmann, P. K., Archinal, B. A., A'Hearn, M. F., et al. 2007. Report of the IAU/IAG Working Group on Cartographic Coordinates and Rotational Elements: 2006. *Celest. Mech. Dyn. Astr.*, **98**, 155–180.

Seifahrt, A., Stürmer, J., Bean, J. L., and Schwab, C. 2018. MAROON–X: A Radial Velocity Spectrograph for the Gemini Observatory. *Page 107026D of:* Evans, C. J., Simard, L., and Takami, H. (eds.), *Ground-Based and Airborne Instrumentation for Astronomy VII*. Society of Photo-Optical Instrumentation Engineers (SPIE) Conference Series, vol. 10702.

Selig, H., Lämmerzahl, C., and Ni, W.-T. 2013. Astrodynamical Space Test of Relativity Using Optical Devices I (ASTROD I)–Mission Overview. *Int. J. Mod. Phys. D*, **22**, 1341003.

Sereno, M., and Jetzer, Ph. 2006. Dark Matter versus Modifications of the Gravitational Inverse-Square Law: Results from Planetary Motion in the Solar System. *Mon. Not. Roy. Astron. Soc.*, **371**, 626–632.

Shao, L. 2016. Testing the Strong Equivalence Principle with the Triple Pulsar PSR J0337 + 1715. *Phys. Rev. D*, **93**, 084023.

Shapiro, I. I. 1990. Solar System Tests of General Relativity: Recent Results and Present Plans. *Pages 313–330 of:* Ashby, N., Bartlett, D. F., and Wyss, W. (eds.), *General Relativity and Gravitation, 1989*. Cambridge University Press.

Shapiro, I. I., Ash, M. E., and Smith, W. B. 1968. Icarus: Further Confirmation of the Relativistic Perihelion Precession. *Phys. Rev. Lett.*, **20**, 1517–1518.

Shapiro, I. I., Smith, W. B., Ash, M. E., and Herrick, S. 1971. General Relativity and the Orbit of Icarus. *Astron J.*, **76**, 588.

Shapiro, I. I., Pettengill, G. H., Ash, M. E., et al. 1972. Mercury's Perihelion Advance: Determination by Radar. *Phys. Rev. Lett.*, **28**, 1594–1597.

Shapiro, I. I., Reasenberg, R. D., Chandler, J. F., and Babcock, R. W. 1988. Measurement of the de Sitter Precession of the Moon: A Relativistic Three-Body Effect. *Phys. Rev. Lett.*, **61**, 2643–2646.

Shapiro, S. L., and Teukolsky, S. A. 1986. *Black Holes, White Dwarfs and Neutron Stars: The Physics of Compact Objects*. Wiley.

Shcherbakov, R. V., Penna, R. F., and McKinney, J. C. 2012. Sagittarius A* Accretion Flow and Black Hole Parameters from General Relativistic Dynamical and Polarized Radiative Modeling. *Astrophys. J.*, **755**, 133.

Silva, H. O., Holgado, A. M., Cárdenas-Avendaño, A., and Yunes, N. 2021. Astrophysical and Theoretical Physics Implications from Multimessenger Neutron Star Observations. *Phys. Rev. Lett.*, **126**, 181101.

Skordis, C. 2011. Modifications of Gravity. *Phil. Trans. R. Soc. A*, **369**, 4962–4975.

Smith, A. M. S., Cabrera, J., Csizmadia, Sz, et al. 2018. K2–137 b: An Earth-Sized Planet in a 4.3-h Orbit around an M-Dwarf. *Mon. Not. Roy. Astron. Soc.*, **474**, 5523–5533.

Soffel, M. H. 1989. *Relativity in Astrometry, Celestial Mechanics and Geodesy*. Springer.

Soffel, M. H., and Brumberg, V. A. 1991. Relativistic Reference Frames Including Time Scales: Questions and Answers. *Celest. Mech. Dyn. Astr.*, **52**, 355–373.

Soffel, M. H., and Han, W.-B. 2019. *Applied General Relativity*. Astronomy and Astrophysics Library. Springer.

Soffel, M. H., Wirrer, R., Schastok, J., Ruder, H., and Schneider, M. 1987. Relativistic Effects in the Motion of Artificial Satellites: The Oblateness of the Central Body I. *Celest. Mech. Dyn. Astr.*, **42**, 81–89.

Soffel, M. H., Klioner, S. A., Petit, G., et al. 2003. The IAU 2000 Resolutions for Astrometry, Celestial Mechanics, and Metrology in the Relativistic Framework: Explanatory Supplement. *Astron J.*, **126**, 2687–2706.

Sotiriou, T. P., and Zhou, S.-Y. 2014. Black Hole Hair in Generalized Scalar-Tensor Gravity: An Explicit Example. *Phys. Rev. D*, **90**, 124063.

Souami, D., and Souchay, J. 2012. The Solar System's Invariable Plane. *Astron. Astrophys.*, **543**, A133.

Sparavigna, A. C. 2016. The Word Satellite, Its Origin from Etruscan and Its Translation into Greek. *PHILICA*, 568.

Stairs, I. H. 2003. Testing General Relativity with Pulsar Timing. *Living Rev. Relativ.*, **6**, 5.

Steffen, J. H., Fabrycky, D. C., Ford, E. B., et al. 2012a. Transit Tming Observations from Kepler III. Confirmation of Four Multiple Planet Systems by a Fourier-Domain Study of Anticorrelated Transit Timing Variations. *Mon. Not. Roy. Astron. Soc.*, **421**, 2342–2354.

Steffen, J. H., Ford, E. B., Rowe, J. F., et al. 2012b. Transit Timing Observations from *Kepler*. VI. Potentially Interesting Candidate Systems from Fourier-Based Statistical Tests. *Astrophys. J.*, **756**, 186.

Stella, L., and Possenti, A. 2009. Lense–Thirring Precession in the Astrophysical Context. *Space Sci. Rev.*, **148**, 105–121.

Stella, L., and Vietri, M. 1998. Lense–Thirring Precession and Quasi-periodic Oscillations in Low-Mass X-Ray Binaries. *Astrophys. J. Lett.*, **492**, L59–L62.

Stepanov, S. 2018. *Relativistic World. Volume 1: Mechanics*. De Gruyter Textbook. De Gruyter.

Stephan, A. P., Wang, J., Cauley, P. W., et al. 2022. Nodal Precession and Tidal Evolution of Two Hot Jupiters: WASP–33 b and KELT–9 b. *Astrophys. J.*, **931**, 111.

Stephani, H. 1990. *General Relativity: An Introduction to the Theory of the Gravitational Field. 2nd edition*. Cambridge University Press.

Sterne, T. E. 1939. Apsidal Motion in Binary Stars. *Mon. Not. Roy. Astron. Soc.*, **99**, 451–462.

Struve, O., and Huang, S. S. 1958. Spectroscopic Binaries. *Pages 243–273 of:* Flügge, S. (ed), *Astrophysik I: Sternoberflächen-Doppelsterne / Astrophysics I: Stellar-Surfaces-Binaries*. Springer.

Stuchlík, Z., and Hledík, S. 1999. Some Properties of the Schwarzschild–de Sitter and Schwarzschild–anti-de Sitter Spacetimes. *Phys. Rev. D*, **60**, 044006.

Stuchlík, Z., Kološ, M., and Tursunov, A. 2021. Penrose Process: Its Variants and Astrophysical Applications. *Universe*, **7**, 416.

Synge, J. L. 1927. On the Geometry of Dynamics. *Phil. Trans. R. Soc. A*, **226**, 31–106.

Synge, J. L. 1960. *Relativity: The General Theory*. North-Holland, Amsterdam.

Taff, L. G. 1985. *Celestial Mechanics: A Computational Guide for the Practitioner*. Wiley.

Takahashi, R. 2004. Shapes and Positions of Black Hole Shadows in Accretion Disks and Spin Parameters of Black Holes. *Astrophys. J.*, **611**, 996–1004.

Tapley, B. D., Schutz, B. E., and Born, G. H. 2004a. *Statistical Orbit Determination*. Elsevier.

Tapley, B. D., Bettadpur, S., Watkins, M. M., and Reigber, C. 2004b. The Gravity Recovery and Climate Experiment: Mission Overview and Early Results. *Geophys. Res. Lett.*, **31**, L09607.

Taratynova, G. P. 1959. Über die Bewegung von künstlichen Satelliten im nicht-zentralen Schwerefeld der Erde unter Berücksichtigung des Luftwiderstandes. *Fortschritte der Phys.*, **7**, 55–64.

Tartaglia, A. 2000a. Detection of the Gravitomagnetic Clock Effect. *Class. Quantum Gravit.*, **17**, 783–792.

Tartaglia, A. 2000b. Geometric Treatment of the Gravitomagnetic Clock Effect. *Gen. Relativ. Gravit.*, **32**, 1745–1756.

Tartaglia, A. 2000c. Influence of the Angular Momentum of Astrophysical Objects on Light and Clocks and Related Measurements. *Class. Quantum Gravit.*, **17**, 2381–2384.

Tartaglia, A. 2002. Angular-Momentum Effects in Weak Gravitational Fields. *Europhys. Lett.*, **60**, 167–173.

Taylor, E. F., and Wheeler, J. A. 1992. *Spacetime Physics: Introduction to Special Relativity; Second Edition*. Freeman.

Teukolsky, S. A. 2015. The Kerr Metric. *Class. Quantum Gravit.*, **32**, 124006.

Teyssandier, P. 1977. Rotating Nonspherical Masses and Their Effects on the Precession of a Gyroscope. *Phys. Rev. D*, **16**, 946–952.

Teyssandier, P. 1978. Rotating Stratified Ellipsoids of Revolution and Their Effects on the Dragging of Inertial Frames. *Phys. Rev. D*, **18**, 1037–1046.

Thebault, P., and Haghighipour, N. 2015. Planet Formation in Binaries. *Pages 309–340 of:* Jin, S., Haghighipour, N., and Ip, W. H. (eds.), *Planetary Exploration and Science: Recent Results and Advances*. Springer.

Thirring, H. 1918. 'Über die Wirkung rotierender ferner Massen in der Einsteinschen Gravitationstheorie'. *Phys. Z.*, **19**, 33–39.

Thirring, H. 1921. Berichtigung zu meiner Arbeit: 'Über die Wirkung rotierender Massen in der Einsteinschen Gravitationstheorie'. *Phys. Z.*, **22**, 29–30.

Thorne, K. S. 1986. Black Holes: The Membrane Viewpoint. *Pages 103–161 of:* Shapiro, S. L., Teukolsky, S. A., and Salpeter, E. E. (eds), *Highlights of Modern Astrophysics: Concepts and Controversies*. Wiley.

Thorne, K. S. 1988. Gravitomagnetism, Jets in Quasars, and the Stanford Gyroscope Experiment. *Pages 573–586 of:* Fairbank, J. D., Deaver, B. S., Jr., Everitt, C. W. F., and Michelson, P. F. (eds.), *Near Zero: New Frontiers of Physics*. Freeman.

Thorne, K. S., and Blandford, R. D. 2021. *Relativity and Cosmology. Volume 5 of Modern Classical Physics*. Vol. 5. Princeton University Press.

Thorne, K. S., MacDonald, D. A., and Price, R. H. (eds.). 1986. *Black Holes: The Membrane Paradigm*. Yale University Press.

Thorsett, S. E., Arzoumanian, Z., and Taylor, J. H. 1993. PSR B1620–26: A Binary Radio Pulsar with a Planetary Companion? *Astrophys. J. Lett.*, **412**, L33–L36.

Thorsett, S. E., Arzoumanian, Z., Camilo, F., and Lyne, A. G. 1999. The Triple Pulsar System PSR B1620–26 in M4. *Astrophys. J.*, **523**, 763–770.

Tino, G. M., Cacciapuoti, L., Capozziello, S., Lambiase, G., and Sorrentino, F. 2020. Precision Gravity Tests and the Einstein Equivalence Principle. *Prog. Part. Nucl. Phys.*, **112**, 103772.

Torres, D. F., and Rea, N. (eds.). 2011. *High-Energy Emission from Pulsars and Their Systems*. Astrophysics and Space Science Proceedings. Springer.

Torres, D. F., Capozziello, S., and Lambiase, G. 2000. Supermassive Boson Star at the Galactic Center? *Phys. Rev. D*, **62**, 104012.

Torretti, R. 1991. The Geometric Structure of the Universe. *Pages 53–73 of:* Agazzi, E., and Cordero, A. (eds.), *Philosophy and the Origin and Evolution of the Universe*. Kluwer.

Touboul, P., Métris, G., Rodrigues, M., et al. 2022a. M I C R O S C O P E Mission: Final Results of the Test of the Equivalence Principle. *Phys. Rev. Lett.*, **129**, 121102.

Touboul, P., Métris, G., Rodrigues, M., et al. 2022b. The MICROSCOPE Space Mission: The First Test of the Equivalence Principle in a Space Laboratory. *Class. Quantum Gravit.*, **39**, 200401.

Triaud, A. H. M. J. 2018. The Rossiter–McLaughlin Effect in Exoplanet Research. *Pages 1375–1401 of:* Deeg, H. J., and Belmonte, J. A. (eds.), *Handbook of Exoplanets*. Springer.

Tsiklauri, D., and Viollier, R. D. 1998. Dark Matter Concentration in the Galactic Center. *Astrophys. J.*, **500**, 591–595.

Tucker, A., and Will, C. M. 2019. Pericenter Advance in General Relativity: Comparison of Approaches at High Post-Newtonian Orders. *Class. Quantum Gravit.*, **36**, 115001.

Turner, M. S. 1979. Influence of a Weak Gravitational Wave on a Bound System of Two Point-Masses. *Astrophys. J.*, **233**, 685–693.

Turyshev, S. G., and Toth, V. T. 2010. The Pioneer Anomaly. *Living Rev. Relativ.*, **13**, 4.

Turyshev, S. G., Shao, M., and Nordtvedt, K. L. 2004. The Laser Astrometric Test of Relativity Mission. *Nucl. Phys. B Proc. Suppl.*, **134**, 171–178.

Turyshev, S. G., Toth, V. T., Kinsella, G., et al. 2012. Support for the Thermal Origin of the Pioneer Anomaly. *Phys. Rev. Lett.*, **108**, 241101.

Turyshev, S. G., Chiow, S.-w., and Yu, N. 2024. Searching for New Physics in the Solar System with Tetrahedral Spacecraft Formations. *Phys. Rev. D*, **109**, 084059.

Vagnozzi, S. 2023. Seven Hints That Early–Time New Physics Alone Is Not Sufficient to Solve the Hubble Tension. *Universe*, **9**, 393.

Vargas dos Santos, M., and Mota, D. F. 2017. Extrasolar Planets as a Probe of Modified Gravity. *Phys. Lett. B*, **769**, 485–490.

Veledina, A., Poutanen, J., and Ingram, A. 2013. A Unified Lense–Thirring Precession Model for Optical and X-Ray Quasi-periodic Oscillations in Black Hole Binaries. *Astrophys. J.*, **778**, 165.

Venkatraman Krishnan, V., Bailes, M., van Straten, W., et al. 2020. Lense–Thirring Frame Dragging Induced by a Fast-Rotating White Dwarf in a Binary Pulsar System. *Science*, **367**, 577–580.

Viollier, R. D., Trautmann, D., and Tupper, G. B. 1993. Supermassive Neutrino Stars and Galactic Nuclei. *Phys. Lett. B*, **306**, 79–85.

Visco, M., and Lucchesi, D. M. 2018. Comprehensive Model for the Spin Evolution of the L A G E O S and L A R E S Satellites. *Phys. Rev. D*, **98**, 044034.

Vladimirov, Yu., Mitskiévic, H., and Horský, J. 1987. *Space Time Gravitation*. Mir.

Voisin, G., Cognard, I., Freire, P. C. C., et al. 2020. An Improved Test of the Strong Equivalence Principle with the Pulsar in a Triple Star System. *Astron. Astrophys.*, **638**, A24.

Vokrouhlický, D., and Farinella, P. 1997. Thermal Force Effects on Slowly Rotating, Spherical Artificial Satellites–II. Earth Infrared Heating. *Planet. Space Sci.*, **45**, 419–425.

Waisberg, I., Dexter, J., Gillessen, S., et al. 2018. What Stellar Orbit Is Needed to Measure the Spin of the Galactic Centre Black Hole from Astrometric Data? *Mon. Not. Roy. Astron. Soc.*, **476**, 3600–3610.

Wald, R. M. 1984. *General Relativity*. The University of Chicago Press.

Wechsler, R. H., and Tinker, J. L. 2018. The Connection between Galaxies and Their Dark Matter Halos. *Annu. Rev. Astron. Astr.*, **56**, 435–487.

Weinberg, S. 1972. *Gravitation and Cosmology: Principles and Applications of the General Theory of Relativity*. Wiley.

Weinberg, S. 1992. *Dreams of a Final Theory*. Pantheon Books.

Weisberg, J. M., and Taylor, J. H. 1984. Observations of Post-Newtonian Timing Effects in the Binary Pulsar PSR 1913+16. *Phys. Rev. Lett.*, **52**, 1348–1350.

Weisberg, J. M., and Taylor, J. H. 2005. The Relativistic Binary Pulsar B1913+16: Thirty Years of Observations and Analysis. *Pages 25–31 of:* Rasio, F. A., and Stairs, I. H. (eds.), *Binary Radio Pulsars*. ASP Conference Series, vol. 328. Astronomical Society of the Pacific.

Welsh, W. F., Orosz, J. A., Carter, J. A., et al. 2012. Transiting Circumbinary Planets Kepler–34 b and Kepler–35 b. *Nature*, **481**, 475–479.

Welsh, W. F., Orosz, J. A., Short, D. R., et al. 2015. Kepler 453 b – The 10th *Kepler* Transiting Circumbinary Planet. *Astrophys. J.*, **809**, 26.

Wex, N. 1995. The Second Post-Newtonian Motion of Compact Binary-Star Systems with Spin. *Class. Quantum Gravit.*, **12**, 983–1005.

Wex, N. 1998. A Timing Formula for Main-Sequence Star Binary Pulsars. *Mon. Not. Roy. Astron. Soc.*, **298**, 67–77.

Wex, N., and Kopeikin, S. M. 1999. Frame Dragging and Other Precessional Effects in Black Hole Pulsar Binaries. *Astrophys. J.*, **514**, 388–401.

Wex, N., and Kramer, M. 2020. Gravity Tests with Radio Pulsars. *Universe*, **6**, 156.

Weyl, H. 1917. Zur Gravitationstheorie. *Ann. Phys.-Berlin*, **359**, 117–145.

White, W. D. 1958. Electromagnetic Analogs for the Gravitational Fields in the Vicinity of a Satellite. *Proc. IRE*, **46**, 920–922.

Will, C. M. 2008. Testing the General Relativistic 'No-Hair' Theorems Using the Galactic Center Black Hole Sagittarius A*. *Astrophys. J. Lett.*, **674**, L25.

Will, C. M. 2011a. Finally, Results from Gravity Probe B. *Physics Magazine*, **4**, 43.

Will, C. M. 2011b. Inaugural Article: On the Unreasonable Effectiveness of the Post-Newtonian Approximation in Gravitational Physics. *Proc. Natl. Acad. Sci.*, **108**, 5938–5945.

Will, C. M. 2014. Incorporating Post-Newtonian Effects in N-Body Dynamics. *Phys. Rev. D*, **89**, 044043.

Will, C. M. 2018. *Theory and Experiment in Gravitational Physics. Second edition.* Cambridge University Press.

Will, C. M., and Yunes, N. 2020. *Is Einstein Still Right?* Oxford Universiy Press.

Williams, J. G., Newhall, X. X., and Dickey, J. O. 1996. Relativity Parameters Determined from Lunar Laser Ranging. *Phys. Rev. D*, **53**, 6730–6739.

Williams, J. G., Turyshev, S. G., and Boggs, D. H. 2004. Progress in Lunar Laser Ranging Tests of Relativistic Gravity. *Phys. Rev. Lett.*, **93**, 261101.

Williams, J. G., Turyshev, S. G., and Boggs, D. H. 2012. Lunar Laser Ranging Tests of the Equivalence Principle. *Class. Quantum Gravit.*, **29**, 184004.

Winn, J. N. 2011. Exoplanets Transits and Occultations. *Pages 55–77 of:* Seager, S. (ed), *Exoplanets*. University of Arizona Press.

Winn, J. N., Noyes, R. W., Holman, M. J., et al. 2005. Measurement of Spin-Orbit Alignment in an Extrasolar Planetary System. *Astrophys. J.*, **631**, 1215–1226.

Wolff, M. 1969. Direct Measurements of the Earth's Gravitational Potential Using a Satellite Pair. *J. Geophys. Res.*, **74**, 5295–5300.

Worku, K., Wang, S., Burt, J., et al. 2022. Revisiting the Full Sets of Orbital Parameters for the XO–3 System: No Evidence for Temporal Variation of the Spin-Orbit Angle. *Astron J.*, **163**, 158.

Worley, A., Krastev, P. G., and Li, B.-A. 2008. Nuclear Constraints on the Moments of Inertia of Neutron Stars. *Astrophys. J.*, **685**, 390–399.

Xie, Y., and Deng, X.-M. 2014. On the (Im)possibility of Testing New Physics in Exoplanets Using Transit Timing Variations: Deviation from Inverse-Square Law of Gravity. *Mon. Not. Roy. Astron. Soc.*, **438**, 1832–1838.

Xu, G. 2008. *Orbits*. Springer.

You, R. J. 1998. The Gravitational Larmor Precession of the Earth's Artificial Satellite Orbital Motion. *Boll. Geod. Sci. Aff.*, **57**, 453–460.

Yu, L., Rodriguez, J. E., Eastman, J. D., et al. 2018. Two Warm, Low-Density Sub-Jovian Planets Orbiting Bright Stars in *K*2 Campaigns 13 and 14. *Astron J.*, **156**, 127.

Yu, Q., Zhang, F., and Lu, Y. 2016. Prospects for Constraining the Spin of the Massive Black Hole at the Galactic Center via the Relativistic Motion of a Surrounding Star. *Astrophys. J.*, **827**, 114.

Yukawa, H. 1935. On the Interaction of Elementary Particles. I. *Proc. Phys.-Math. Soc. Jpn.*, **17**, 48–57.

Zakharov, A. F., Nucita, A. A., de Paolis, F., and Ingrosso, G. 2007. Apoastron Shift Constraints on Dark Matter Distribution at the Galactic Center. *Phys. Rev. D*, **76**, 062001.

Zakharov, A. F., de Paolis, F., Ingrosso, G., and Nucita, A. A. 2010. Constraints on Parameters of Dark Matter and Black Hole in the Galactic Center. *Phys. At. Nucl.*, **73**, 1870–1877.

Zee, A. 2013. *Einstein Gravity in a Nutshell*. Princeton University Press.

Zel'Dovich, Ya. B. 1971. Generation of Waves by a Rotating Body. *J. Exp. Theor. Phys.*, **14**, 180.

Zhang, F., and Iorio, L. 2017. On the Newtonian and Spin–Induced Perturbations Felt by the Stars Orbiting around the Massive Black Hole in the Galactic Center. *Astrophys. J.*, **834**, 198.

Zhang, F., Lu, Y., and Yu, Q. 2015. On Testing the Kerr Metric of the Massive Black Hole in the Galactic Center via Stellar Orbital Motion: Full General Relativistic Treatment. *Astrophys. J.*, **809**, 127.

Zhao, S.-S., and Xie, Y. 2013. Parametrized Post-Newtonian Secular Transit Timing Variations for Exoplanets. *Res. Astron. Astrophys.*, **13**, 1231–1239.

Zhao, X.-F. 2016. On the Moment of Inertia of PSR J0348+0432. *Chin. J. Phys.*, **54**, 839–844.

Zhongolovich, I. D. 1960. Certain Formulas Related to the Motion of a Material Point within the Gravitational Field of an Ellipsoid of Revolution. *Byul. Inst. Teor. Astron.*, **7**, 521–536.

Zhongolovich, I. D. 1966. On the Use of the Results Obtained from Synchronous Observations of the Artificial Satellites of the Earth from the INTEROBS Programme for Scientific Purposes. *Pages 1–5 of:* Kovalevsky, J. (ed.), *Trajectories of Artificial Celestial Bodies as Determined from Observations/Trajectoires des Corps Celestes Artificiels Déterminées D'après les Observations*. Springer.

Zuber, M. T., Smith, D. E., Lehman, D. H., et al. 2013. Gravity Recovery and Interior Laboratory (GRAIL): Mapping the Lunar Interior from Crust to Core. *Space Sci. Rev.*, **178**, 3–24.

Author Index

Bayes, T., 8
Bessel, F. W., 18, 168, 169, 182
Blandford, R. D., 87
Brumberg, V. A., 147

Christoffel, E. B., 3, 182
Ciufolini, I., 223
Coriolis, G.-G., 4, 15

Damour, T., 73, 80, 81, 188, 189
Darwin, G. H., 84, 183
de Sitter, W., 15, 155, 160, 166
Deruelle, N., 73, 80, 81, 188, 189

Einasto, J., 172
Einstein, A., 1–5, 54, 88, 191, 227
Euler, L., 17, 22

Fokker, A. D., 15, 155, 160

Galilei, G., 2, 88
Gauss, C. F., 22, 80

Hamilton, W. R., 73

Jacobi, C., 5, 73

Kepler, J., 7, 8, 11–13, 16, 18–31, 34, 36, 38, 48, 55, 58, 59, 63–66, 73, 74, 78, 80–82, 88, 92, 97–99, 117, 126, 127, 129, 131, 144, 145, 150, 154, 160, 185–187, 189, 191–193, 197, 199, 216
Kerr, R. P., 86, 91, 125, 181, 227
Kottler, F., 166
Kronecker, L., 158, 182, 190

Lagrange, J.-L., 13, 26, 32, 147
Laplace, P. S., 32, 58, 59, 97, 98, 129, 131, 188
Legendre, A.-M., 119, 123, 182
Lense, J., 7, 88
Lenz, W., 58, 59, 97, 98, 129, 131, 188
Levi-Civita, T., 182, 191
Lorentz, H. A., 2, 4, 10, 13, 165, 175
Love, A. E. H., 85, 183

Maxwell, J. C., 87
McLaughlin, D. B., 92
Messier, C., 6
Minkowski, H., 182, 190

Newton, I., 1–7, 9–13, 15, 16, 26, 31, 55, 104, 122, 123, 146, 147, 150, 155, 156, 158, 165, 168, 180, 182, 183, 185, 193
Nordström, G., 165
Nordtvedt, K. L., 3, 225

Panhans, M., 119
Penrose, R., 87
Poisson, S.-D., 5, 173
Pugh, G. E., 7, 15, 90, 155, 160

Radau, R., 84, 183
Ricci–Curbastro, G., 4, 5, 182
Riemann, G. F. B., 1, 2, 4, 5, 182
Rossby, C.-G. A., 92
Rossiter, R. A., 92
Runge, C., 58, 59, 97, 98, 129, 131, 188

Schiff, L., 7, 15, 90, 155, 160
Schwarzschild, K., 68, 166, 181
Shapiro, I. I., 225
Soffel, M. H., 119, 147

Thirring, H., 7, 88

Yukawa, H., 13, 165, 168

Znajek, R., 87

Subject Index

1960 ε 3 (Earth satellite), 33
1pN Lense–Thirring
acceleration, 12, 31, 88, 90, 98, 101, 103, 119, 223
acceleration, normal component, 90
acceleration, radial component, 90
acceleration, transverse component, 90
correction to the draconitic period, 99
correction to the sidereal period, 100
dec., instantaneous shift, 115, 116
dec., net shift per orbit, 115, 229
effect, 7, 88–90, 93, 96, 231–233
equations of motion, numerical integrations, 98, 99, 101–103, 118
full width at half maximum transit duration, net shift per orbit, 111
inclination, net shift per orbit, 106, 228, 229
inclination, precession, 89, 217, 227
ingress/egress transit duration, net shift per orbit, 110
instantaneous orbital shifts, 92
instantaneous orbital shifts, 13, 92, 197
longitude of the pericentre, precession, 93
mean anomaly at epoch, net shift per orbit, 98
microquasar, precession, 89
node, net shift per orbit, 228, 229
node, precession, 219, 227
orbital precessions, equatorial orbit, 94
orbital precessions, general case, 93, 97, 100, 102, 217–219
orbital precessions, polar orbit, 95, 231
periastron, precession, 89, 224
pericentre, net shift per orbit, 228, 229
perihelion, precession, 89, 220
perturbed orbit, 102
RA, instantaneous shift, 115, 116
RA, net shift per orbit, 115, 229
radial velocity curve, instantaneous shift, 107–109
radial velocity curve, net shift per orbit, 108, 109
radial velocity semiamplitude, net shift per orbit, 106, 109, 110, 229
range rate, instantaneous shift, 116–118
range, instantaneous shift, 116–118
semilatus rectum, instantaneous shift, 106
sky-projected spin-orbit angle, precession, 113, 114
time of inferior conjunction, net shift per orbit, 111, 113
total transit duration, net shift per orbit, 110
transit characteristic timescales, net shifts per orbit, 112
variation of the times of arrival, instantaneous shift, 114
variation of the times of arrival, net shift per orbit, 109, 114
1pN graviteolectric test particle
mean anomaly at epoch, net shift per orbit, 229
pericentre, net shift per orbit, 229
1pN gravitoelectric
feature of motion, 12
orbital effects, 16, 64
1pN gravitoelectric test particle
acceleration, 31, 54, 57–61, 146
acceleration, normal component, 55
acceleration, radial component, 55
acceleration, transverse component, 55
dec., instantaneous shift, 62
dec., net shift per orbit, 229
equations of motion, 59–61, 63
equations of motion, numerical integrations, 58–61, 63
instantaneous orbital shifts, 55
mean anomaly at epoch, net shift per orbit, 229
mean anomaly at epoch, precession, 220
mean anomaly, instantaneous shift, 57, 63
net orbital shifts, 154
orbital precessions, 56, 58, 60, 217–219
pericentre, net shift per orbit, 228
perihelion, precession, 7, 54, 220
RA, instantaneous shift, 62
RA, net shift per orbit, 229
1pN gravitoelectric two-body
acceleration, 64, 73
acceleration, normal component, 65, 75
acceleration, radial component, 65, 75
acceleration, transverse component, 65, 75

eccentricity, instantaneous shift, 76
full width at half maximum transit duration, net shift per orbit, 70
ingress/egress transit duration, net shift per orbit, 70
instantaneous orbital shifts, 65, 74
mean anomaly, instantaneous shift, 66
net orbital shifts, 66, 78
orbital precessions, 224
periastron, precession, 224
pericentre, fractional advance, 80, 83
pericentre, instantaneous shift, 74
radial velocity curve, instantaneous shift, 68
radial velocity curve, net shift per orbit, 68
radial velocity semiamplitude, net shift per orbit, 68
range rate, instantaneous shift, 63
range, instantaneous shift, 63
semimajor axis, instantaneous shift, 76
time of inferior conjunction, net shift per orbit, 71, 72
total transit duration, net shift per orbit, 69
transit characteristic timescales, net orbital shifts, 71
transit characteristic timescales, net shifts per orbit, 68
variation of the times of arrival, instantaneous shift, 72
variation of the times of arrival, net shift per orbit, 72

1pN gravitomagnetic
feature of motion, 12, 88
orbital effects, 16, 91, 227
orbital precessions, 88, 94

1pN gravitomagnetic spin octupole
acceleration, 119
acceleration, normal component, 120
acceleration, radial component, 120
acceleration, transverse component, 120
orbital precessions, equatorial orbit, 121
orbital precessions, general case, 120
orbital precessions, polar orbit, 122

1pN quadrupole
acceleration, 147
acceleration, normal component, 147
acceleration, radial component, 147
acceleration, transverse component, 147
direct orbital precessions, 148
direct orbital precessions, equatorial orbit, 149
direct orbital precessions, general case, 148
direct orbital precessions, polar orbit, 149
mixed net orbital shifts, total, equatorial orbit, 151
mixed net orbital shifts, total, general case, 13, 150, 203
mixed net orbital shifts, total, polar orbit, 152
mixed orbital precessions, total, 151, 154
orbital precessions, total, 154

2pN
acceleration, 73
acceleration, normal component, 74
acceleration, radial component, 74
acceleration, transverse component, 74
direct, net orbital shifts, 74
mixed I, net orbital shifts, 75
mixed II, net orbital shifts, 76
mixed, total net orbital shifts, 77
net orbital shifts, total, 73, 77, 78
orbital effects, 12
orbital precessions, 225
orbital precessions, total, 78
periastron, precession, 73, 224
pericentre, fractional advance, 80, 83
pericentre, net shift per orbit, 74, 80

3rd body
disturbing function, gravitoelectric, 160
disturbing function, gravitomagnetic, 162
mass, 155, 156, 160–162, 189
spin angular momentum, 155, 162, 189, 226
spin angular momentum unit vector, 189

3rd body, orbit around
argument of pericentre, 189, 225, 226
de Sitter, orbital precessions, 161, 232
distance, 161, 162, 189
eccentricity, 189, 225
gravitomagnetic, orbital precessions, 163
inclination of the orbital plane, 189
Keplerian mean motion, 156, 189
longitude of the ascending node, 189
orbital angular momentum, 161, 217, 226
orbital angular momentum per unit mass, 160, 189
orbital angular momentum unit vector, 189
orbital period, 155, 161, 162, 189, 225
semimajor axis, 189
true anomaly, 189

acceleration, centripetal, 104
accretion, disk, 87, 90
activity indices
photometric time series, 85
spectroscopic time series, 85
affine, parameter, 3, 182
altimetry, 170
angle, broken, *see* angle, dogleg
angle, dogleg, 17, 35
anomalistic period, *see* orbital period, anomalistic
anomaly
eccentric, 18, 157, 166, 168, 169, 173, 187
mean, 17, 18, 63, 98, 116, 144, 186
mean, at epoch, 8, 17, 26, 31, 94, 98, 186, 220, 229
mean, at epoch, net shift per orbit, 228
mean, at epoch, precession, 26, 31
mean, instantaneous shift, 36, 37, 52
true, 17, 18, 21, 30, 36, 46, 186
true, at epoch, 21, 25, 28–30, 37, 56, 74, 83, 127, 186, 189, 225
true, at midtransit, 46
true, instantaneous shift, 36

apocentre
height, 231
position of, 18, 55, 97
approximation, weak-field and slow-motion, 5, 87, 156, 175
apsidal, constant, 85, 183
apsidal period, *see* orbital period, anomalistic
apsides
line of, 17, 27, 30, 33, 58, 59, 97, 98, 129–131
position of, 55
Astrodynamical Space Test of Relativity using Optical Devices (ASTROD) (interplanetary probe), 233
astrodynamics, 11
astrometry, relativistic, 11
astronomy, 11
atoms, 87
attraction, gravitational, 3

B1620–26, triple system, 96
barycentric, radial velocity, *see* radial velocity, spectroscopic binary, systemic
Bayesian, approach, 8
BepiColombo (interplanetary probe), 51, 52, 63, 89, 118, 144
Bessel
function of the first kind of order s, 18, 182
modified function of the first kind of order s, 182
modified function of the first kind of order $s = 0$, 168
modified function of the first kind of order $s = 1$, 169
binary, 6, 7, 64, 73, 84, 94, 96, 107, 109, 125, 136, 166
angular momenta, 94, 107, 115
angular momentum, body A, 94, 96, 107, 184
angular momentum, body B, 94, 107, 184
companion, 38, 48, 72, 85, 96, 107, 114, 124, 136, 141, 142, 184, 225
companion, mass, 38, 107, 136, 184
companion, standard gravitational parameter, 184
equatorial radius, body A, 185
equatorial radius, body B, 185
magnitude of the angular momentum, body A, 184
magnitude of the angular momentum, body B, 184
mass, body A, 7, 37, 64, 73, 94, 107, 125, 136, 184
mass, body B, 7, 37, 64, 73, 94, 107, 125, 136, 184
masses, 7, 37, 64, 73, 94, 107, 115, 125, 136, 141, 161, 166
normalized mass, body A, 80, 184
normalized mass, body B, 80, 184
pulsar-white dwarf, minimum distance, 72, 115
quadrupole mass moment (dimensional), body A, 125, 136, 185
quadrupole mass moment (dimensional), body B, 125, 136, 185
quadrupole mass moment (dimensionless), body A, 185
quadrupole mass moment (dimensionless), body B, 185
quadrupole mass moments, 125, 136, 141
reduced mass, 96, 184
spectroscopic, 37
spectroscopic, double-lined (SB2), 37
spectroscopic, light curve, 37
spectroscopic, single-lined (SB1), 37
spin axes, 11, 141
spin axis, body A, 184
spin axis, body B, 184
standard gravitational parameter, 64, 184
symmetric mass ratio, 65, 184
total mass, 38, 64, 80, 107, 184
black hole, 6, 48, 62, 68, 69, 74, 87, 124, 227
accretion disk, 92
horizon, 86
mass, 181
Schwarzschild radius, 68, 181
singularity, naked, 86, 227
standard gravitational parameter, 181
Blandford–Znajek effect, 87
blazar, 74
body
characteristic size, 182
fluid, 85
fluid, centre, 85
fluid, homogeneous, 85
fluid, mass concentration, 85
hydrostatic equilibrium, 84, 85, 124, 183
massive, 3, 13, 84, 190, 192
massive, spherically symmetric, 191
boson stars, 227
bosons, 227

causality violations, 86
Celestial
Equator, 218, 219, 231
Sphere, 38
celestial mechanics, 11
perturbative methods, 19
relativistic, 11
censorship, cosmic conjecture, 86
centrifugal quadrupole parameter, 124, 183
Christoffel, symbols of the second kind, 3, 182
Chronos, personification of time in pre-Socratic philosophy, 2
clock, 87
gravitomagnetic effect, 91, 103, 104
clouds, 41
collapse, gravitational, 86
conditions, initial, 9, 83
coordinate, system, 5
coordinates, 5
harmonic, 11, 190
coordinates, spatially isotropic, *see* coordinates, harmonic
correlations, matrix, 8
cosmic expansion, accelerated, 10
cosmological constant, 165, 166, 168

counter orbiting particles, 91, 103, 104
covariance matrix, 8
curvature of spacetime, *see* spacetime, curvature of

Damour-Deruelle
  mean motion, 81, 188
  parametrization, 73, 80, 81, 189
  proper time eccentricity, 80, 189
dark matter, 10, 165, 172, 173
  density profile, 165
  density profile, exponential, 13, 172
  density profile, power-law, 13, 173
  density profile, scale length, 181
  density profile, scaling parameter, 181
  exponential density profile, orbital precessions, 173
  power-law density profile, exponent, 182
  power-law density profile, mean anomaly at epoch precession, 213
  power-law density profile, orbital precessions, 13, 174, 212
  power-law density profile, pericentre precession, 212
dark matter, nonbaryonic, 171
Darwin–Radau, relation, 84, 183
  dimensionless parameter, 84, 183
data centres, 11
data reduction, 170
data reduction software, 223
de Sitter–Fokker perigee precession, 162
de Sitter–Fokker spin precession, 15, 48, 155, 160
dec., of the north pole of rotation, 94, 95, 121, 122, 128, 149, 151, 183
declination (dec.), 11, 12, 36, 49, 185, 229
DI Herculis (binary star), 85
differential geometry, *see* geometry, differential
disturbing function, 26, 182, 190
  dark matter, exponential density profile:, 173
  dark matter, power-law, density profile, 173, 212
  logarithmic, 169
draconitic period, *see* orbital period, draconitic
drag-free, spacecraft, 230, 232, 233
dragon, 31

Earth, 2, 7, 8, 13, 16, 33, 39, 46, 48–50, 54, 59–61, 63, 64, 88, 90, 98, 101, 103, 116, 131, 135, 144, 146, 164, 196, 219, 223, 225, 230, 233
  angular momentum, 7, 180
  angular momentum per unit mass, magnitude, 215
  angular momentum, magnitude, 180, 215
  dec. of the north pole of rotation, 215
  ellipticity, 215
  equatorial plane, 230, 232
  geopotential, 223, 224, 232
  geopotential, even zonal harmonics, 223, 224, 232
  geopotential, odd zonal harmonics, 232
  mass, 180
  physical parameters, numerical values, 215, 223
  quadrupole mass moment (dimensionless), 180, 230
  RA of the north pole of rotation, 215
  radius, equatorial, 180, 215
  radius, polar, 215
  satellites, 33, 49, 88, 232
  satellites, geodetic, 7, 54, 88, 217, 222, 223
  spin axis, 181
  standard gravitational parameter, 180
eccentricity, 7, 16, 20, 31, 63, 68, 69, 72, 80, 81, 83, 97, 115, 140, 144, 186, 189
  instantaneous shift, 37
  rate of change, 21
Echelle SPectrograph for Rocky Exoplanets and Stable Spectroscopic Observations (ESPRESSO), 39
eclipse, 31
  primary, 40
  secondary, 40
ecliptic, plane of, 16, 231
Einasto, mass density profile, 172
Einstein, field equations, 4–6, 11, 227
Einstein, gravitational constant, 5, 181
Einstein, summation convention, 191
ejecta, angle, 89
electric
  charges, 87
  currents, 87
electromagnetic waves, *see* waves, electromagnetic
electromagnetism, 87
Elliptical Uranian Relativity Orbiter (EURO) (interplanetary probe), 14, 230, 231
ellipticity, 119, 183
ELXIS (Earth satellite), 14, 230, 232
energy, 4
  density, 5
  potential, of interaction, 192
  rest, 4
energy-momentum, tensor, 4, 181
Ephemeris of Planets and the Moon (EPM), 220
equation of geodesics, 3
equations of motion, 3, 6, 11, 35
equator, 104, 105
Equivalence Principle (EP), 2–4
  strong, 3, 226
  weak, 2
Euler angles, 17
European Southern Observatory (ESO), 40
European Space Agency (ESA), 233
even zonal harmonic, degree $\ell = 2$, *see* quadrupole mass moment (dimensionless)
Event Horizon Telescope (EHT), 6, 92
exoplanets, 16, 37–39, 64, 68, 69, 89, 92, 107–109, 112, 136
  angular momentum, 108, 109, 141
  circumbinary, 96
  eccentricity, 68, 69, 71, 112, 141
  mass, 69, 108, 109, 136, 137
  orbital angular momentum, 141
  orbital frequency, 108, 109, 136, 137

exoplanets (cont.)
orbital period, 68, 69, 108, 109, 136
quadrupole mass moment (dimensionless), 136, 137
radius, 42, 46, 69, 108, 109, 136, 137
rotational frequency, 108, 109, 136, 137
sky–projected spin–orbit angle, precession, 47, 113, 141, 142
sky-projected orbital angular momentum, 46, 92
sky-projected spin-orbit angle, 92
spin axis azimuth in the plane of the sky, 92
spin-orbit, misalignment, 108
star–planet, distance, 108, 109, 136, 137
star–planet, minimum distance, 68, 71, 72, 112, 140, 141
exoplanets, habitable, 39
exoplanets, host star, 71, 92, 107, 112, 113, 140–142
angular momentum, 47, 108, 109, 113, 141
azimuthal angle of the spin axis in the plane of the sky, 47
equatorial radius, 42, 46, 141
mass, 108, 109, 137
quadrupole mass moment (dimensional), 142
quadrupole mass moment (dimensionless), 137, 141
radius, 108, 109, 137
sky-projected angular momentum, 47, 92
spin axis, 92, 113, 141, 142
spin axis azimuth in the plane of the sky, 113
spin axis tilt to the line of sight, 92, 113, 142
exoplanets, transiting, 11, 12, 36, 40, 41, 46, 68, 71, 72, 92, 109, 112, 113, 137, 140–142
characteristic timescales, 11, 12, 40, 71, 72, 113, 141
disk, 40
impact parameter, 42, 183
planet to star radius ratio, 42, 185
sky-projected spin-obit angle, 12, 46
sky-projected spin-orbit angle, 113, 141
exotic compact objects (ECOs), 227
explanets
longitude of the ascending node, 47
Extreme Precision Radial Velocity (EPRV), 39
EXtreme PREcision Spectrograph (EXPRES), 39

fermion balls, 227
fermions, 227
field
1pN gravitoelectric, 48, 63, 217–220, 232, 233
1pN gravitomagnetic, 84, 87–90, 96, 98, 116, 164, 217–220, 223, 224
1pN gravitomagnetic, Lorentz-violating, 175, 184
1pN gravitomagnetic, matter ring, 96, 217, 226
1pN gravitomagnetic, spin octupole, 119, 184
gravitational, 2, 3, 5–7, 11, 15, 84, 88, 155
gravitational, external, 3, 226
pK quadrupolar, 217, 219, 220
field effective theory, 10
free fall, 2, 3, *see* fall, free
full-width at half-maximum (FWHM), cross-correlation function (CCF), 85

Galactic Centre, 6, 14, 20, 38, 49, 55, 172, 217, 227
galaxies, 10, 171
halos, 172, 173
Galaxy (Milky Way), 97
Gauss
equations for the variations of the orbital elements, 22, 80, 83
general theory of relativity (GTR), 1–6, 9, 11, 37, 54, 65, 82, 87, 89, 156, 165, 190, 224, 225, 227, 233
generalized hypergeometric function, 174, 182
geodesic deviation equation, 5
geodesics, equation of, *see* equation, of geodesics
geodetic precession, spin, *see* de Sitter–Fokker, precession, spin
geometry, differential, 2
GRACE Follow–On (GRACE–FO) (Earth satellite), 50
grava-stars, 227
gravitation, 1–5, 11
gravitational attraction, *see* attraction, gravitational
gravitational field, *see* field, gravitational
gravitational potentials, *see* potentials, gravitational
gravitational waves, *see* waves, gravitational
gravitoelectromagnetism, paradigm, 87
gravitomagnetic field, *see* field, 1pN gravitomagnetic
gravitomagnetic precessions, *see* 1pN Lense–Thirring, orbital precessions, general case
gravitomagnetism, 4, 87, 89
gravity
absence of, 3
field, mapping, 170
models of, modified, 10, 13, 16, 31, 37, 82, 169
Gravity Probe B (GP–B) (Earth satellite), 7, 90
Gravity Recovery and Climate Experiment (GRACE) (Earth satellite), 50
Gravity Recovery and Interior Laboratory (GRAIL) (Moon satellite), 50
GRAVSAT (Earth satellite), 50
Greeks, ancient, 31
gyroscopes, 7, 87, 90

Hēlios, Greek deity of Sun, 54
H1743–322 (microquasar), 90
Hamilton–Jacobi method, 73, 80
HD 209458 (star), 92
HD 286123 (star), 40, 41
helioseismology, 85
hierarchical triple system, 96, 155, 165, 225
distant companion, 96, 97, 155, 165, 225, 227
inner binary, 155, 165, 217, 225, 226
Highly Elliptical Relativity Orbiter (HERO) (Earth satellite), 14, 230

Horizons System, NASA JPL, 63, 64, 117, 144, 145, 221
Hubble parameter tension, 10

Iūnō, Roman deity, 13
Icarus (asteroid), 7, 54
imaginary unit, 87
in-plane acceleration, 90
inclination
  of the orbital plane, 7, 16, 20, 106, 133, 186, 217, 228, 231, 232
  of the orbital plane, critical, 230
inclination, net shift per orbit, 228
inclination, precession, 22, 218
inertial forces, 15
  centrifugal, 15
  Coriolis, 4, 15
inner planets of the solar system, *see* solar system, planets, inner
interactions, fundamental, 1
International Astronomical Union (IAU), 11
International Celestial Reference Frame (ICRF), 63, 64, 117, 144, 145, 214, 215, 217, 220, 221, 223
inverse-square, law, *see* Newton, inverse-square law, acceleration
IORIO (interplanetary probe), 14, 230, 231
iron line, 90

J2000.0, reference epoch, 49, 214, 215, 218, 219
Jacobi equation, *see* geodesic, deviation, equation
Jet Propulsion Laboratory (JPL), 63, 64, 117, 144, 145, 221
Juno (interplanetary probe), 13, 88, 217, 220–222
  node period, 221
  orbital parameters, 221
  perijove period, 221
Jupiter, 13, 40, 68, 69, 71, 85, 88, 90, 96, 107–109, 112, 113, 136, 137, 140, 141, 146, 217, 220, 221, 231
  angular momentum, 181
  angular momentum, magnitude, 181
  dec. of the north pole of rotation, 181, 215
  ellipticity, 215
  mass, 181
  physical parameters, numerical values, 215, 221
  quadrupole mass moment (dimensionless), 181, 221
  RA of the north pole of rotation, 181, 215
  radius, equatorial, 181, 215
  radius, polar, 215
  spin axis, 181
  standard gravitational parameter, 181

K2–137 (star), 40, 107, 136
Kelt–9 (star), 47, 107, 113, 114, 141, 142
Kepler
  mean motion, 185
  orbital elements, 7, 8, 11, 13, 16, 18–22, 25, 27, 29, 30, 36, 58, 59, 63, 64, 73, 78, 80, 82, 88, 117, 120, 126, 129, 131, 144, 145, 148, 150, 160, 197, 199
Kepler mean motion, *see* mean motion, Keplerian
Kepler orbital period, *see* orbital period, Keplerian
Kepler third law, 216
Kepler, ellipse, 16, 19, 20, 26–28, 30, 31, 34, 74, 191–193
  osculating, 19–25, 31, 58, 60, 97, 100, 102, 130, 132, 134
Kepler–13 (star), 47
Kerr black hole, 86, 87, 91, 125, 227
  angular momentum, 86, 89, 115, 181, 227
  angular momentum, magnitude, 86, 181
  even spin moments, 87
  mass, 227
  mass multipole moments, 86, 181
  multipole moments, 227
  odd mass moments, 87
  quadrupole mass moment (dimensional), 125, 143, 181
  shadow, 6, 92
  spacetime, 91
  spin axis, 181
  spin axis azimuth in the plane of the sky, 115, 181
  spin axis tilt to the line of sight, 181
  spin dipole moment, 86
  spin multipole moments, 86, 181
  spin parameter, 86, 181
Kronecker delta, 158, 182, 190
Kronos (Greek pre-Olympian deity), 2

LAGEOS (Earth satellite), 14, 217, 222, 223
  node period, 222
  orbital period, 222
  perigee period, 222
  retroreflectors, 222
LAGEOS–2 (Earth satellite), 222
Lagrange equations of motion, 147
Lagrange planetary equations, 13, 26
Lagrangian per unit mass, 182
  1pN Lense–Thirring, 192
  1pN gravitoelectric test particle, 147, 191
  1pN gravitoelectric, test particle, 191
  Newtonian, quadrupole mass moment (dimensionless), 193
  Newtonian, spherical body, 26, 191
  pK, 13, 26, 27, 190, 193
Laplace–Lagrange, first parameter, 32, 186
Laplace–Lagrange, second parameter, 32, 186
Laplace–Runge–Lenz
  unit vector, 58, 129, 188
  vector, 32, 59, 97, 98, 188
LARES (Earth satellite), 222, 230
LARES–2 (Earth satellite), 222
Laser Astrometric Test Of Relativity (LATOR) (interplanetary probe), 233
Laser Interferometer Gravitational-wave Observatory (LIGO), 6

laser, pulses, 222
laser-ranging stations, 11
latitude, argument of, 17, 35, 106, 186
latitude, argument of
  at epoch, 186, 227, 229
latitude, argument of
  at epoch, 229
least-square approach, 8, 146
Legendre
  polynomial of degree ℓ, 123, 182
  polynomial of degree $\ell = 3$, 119
LEnse–Thirring Sun–Geo Orbiter (LETSGO) (interplanetary probe), 14, 230, 231
lensing, gravitational, 92
Levi-Civita 3-dimensional symbol, 182, 191
line
  of sight, 113, 142
  straight, 3
line of sight, *see* sight, line of
logarithmic potential, 169
  characteristic length scale, 169, 184
  mean anomaly at epoch, precession, 169
  orbital precessions, 169
  pericentre, precession, 169
longitude, true, 17, 35, 101, 133, 186
  at epoch, 186
Lorentz invariance, *see* Lorentz, symmetry
Lorentz, symmetry, 4, 10
  spontaneous breaking, 175
  violations, 10, 13, 165, 175
Lorentz-violating coefficients, 175
  vector, 184
Lorentz-violating gravitomagnetic
  acceleration, 175
  acceleration, normal component, 175
  acceleration, radial component, 175
  acceleration, transverse component, 175
Lorentz-violating off-diagonal coefficients
  vector, 175
Love, number, 85, 183
LS I+61°303 (microquasar), 89
Lunar Laser Ranging (LLR), 2, 6, 162

M dwarf Advanced Radial velocity Observer Of Neighbouring eXoplanets (MAROON–X), 39
magnetohydrodynamical general relativistic simulations, 92
manifold
  differentiable, 2
    Lorentzian, 1, 2, 181
    pseudo-Riemannian, 2
    Riemannian, 2
  number of dimensions, 181
Mars, 88, 218, 220
Mars Global Surveyor (MGS) (interplanetary probe), 89
mass
  central body, 7, 9, 15, 19, 20, 54, 58, 60, 182, 190, 192
  density, 4, 119, 181
    profile, 172
  monopole moment, 7, 16, 54, 155, 156, 160–162, 168, 175, 221
  multipole moments, 7, 16
  point approximation, 85
  quadrupole moment, 16, 221
mass-energy
  currents, 87, 190
  distribution, 190
material bodies
  composition, 4
  mass, 4
matter, 4
Maxwell equations, 87
mean anomaly at epoch, precession, 20, 98, 219
mean motion
  instantaneous shift, 37
  Keplerian, 16, 22–25, 38, 80, 156
Mercury (planet), 7, 51, 52, 54, 63, 64, 116, 117, 144, 145, 217, 219, 220, 232
Mercury (Roman deity), 89
MESSENGER (interplanetary probe), 89
Messier M87 (galaxy), 6, 90
metric tensor signature, 2
MicroSCOPE (Earth satellite), 2
midtransit, 42
Modified General Relativity (MGR), 10
MOdified Gravity (MOG), 10
MOdified Newtonian Dynamics (MOND), 10
modulation, quasi-periodic, 89
moment of inertia, 84, 183
Moon, 2, 3, 17, 31, 48, 50, 162, 164
  orbital angular momentum, 164

NASA Exoplanet Archive, 41
National Aeronautics and Space Administration (NASA), 63, 64, 117, 144, 145, 221
near-infrared emission, 92
NEID, 39
net shifts per orbit
  1st order, 25, 28
  2nd order, 27, 28
  mixed, 29
neutron star, 6, 47, 48, 86, 124, 224–226
  angular momentum, 86
  equation of state (EOS), 224
  moment of inertia, 86, 216, 224
  quadrupole mass moment (dimensional), 124, 216
Newton
  dynamics, laws of, 1
  gravitational constant, 5, 15, 180
  gravitational force, 1–3

inverse-square law, 165
acceleration, 11, 15, 104, 165, 185
Universal Gravitation, theory of, 1, 2, 5
Newton potential, *see* potential, Newtonian, arbitrary mass distribution
no-hair theorems, 86, 227
node
ascending, longitude of, 7, 17, 19, 31, 61, 99, 131, 186, 218, 223, 228, 231
ascending, position of, 59–61, 101, 133, 135
descending, longitude of, 17, 186
node, net shift per orbit, 228
node, precession, 20, 23, 113, 141, 218
nodes
line of, 17, 19, 33, 36, 42, 59–61, 92, 99–101, 106, 131–133
line of, unit vector, 18, 59, 60, 99, 101, 131, 133, 187
of the orbit, 31, 36
non-gravitational forces, 2, 16, 171, 222, 232
nonconservative forces, *see* non-gravitational forces
Nordtvedt, effect, 3, 226
normal unit vector, *see* orbital angular momentum, unit vector
normalized moment of inertia (NMoI), 84, 183

oblateness, 16, 133, 136–138, 146
observables, 8, 12, 25, 36, 37
calculated values, 8
instantaneous shifts, 1st order, 36
measured values, 8
net shifts per orbit, 1st order, 37
observatories, astronomical, 11
off-diagonal gravitomagnetic metric tensor components, 87, 190, 192
OJ 287 (blazar), 74
once-per-revolution
acceleration, 169
acceleration, cosine coefficient of the normal component, 170, 188
acceleration, cosine coefficient of the radial component, 170, 188
acceleration, cosine coefficient of the transverse component, 170, 188
acceleration, normal component, 170
acceleration, orbital precessions, 170
acceleration, radial component, 170
acceleration, sine coefficient of the normal component, 170, 188
acceleration, sine coefficient of the radial component, 170, 188
acceleration, sine coefficient of the transverse component, 170, 188
acceleration, transverse component, 170
OPTIS (Earth satellite), 233
orbit, circular, 18, 20, 41, 55, 70, 90, 99, 103–111, 136–140, 172, 187, 232
polar angle, 104–106
orbit, equatorial, 90, 99–101, 104–106, 125, 133
orbit, polar, 20, 90, 99, 125, 231, 232
orbit, prograde, 90, 95, 104, 105
orbit, retrograde, 91, 95, 105, 106
orbit, rotational sense, 90, 91, 104, 105
orbital angular momentum, 19, 24, 96, 107, 112, 114, 115, 136, 141, 142, 161, 162, 187, 192, 195
unit vector, 24, 55, 74, 90, 120, 125, 136, 147, 188
orbital effects, mixed, 28
orbital effects, post-Keplerian (pK), 11, 217–219, 227
orbital effects, post-Newtonian (pN), 14, 230
orbital element, non-singular
$\widetilde{k}$, 35, 186
$\widetilde{q}$, 35, 186
k, 106
q, 106
orbital elements, non-osculating, 8, 21
orbital frequency, *see* mean motion, Keplerian
orbital motions, 172
orbital period, anomalistic, 12, 28–30, 186
1pN Lense–Thirring, 97, 98
1pN gravitoelectric test particle, 57, 58, 154
1pN gravitoelectric two-body, 67, 78
quadrupole mass moment (dimensionless), 129, 154
orbital period, Keplerian, 16, 18, 21–25, 28, 30, 40, 42, 48, 58–60, 63, 68, 69, 78, 97, 100, 102, 115, 135, 137, 144, 154, 155, 168, 173, 185, 221, 222, 225
orbital period, sidereal, 12, 33, 35, 36, 100, 133, 186
orbital period, draconitic, 12, 31, 33, 34, 36, 99, 100, 186
1pN Lense–Thirring, 98, 100
1pN gravitoelectric test particle, 58, 61
1pN gravitoelectric two-body, 67, 68
quadrupole mass moment (dimensionless), 130, 132, 133
orbital precessions, *see* precessions, orbital
1st order, 25
mixed, 29
orbital shifts, instantaneous
1st order, 11, 25, 28, 29, 36
2nd order, 11, 27, 28
oscillations, quasi-periodic, 89
out-of-plane acceleration, component of, *see* post-Keplerian (pK) acceleration, normal component
out-of-plane unit vector, *see* orbital angular momentum, unit vector

parallel (Earth line of latitude), 33
parallel transport, 15, 156
parameters, solving for, 8, 74, 146
parametrized post-Newtonian (PPN) formalism, 9
β_{PPN} parameter, 9, 182
γ_{PPN} parameter, 9, 182
Penrose process, 87

periastron
advance, 7, 66
advance, fractional, 74
distance, 141
net shift per orbit, 228
pericentre
advance, 66, 122, 150
argument of, 7, 17, 31, 80, 186, 219, 228, 231
longitude of, 17, 93, 186, 220
passage at, 28, 31, 98
position, 31
position of, 17, 18, 26, 30, 46, 55, 58, 97, 98, 130
precession, 20, 24
precession, in plane, 94
time of passage at, 17, 26, 31, 98
perigee
argument of, 162, 230
height, 230
perihelion, precession, 172, 219
perijove, 222
perinigricon
advance, 7
distance, 62, 68, 115, 143
perturbations, orbital secular, 20
Pioneer, anomaly, 171
Pisa, Leaning Tower of, 2
plane
equatorial, 16, 34, 94, 103, 104, 121, 127, 149, 151, 187, 195, 196
fundamental, of the reference frame, 16, 17, 33, 35, 36, 38, 42, 48, 49, 61, 92, 99–104, 134, 135, 225
of the orbit, 17, 19, 21, 24, 34–36, 55, 58, 90, 94, 95, 99–101, 104–106, 121, 122, 125, 127, 128, 149, 151, 152, 195, 225, 230, 232
of the sky, 17, 42, 47, 48, 91, 92, 113, 181, 225
planet
gaseous giant, 68, 69, 84, 108, 109, 113, 136, 137, 141
gravity field, multipoles, 231
Planetary Laser Ranging (PLR), 6
planets, rocky, *see* solar system, planets, inner
platinum, 2
Poisson, equation, 5, 173
polarimetric observations, 92
polarized submillimetre infrared observations, 92
position unit vector, *see* radial unit vector
position, vector, 16, 18, 50, 51, 54, 58–60, 88, 97–99, 101, 129, 131, 133, 187
projection onto the primary's spin angular momentum, 88, 147, 187
shift, 51, 187
shift, normal component, 51, 188
shift, radial component, 51, 188
shift, transverse component, 51, 188
post-fit residuals, *see* residuals, post-fit
post-Keplerian (pK)
anomalistic correction to the Keplerian orbital period, 186
draconitic correction to the Keplerian orbital period, 186
perturbed orbital period, 28
perturbed trajectory, 19, 20, 27, 31
potential, 26
potential, spherically symmetric, 27
radial velocity, instantaneous shift, 38
sidereal correction to the Keplerian orbital period, 186
post-Keplerian (pK) acceleration, 11, 15, 16, 19, 25, 28, 30, 32–35, 47, 51, 123, 146, 185
r^{-1}, 169
constant, 171
constant, normal component, 171
constant, orbital precessions, 171
constant, radial component, 171
constant, transverse component, 171
normal component, 23, 188
radial component, 23, 31, 188
transverse component, 23, 188
post-Keplerian (pK) orbital effects, *see* orbital effects, post-Keplerian (pK)
post-Newtonian (pN)
corrections to the Minkowski metric tensor, 182, 190
spin precessions, 13, 15
post-Newtonian (pN), approximation, 6, 190
1pN order, 7, 83, 84, 119, 146, 191
2pN order, 7, 73, 191
potential, energy
interaction, 181
spin-spin, 162
two-body, 161
potential, gravitomagnetic, spin octupole, 119, 184
potential, Newtonian
arbitrary mass distribution, 4, 5, 181, 182, 190
axisymmetric body, 16, 122, 123, 147, 183
spherical body, 123, 182
potentials, gravitational, 5
potentials, metric tensor components, 4
power-law potential $n = -2$
mean anomaly at epoch, precession, 168
pericentre, precession, 168
power-law potential $n = 2$
mean anomaly at epoch, precession, 167
pericentre, precession, 167
power-law potential $n = 3$
mean anomaly at epoch, precession, 167
pericentre, precession, 167
power-law potential n generic
mean anomaly at epoch, precession, 166
pericentre, precession, 166
power-law, potential, 165
dimensional strength parameter, 166, 184
disturbing function, 166

precession of the inclination, *see* inclination, precession
precession of the node, *see* node, precession
precession of the pericentre, *see* pericentre, precession
precessions, orbital, 13
  1st order, 26
  2nd order, 28, 30
precise orbit determination (POD), 170
pressure, 5
probe, 146, 221, 222, 232
PSR B1913+16 (binary pulsar), 225
PSR J0337+1715 (triple pulsar), 2, 96, 225, 226
PSR J0337+1715 (triple pulsar), inner binary, 3, 96
PSR J0737–3039 (binary pulsar), 13, 14, 48, 89, 216, 217, 225
  physical parameters, numerical values, 225
PSR J0737–3039A (pulsar), 48, 86
  angular momentum, 86
  rotational period, 86
PSR J0737–3039B (pulsar), 48, 86
  angular momentum, 48, 86
  rotational period, 86
PSR J0737–3039 (binary pulsar), 214
PSR J1141–6545 (binary pulsar), 72, 85, 89, 114, 115, 124, 141, 142
Pugh–Schiff spin precession, 7, 15, 90, 155, 160
pulsar
  emission, 47
  millisecond, 96
  triple, 14, 96, 97, 217, 226, 227
pulsar, binary, 2, 7, 11, 12, 36, 47, 48, 64, 66, 72–74, 85, 89, 96, 114, 124, 141, 142, 146, 225
  barycentre, 48
  barycentric orbit, 48, 226
  barycentric semimajor axis, 48
  radio pulses, 47, 48
  semimajor axis, 216
  times of arrival (TOAs), 47
  variation of the times of arrival, 11, 48

quadrupole mass moment (dimensional), 124, 183, 225
  acceleration, 125
  dec., instantaneous shift, 143
  dec., net shift per orbit, 229
  inclination, net shift per orbit, 228, 229
  mean anomaly at epoch, net shift per orbit, 229
  node, net shift per orbit, 228, 229
  orbital precessions, 227
  pericentre, net shift per orbit, 228
  perinigricon, net shift per orbit, 229
  RA, instantaneous shift, 143
  RA, net shift per orbit, 229
  radial velocity semiamplitude, net shift per orbit, 137, 229
  variation of the times of arrival, instantaneous shift, 141, 142
  variation of the times of arrival, net shift per orbit, 141
quadrupole mass moment (dimensionless), 11, 12, 16, 31, 84, 123, 124, 129, 130, 132, 134, 137, 150, 183, 193
  acceleration, 13, 123, 146
  acceleration, normal component, 125
  acceleration, radial component, 125
  acceleration, transverse component, 125
  correction to the sidereal period, 133
  dec., net shift per orbit, 143
  dec., instantaneous shift, 142
  equations of motion, numerical integrations, 129–131, 133, 135, 145
  full width at half maximum transit duration, net shift per orbit, 139
  inclination, net shift per orbit, 136
  inclination, precession, 218
  ingress/egress transit duration, net shift per orbit, 138
  instantaneous orbital shifts, 13, 126, 199
  mean anomaly at epoch, precession, 220
  mean anomaly, instantaneous shift, 129, 144
  net orbital shifts, equatorial orbit case, 128
  net orbital shifts, general case, 127, 154
  net orbital shifts, polar orbit case, 128
  node, precession, 219
  pericentre, precession, 220
  RA, instantaneous shift, 142
  RA, net shift per orbit, 143
  radial velocity curve, instantaneous shift, 136, 137
  radial velocity semiamplitude, net shift per orbit, 136
  range rate, instantaneous shift, 144, 145
  range, instantaneous shift, 144
  sky-projected spin-orbit angle, precession, 141
  time of inferior conjunction, net shift per orbit, 140, 141
  total transit duration, net shift per orbit, 138
  transit characteristic timescales, net shifts per orbit, 137, 140
quantum gravity, theory of, 10
quantum regime, 3

RA of the north pole of rotation, 94, 95, 121, 122, 128, 149, 151, 183
radar ranging, 54
radial unit vector, 24, 55, 74, 90, 120, 125, 136, 147
radial velocity, spectroscopic binary, 11, 12, 36–39, 187
  accuracy, 39
  curve, 37, 68, 106, 135
  semiamplitude, 39, 68, 106, 135, 187, 229
  systemic, 38, 187

radiation
gamma, 47
optical, 47
radio, 47, 225
X, 47
radius, circular orbit, 104, 187
radius, equatorial, 119, 182
radius, polar, 119, 183
rate of change of the eccentricity, *see* eccentricity, rate of change
rate of change of the semimajor axis, *see* semimajor axis, rate of change
redshift, gravitational, 225
reference
x axis, 16, 17, 33, 42, 47, 91, 92
y axis, 92
z axis, 17, 38, 42, 48, 92, 93, 104, 125, 147, 148, 150
direction, 16, 33, 36, 47, 102, 104, 106, 135
reference frame, 13, 16, 38, 94, 121, 128, 147–151, 155, 160, 196
accelerated, 4, 5
dynamically non-rotating, 15, 155, 160, 232
inertial, 2, 15
kinematically non-rotating, 15, 155
kinematically rotating, 160, 232
rotating, 4
relativistic
jets, 87
multipoles, 120
remote sensing, 170
residuals, post-fit, 8, 9, 146, 170, 223
Ricci
spacetime curvature tensor, 4, 182
tensor trace, 5, 182
Riemann, spacetime curvature tensor, 4, 5, 156, 182
right ascension (RA), 11, 12, 34, 36, 49, 185, 229
ring, massive, 96, 165, 226, 227
potential expansion, 165, 166, 226
quadrupole, 165, 166, 226
Rossby wave-induced spiral pattern, 92
Rossiter–McLaughlin effect, 47, 92
rotation, rigid, 119
rotational
frequency, 85, 107, 183
period, 86, 183

S star, 20, 37–39, 49, 62, 68, 69, 74, 89, 91, 115, 116, 143, 172
S2 (star), 7, 39, 54
S4716 (star), 14, 49, 109, 110, 137, 138, 217, 227–229
satellite, 15, 19, 20, 27, 30, 88, 94, 95, 121, 122, 127, 128, 149, 151, 152
data reductions, 169
Satellite Laser Ranging (SLR), 6, 7, 88, 222, 223
stations, 223
Saturn, 88
Saturn (Roman deity), 2
Scalar Tensor Vector Gravity (STVG), 10
self-energy, gravitational, 226
self-gravity, 2, 3
semilatus rectum, 16, 31, 106, 186
semimajor axis, 7, 16, 20, 37, 38, 40, 48, 59–61, 81, 83, 96, 98, 101, 103, 131, 133, 135, 185, 189, 222, 230
instantaneous shift, 37
rate of change, 21, 230
Sgr A*, 6, 7, 20, 38, 54, 62, 91, 92, 107, 109, 115, 136, 142, 227
Shapiro, delay, 225
sidereal period, *see* orbital period, sidereal
sight, line of, 17, 36, 38, 42, 48, 85, 91, 92, 181, 225
SINgle Faint Object Near–IR Investigation (SINFONI), 40
sky-projected spin-orbit angle, 11, 183
SLALOM (Earth satellite), 50
solar system, 6, 7, 11, 13, 16, 50, 88, 89, 172, 214, 217, 230, 233
ice giants, 220
moons, 88
planets, 13, 88, 217–220
planets, inner, 7, 54, 88
planets, outer, 88
spacecraft, *see* probe
spacetime, 2–5, 13, 15, 156, 160, 166
curvature of, 1, 3, 4
geodesic, 3
metric inverse tensor, 3, 182
metric tensor, 3, 4, 13, 27, 87, 165, 182, 190
Special Theory of Relativity (STR), 1, 2, 4
spectral energy distribution (SED), 85
spectrum, electromagnetic, 47
speed of light in vacuum, 2, 5, 6, 48, 180
spheroid, oblate, 119
spin
dipole moment, 7, 12, 16, 119, 221
multipole moments, 7, 16
octupole moment, 12, 16, 221
spin angular momentum, 7, 19, 20, 84, 88, 94, 96, 99, 102, 107–109, 112, 114, 129, 131, 135, 136, 142, 147, 183, 190, 192, 195, 196, 231
magnitude, 183
unit vector, 88, 90, 91, 93, 95, 97, 99, 100, 102, 104, 122, 125, 148–150, 183
spin axis, *see* spin angular momentum, unit vector
spin-orbit coefficients
equatorial orbit, 195
general case, 194
polar orbit, 195
spin aligned with the *z* axis, 196
spin-orbit coefficients are defined in Appendix D, 13
standard gravitational parameter, 15, 182
Standard Model Extension (SME), 10, 175
Standard Model of elementary particles and fields, 10
star, main sequence, 37–40, 48, 68, 69, 72, 84, 85, 87, 89, 96, 107–109, 124, 136, 137, 146

activity indices, 85
disk, 40
equatorial rotational velocity, 85, 183
mask, 85
projected rotational velocity, 85, 183
quadrupole mass moment, 107
radius, equatorial, 85
reduced spectrum, 85
rotational period, 85
spin axis, 85
spin axis azimuth in the plane of the sky, 183
spin axis tilt to the line of sight, 85, 183
spots, 41
stress-energy tensor, *see* energy-momentum, tensor
stresses, internal, 4
strong, gravitational regime, 5, 6, 89
sub-satellite, point, 33
time of passage of, 33
Sun, 2, 7, 13, 40, 63, 68, 69, 71, 90, 107–109, 112, 116, 136, 137, 140, 162, 164, 217–220, 232, 233
angular momentum, 89, 117, 180
angular momentum, magnitude, 85, 117, 180
dec. of the north pole of rotation, 117, 144, 145, 180, 214
ellipticity, 214
mass, 63, 64, 180
physical parameters, numerical values, 214, 217–219
quadrupole mass moment (dimensionless), 107, 144, 145, 180, 217–220, 232
RA of the north pole of rotation, 117, 144, 145, 180, 214
radius, equatorial, 180, 214, 232
radius, polar, 214
spin axis, 144, 145, 180
standard gravitational parameter, 180
supermassive black hole in M87, 6, 90
supermassive black hole in Sgr A*, 6, 7, 20, 38, 54, 89, 107, 110, 115, 116, 136, 138, 143, 227–229
angular momentum, 107
mass, 107, 110, 116, 136, 138, 143, 228, 229
quadrupole mass moment (dimensional), 136
spin axis, 91, 92, 109, 110, 115, 116, 137, 138, 143, 144, 227–229
spin axis azimuth in the plane of the sky, 91, 109, 110, 116, 137, 138, 143, 144
spin axis tilt to the line of sight, 91, 92, 109, 110, 115, 116, 137, 138, 143
spin parameter, 107, 110, 116, 136, 138, 143, 227–229
supermassive black hole in Sgr A*, 229
mass, 229
spin axis, 229
spin parameter, 229
superradiance, 87
symmetry
axial, 123
spherical, departures from, 16
symmetry axis, 121, 123
systematic errors, 41

temporal coordinate, 182
test particle, 3, 4, 11, 13, 15–20, 24, 28, 30, 31, 33, 35, 49, 58, 59, 61, 62, 64, 65, 69, 84, 90, 97–99, 102, 104–106, 108, 115, 119, 129, 131, 135, 137, 142, 155, 156, 161, 175, 192
tetrad components, 156
TianQuin (interplanetary probe), 233
tidal
angular momenta alignment, 108, 109, 137
effects, 15, 107, 155
forces, 3, 107, 136
matrix, 156
matrix, 1pN gravitoelectric, 158
matrix, 1pN gravitomagnetic, 159
matrix, Newtonian, 49, 158
orbit circularization, 108, 109, 137
orbital effects, 13
spin-orbit synchronization, 108, 109, 136, 137
tidal acceleration, 156, 172
normal component, 156
radial component, 156
transverse component, 156
tidal effects, 84, 108, 109, 137
tidal quadrupole parameter, 124, 183
tide, K_1, 232
tide-raising companion, 124
mass, 124, 183
semimajor axis, 124, 183
tides, 3, 108, 109, 136, 137
Time, *see* Chronos
time
coordinate, 11, 182, 191
inferior conjunction of, 40, 46, 71, 72, 112, 141, 185
initial instant, 17, 25, 58, 60, 83, 97, 100, 102, 127, 130, 132, 134, 186
proper, 3, 11, 182
standards, 41
timelike curves, closed, 86
times of arrival (TOAs), 12, 226
titanium, 2
Titans (Greek pre-Olympian gods), 2
transit
egress duration, 44
flux, 40
ingress duration, 43
ingress/egress duration, 40, 44, 185
primary, 40, 44
primary, full width at half maximum duration, 40, 44, 185
secondary, 40
total duration, 40–43, 185
transverse unit vector, 24, 55, 74, 90, 120, 125, 136, 147

two-body
distance, 1, 16, 19, 31, 109, 125, 161, 162, 175, 187, 192
range, 12, 50, 52, 185, 232
range rate, 12, 50, 52, 185
range rate shift, 51, 185, 231
range shift, 51, 185, 232
two-body system, 88
restricted, 9, 11, 13, 15, 155, 156, 160–162
restricted, primary, 7, 13, 15, 19, 20, 24, 31, 49, 50, 58, 60, 62, 88, 90, 99, 102, 103, 115, 121, 123, 129–132, 134, 135, 142, 146, 155, 168, 172, 187, 193
restricted, primary, 11, 102, 123

unit vector
in the orbital plane normal to the line of nodes, 19, 187
in the orbital plane along the line of nodes, *see* node, ascending, unit vector
perpendicular to the orbital plane, *see* angular momentum, orbital, unit vector
range rate shift, auxiliary, 51, 185
Uranus, 88, 231

velocity, vector, 18, 50, 51, 147, 175, 187, 191–193
magnitude, 187
projection onto the position vector, 54, 187
projection onto the primary's spin angular momentum, 187
shift, 51, 187
shift, normal component, 51, 188
shift, radial component, 51, 188
shift, transverse component, 51, 188
Vernal Equinox
at some reference epoch, 16, 180, 232
Very Large Telescope (VLT), 40
Virgo (laser interferometer), 6

WASP–33 (star), 47, 107
waves
electromagnetic, 3–5, 87
gravitational, 5, 6, 156, 157, 233
gravitational, frequency, 157, 233
weak-field and slow-motion approximation, *see* approximation, weak-field and slow-motion
weight force, 2
white dwarf, 3, 48, 85, 124, 225, 226
angular momentum, 86
binary, 2, 72, 89, 96, 114, 141, 142, 225
mass, 86
moment of inertia, 85
quadrupole mass moment (dimensional), 124, 142
radius, 86
rotational frequency, 85
triple, 96
worldline, 2, 3
geodesic, 15, 156, 160
wormholes, 228

X-ray
binaries, 89
lightcurve, 92
XMM–Newton (spaceborne mission), 92
XO-3 (star), 47

Yukawa potential, 165, 168
dimensional strength intensity, 168, 184
dimensionless strength intensity, 168, 184
disturbing function, 168
mean anomaly at epoch, precession, 168
orbital precessions, 168
pericentre, precession, 168
scale distance, 168, 184

Zeus (Greek deity), 2

For EU product safety concerns, contact us at Calle de José Abascal, 56–1°, 28003 Madrid, Spain or eugpsr@cambridge.org.

www.ingramcontent.com/pod-product-compliance
Lightning Source LLC
LaVergne TN
LVHW062208200926
848484LV00006B/1346